D1224002

Flowering Plants

Flowering Plants

Evolution above the Species Level

G. Ledyard Stebbins

The Belknap Press of Harvard University Press
Cambridge, Massachusetts, 1974

Library of Congress Catalog Card Number 73-86939
ISBN 0-674-30685-6
Printed in the United States of America

To Barbara

without whose devotion this book
would never have appeared

Preface

The present book has five objectives: *first,* to explore and test as much as possible the hypothesis that the origin of genera, families, and orders of flowering plants, as well as the origin of the angiosperms themselves, has resulted from the same kinds of processes that are responsible for evolutionary change and adaptive radiation at the level of populations and species; *second,* assuming that this hypothesis is correct, to determine and to explain the changes in emphasis that must be made when the well-known processes of population genetics and microevolution—mutation, genetic recombination, natural selection, reproductive isolation—are considered in the context of the long periods of time and the great changes in the earth's habitats that have taken place during and since the origin of the major groups of flowering plants; *third,* to describe and analyze as far as possible the kinds of adaptive radiations that could have given rise to the kinds of morphological and anatomical differences that are used by taxonomists to distinguish between families, orders, and other higher categories; *fourth,* to determine the kinds of ecological conditions and environmental changes that would have been most likely to have given rise to the recognized differences between major categories; *fifth,* to examine in this context current theories about the origin and early differentiation of angiosperms, and to develop a hypothesis about angiosperm origins that will be logical and consistent with the evolutionary principles developed in the first part of the book.

I recognize at the outset that these objectives are very difficult to attain, and might be considered impractical or even idealistic given the present state of our knowledge. The principal barrier is the extreme poverty of the fossil record of angiosperms. With respect to the all-important reproductive characteristics, the record is so fragmentary that it is essentially worthless. With respect to wood, leaves, and pollen, it fails to provide us with even a partial record of evolutionary continuity. Moreover, it is even more biased than is the record of animal phyla in favor of plants that grew in moist lowland habitats. Taken at face value, therefore, the fossil record of angiosperms can be more misleading than enlightening as a means of interpreting the major trends of their evolution. Given the present state of our knowledge, or ignorance, all that the botanist can do is to make logical inferences from such data as can be obtained from studying both living plants and the available fossils, and hope that at some future time facts will become available that will enable us to reaffirm or to reject the hypotheses that have been constructed on the basis of these inferences.

This being the case, the botanist who is accustomed to formulating only hypotheses that can be tested immediately by means of carefully planned experiments and thorough, systematic observations may justly ask: Why bother with these problems at all? Wouldn't it be better to forget about them for the present, and to direct all of our attention to problems and questions to which we can expect to get firm and clear answers in the reasonably near future? My answer to these questions is that they are well justified. I do not believe that anybody can quarrel with the point of view which maintains that all questions concerning the evolutionary history of organisms, in the absence of a significant fossil record, are pointless guessing games. The biologist who enters this field must resign himself to the fact that he can never achieve certainty. His end point must always be a judgment as to which of several hypotheses appears to be most plausible on the basis of presently available facts.

My excuses for entering this field are two. In the first place, speculations about angiosperm phylogeny have been made by scores of botanists during the past century. Hypotheses that embody these speculations are widely known and taught, and have acquired a credibility and illusion of certainty that is not justified by the facts and assumptions upon which they are based. For the most part, these hypotheses are based partly upon a philosophy of taxonomy in which separations and clear distinctions between major categories are given a higher priority than are evidences of continuities between them and partly upon an idealistic morphology in which anatomical and structural features, such as the various organs of the plant, are regarded as fundamental units rather than the products of complex, interacting processes of gene action. One of my major objectives is to find out to what extent both phylogeny itself and the methods of gaining new information about evolutionary history will be modified if botanists shift their major emphasis away from traditional taxonomy and idealistic morphology, and toward population and developmental genetics, comparative developmental physiology, and an ecological viewpoint that places primary emphasis upon interactions between populations and their environment.

Second, the problem of angiosperm origins has a fatal fascination for anyone who has spent a lifetime learning to recognize and to see the relationships between the diverse forms of plant life that exist on this earth today. A botanist who chooses this field does not expect to achieve either riches or scientific glory in terms of the international awards that go to leaders in more prestigious fields of science.

My motivation, like that of my colleagues and associates in this field, has been largely curiosity and wonder. Consequently, after fifty years of observation, experimentation, and reading hundreds of books and journal articles, I cannot escape the urge to put my speculative conclusions in print, however tentative they may be. I recognize that many of them may have to be modified as more facts become available. Nevertheless, I hope that my readers will judge them not on the basis of preconceived notions or the agreement or divergence of my ideas from already published hypotheses, but upon the facts and the logic of the inferences upon which each one of them individually is based. Most important of all, I hope that some of my readers, particularly those who belong to the younger generation of botanists, will be stimulated to search for new ways of observation, experimentation, and synthesis, by means of which both my hypotheses and those of other botanists may be tested and compared, so that a truer and more reliable picture may eventually emerge.

The plan of this book is as follows. In the first two chapters, I state my reasons for believing that the processes responsible for the origin of genera, families, and other higher categories are essentially the same as those that give rise to the divergence of populations and the origin of species. Transspecific evolution is basically a projection into the paleontological time scale of the population–environment interactions that are responsible for change at lower levels. The unifying concept that brings together all of evolution is adaptive radiation, which can be recognized at least to some extent at all levels; the divergence of races within a species, of species within a genus, of genera within a family, of families within an order, and finally of the two major subdivisions of angiosperms, dicotyledons and monocotyledons, from each other. Nevertheless, when the historical dimension of evolution is added to the analysis of processes that take place in contemporary populations, three major shifts in emphasis are required. First, the sequential or epigenetic nature of natural selection must be considered. I have expressed this as the hypothesis of evolutionary canalization, which is discussed in Chapter 2. Second, we must recognize that some evolutionary lines can be highly successful in a narrow range of environments and so may be restricted to a few widespread and common forms, while other lines become not only highly successful but also strongly diversified with respect to both the habitats that they occupy and their degree of morphological divergence. This leads to the distinction between evolutionary success and evolutionary opportunity. Finally, the environmental changes that take place over long periods of time inevitably cause numerous extinctions of forms that are unable to become adaptively modified to new conditions, and bring about radical alterations in the geographic and ecological distributions of those evolutionary lines that do survive. Extinction of intermediate forms is the principal reason for the clear delimitation of categories that is often possible in relatively ancient groups. Because of extinctions and of changes in distributional patterns, modern patterns of geographic distribution are poor and often misleading guides to the place of origin of a group. In addition, the extinct ancestors of a group can rarely if ever be reconstructed by extrapolation from the comparative morphology of contemporary forms.

In Chapters 3 to 5, the recognized and hypothetical adaptive nature of differences between angiosperm groups is reviewed. Because of the diffuse nature and sedentary condition of higher plants, three kinds of adaptive syndromes must be clearly distinguished from one another: (1) those associated with adaptation of the individual plant to its immediate surroundings; (2) those associated with pollination and fertilization; (3) those associated with seed development, dispersal, and seedling establishment. In general, though with numerous exceptions, syndromes of the first kind are the most significant in separating races and species that belong to groups having relatively unspecialized adaptations for pollination; those of the second kind are most significant in separating species, genera, and to a lesser extent families; the importance of the third kind relative to the other two increases as we pass from the race to the species, genus, family, order, and class.

Chapters 6 to 9 are designed to provide a transition from the discussion of factors that control phylogenetic differentiation to the construction of hypotheses about phylogenetic trends themselves. Chapter 6 is a review of those aspects of genetics and morphogenesis that are necessary for understanding how natural selection can act to produce evolutionary diversification and canalization at evolutionary levels. Chapter 7 presents the postulates that form the basis of the hypotheses about phylogeny that appear in the remaining chapters. Chapter 8 discusses ecological factors that affect diversification. The hypothesis is advanced that the present concentration of archaic and apparently primitive forms in moist tropical regions is due not to their origin under these conditions but to the relatively low rate of extinction that prevails there. In Chapter 9, individual examples are described which suggest that adaptive radiation has frequently occurred from common ancestors that lived in marginal regions to derivative forms, some of which are adapted to more mesic habitats, others to more xeric or arctic-alpine habitats. In addition, the hypothesis is advanced that new adaptive syndromes for exploiting more efficiently a particular environment are often and perhaps usually evolved in an indirect fashion, through two more successive adaptive radiations that include entrance into and return from regions having environments that are very different from the original one.

Chapters 10 through 13 contain hypotheses about the ancestry and origin of the angiosperms, the reasons for believing that no single modern group can be regarded as comparable to the original ancestors of the class, the major lines of initial divergence, and a special consideration of the origin and evolution of the monocotyledons. Chapter 14 looks to the future, and suggests kinds of research that might strengthen or overthrow the hypotheses that have been presented herein.

The first serious writing on the material encompassed in this book was the preparation of chapter 13 of my *Variation and Evolution in Plants* (1950). This was followed by a number of articles written during the 1950's, when I became increasingly aware of the possibility that adaptive radiation in angiosperms has not taken place chiefly from more mesic to more arid situations, and that the reductional series that have captured so much of the attention of comparative morphologists represent only half of the true picture. At the time of the Prather Lectures that I delivered at Harvard in 1958, I conducted seminars on the topic of macroevolution in angiosperms, and realized that at that time my ideas were not ready for a full-dress treatment.

Since then, the research done in my laboratory on genetics and morphogenesis has strengthened my ideas about the basis of morphological trends, and some modern anatomists, particularly Sherwin Carlquist, have in my opinion successfully challenged some of the previously held tenets of comparative morphology and anatomy. At the same time, additional research at the subspecific level has made more plausible than ever the hypothesis that all continued evolutionary trends are under the guidance of natural selection. Consequently, our understanding of evolution in the angiosperms will be brought into harmony with knowledge derived from studies of other kinds of organisms only if an adaptive basis for their major evolutionary trends can be found.

During the same period, three different authors, the late J. B. Hutchinson, A. L. Takhtajan, and Arthur Cronquist, have dealt with problems of angiosperm phylogeny and have reached conclusions somewhat divergent from mine. The publication of their books has been another reason for preparing the present volume, even though many of the ideas that it contains might have been more convincing if better evidence for them were available. This book should be regarded as a progress report rather than a final exposition of the nature of angiosperm evolution.

During the twenty-five years over which the ideas presented here were developed, I have had conversations with many botanists, a number of whom have reviewed either preliminary drafts of parts of the manuscript or final versions of some of the chapters. I am most grateful for their comments, both positive and negative. I should like to acknowledge particularly the advice and suggestions of Francisco Ayala, Daniel Axelrod, Herbert G. Baker, B. L. Burtt, Richard Cowan, Sherwin Carlquist, Arthur Cronquist, Elizabeth Cutter, Theodosius Dobzhansky, the late Carl Epling, F. Ehrendorfer, L. D. Gottlieb, E. R. Gifford,

John Madison, H. Meusel, H. Merxmüller, Philip Smith, Tony Swain, A. L. Takhtajan, R. Thorne, and G. L. Webster. Many thanks go to my wife, Barbara, for assistance with the illustrations as well as with editing and preparation of the literature list and index; to Joanne Lina for typing; and to Meg Hehner, Cheryl Hendrickson, and Cheri Hinckley for preparation of the illustrations. These acknowledgments should by no means be taken as indications that any of the botanists named agree with the ideas presented. I take full responsibility for them.

Contents

Tables

Figures

Part I
Factors That Determine Evolutionary Trends

1 / The Basic Processes of Evolution

The study of organic evolution has several aspects. The first is the analysis of evolutionary processes, particularly those that can be followed or reproduced experimentally. Next comes the description of differences between contemporary forms that are related to one another closely enough that the evolutionary processes which gave rise to them can be inferred without reference to events that occurred in the remote past. Finally, the evolutionist may attempt to reconstruct evolutionary history, either by interpreting the fossil record, by making inferences from comparisons between contemporary forms, or, preferably, by combining these two methods. This study of evolutionary history or trends is often designated macroevolution, phylogeny, or transspecific evolution. The last term is used in the present volume.

In recent years, largely through the development of population genetics, evolutionary studies of contemporary organisms have made spectacular advances (Dobzhansky 1970, Mayr 1970, Grant 1963, 1971). The differentiation of populations and the origin of some kinds of species, particularly polyploids, have been both followed in nature and simulated in models. Consequently, most evolutionists believe that five basic processes can explain evolutionary change at the level of populations and species: mutation (in the broadest sense, including chromosomal changes), genetic recombination, natural selection, chance fixation of genes, and reproductive isolation. Disagreements now exist chiefly with respect to the relative importance of these five processes.

The purpose of the present chapter is to review those aspects of the synthetic theory of evolutionary processes that are particularly relevant to problems of macroevolutionary trends in angiosperms.

The operation throughout the course of evolution of these five processes mentioned above is generally recognized. Nevertheless, some evolutionists believe that additional basic processes have operated. In particular, interactions between mutation, recombination, and selection appear, at first sight, to be insufficient to account for directional trends. This topic is taken up in the next chapter, in which a case is made for the sufficiency of the five processes mentioned above.

Even if one accepts the proposition that these five processes are sufficient to explain transspecific evolution, one can only conclude that, in considering trends of evolution above the species level, certain aspects of them must receive particular attention, as follows.

In considering genetic recombination with respect to higher animals and plants, primary emphasis must be placed upon the significance of *gene pools,* or stores of genetic variability that exist in all populations, even those of largely self-fertilizing species (Allard 1965, Allard and Kahler 1972). The richness of such gene pools, particularly with respect to variation that cannot be discovered by macroscopic examination of phenotypes, is becoming increasingly evident. Extensive chromosomal polymorphism, which has long been known in organisms such as *Drosophila* (Dobzhansky 1970) and *Trillium* (Stebbins 1971), has now been revealed in a variety of organisms, including mice and men (Sparks and Arakaki 1971, Waterbury 1972, Craig-Holmes and Shaw 1971). Biochemical polymorphism, especially for proteins that can be differentiated by means of electrophoresis, is far more extensive than any geneticist had imagined it to be before this technique came into general use (Dobzhansky 1970, Allard and Kahler 1972). It extends even to bradytelic species like the horseshoe crab (*Limulus*), which have remained phenotypically constant or nearly so for hundreds of millions of years (Selander *et al.* 1970).

Nevertheless, gene pools are finite. To imagine that a population could respond by natural selection to any kind of environment to which it might be subjected would be a serious error. Several different species have evolved ecotypes that are adapted to such extreme habitats as sea cliffs and mine workings (Bradshaw 1971), but an equal or larger number of species that have been exposed just as much to these conditions have not done so.

Consequently, if a particular change in the environment takes place, a population may respond to it either by evolving in a new direction, that is, by carrying out an adaptive shift, or by becoming extinct. Which response will take place depends largely upon the composition of the gene pool at the time when the environmental change appears.

The role of mutations in determining transspecific evolutionary change depends upon the basis of the entire spectrum of effects that they produce, upon the adaptive advantage or disadvantage of mutant individuals, and upon the role of the mutant gene in population-environment interactions. In a recent review of mutations in flowering plants, chiefly *Pisum, Vicia, Antirrhinum, Hordeum,* and other cultivated plants, Gottschalk (1971) has reemphasized the fact, long known to geneticists, that although the great majority of mutations reduce fitness, a small percentage of them do not, and a still smaller percentage may increase fitness under certain environmental conditions.

Nevertheless, a broader consideration of Gottschalk's examples shows that even those mutations that he believed to have evolutionary significance have rarely if ever been significant in determining evolutionary trends. As he himself recognizes, many of them produce morphological changes that are not duplicated at all among the thousands of living species of the families Leguminosae and Scrophulariaceae, to which, respectively, the genera *Pisum* and *Antirrhinum* belong. Others simulate the morphology of certain members of the families concerned, but in every case the natural counterparts are distantly related to the cultivated species in which the mutations have occurred.

For instance, one of the mutations in *Pisum* converts the androecium from the normal diadelphous condition of nine "fused" stamens and one that is distinct to that of ten distinct stamens. As Gottschalk points out, the latter condition exists in the tribe Sophoreae of the subfamily Papilionoideae, as well as in many genera belonging to the subfamilies Caesalpinoideae and Mimosoideae. In these groups it is probably the original condition, from which the diadelphous condition was derived. There is no indication that the reversion from the diadelphous condition to free stamens has ever been established in a natural species of Papilionoideae. Since thousands of diadelphous species possessing this condition have existed or have succeeded each other for millions of years, any mutation that occurs often enough that it can be detected in the relatively small number of plants that a single investigator can grow in his garden must have occurred thousands or even millions of times in one species or another during the evolutionary history of the diadelphous Leguminosae. Consequently, the failure of this reversion to have become established in a population, in spite of the number of times that it must have occurred, suggests strongly that this condition reduces fitness when it appears in the genetic background of the advanced diadelphous tribes.

The same can be said of the mutations and combinations of mutations that, in *Pisum,* alter the leaf structure from simply pinnate to doubly pinnate. Although doubly pinnate leaves are predominant among the species of the subfamilies Mimosoideae and Caesalpinoideae, they do not occur in any species of the tribe Vicieae or any others of the advanced tribes of herbaceous Papilionoideae. Evidently, for some reason that we do not understand, doubly pinnate leaves are sufficiently incompatible with the overall architecture of the plant in these herbaceous Leguminosae that they reduce fitness under natural conditions and cannot become established.

The same reasoning can be applied to the evolutionary significance, or lack of significance, of the mutation *hyperandra* of

Matthiola incana (family Cruciferae) that increases the stamen number from six to ten. Six stamens are a constant number throughout the Cruciferae, whereas flowers having ten stamens do not exist in any natural population of this very large family.

Of possibly greater significance is the *transcendens* mutation of *Antirrhinum majus,* which reduces the number of stamens from four to two and gives rise to a floral organization simulating that of the genus *Ixianthes,* a distantly related member of the family Scrophulariaceae, to which *Antirrhinum* belongs (Fig. 1-1). The same argument can be applied to this mutation as to those of *Pisum:* the absence of this condition from species of both *Antirrhinum* and the closely related, much larger genus *Linaria* suggests that the condition of two stamens is sufficiently incompatible with the overall floral architecture of these genera that it reduces fitness. Nevertheless, mutations of this kind must be occurring from time to time in various species of the family. In the ancestors of *Ixianthes,* a *transcendens*-like mutation apparently became associated with a genetic background and an environmental situation that was suitable for its establishment.

Antirrhinum, normal

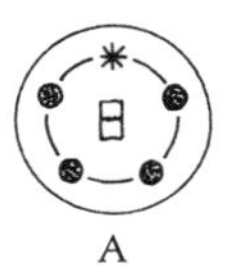

A

Antirrhinum, mut. transcendens

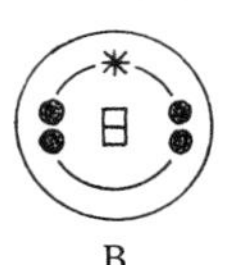

B

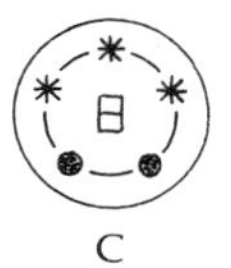

C

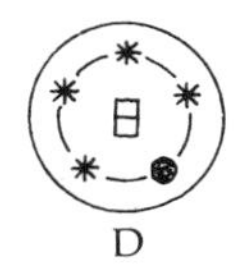

D

Ixianthes

E

Fig. 1-1. Floral diagrams, showing idealized cross-sections of the corolla (*outer circle*), stamens (*black disks*), and carpels (*central squares*) of the following phenotypes: (*A*) *Antirrhinum majus,* normal phenotype; (*B, C, D*) *A. majus,* phenotypes of three different *transcendens* mutants; (*E*) the genus *Ixianthes.* An asterisk (*) indicates the absence of an organ where it might be expected to occur. (From Gottschalk 1971.)

Gottschalk's (1971) review of mutations that affect floral structure is enlightening in another respect. He points out (p. 163) that the normal stability of carpel number in the great majority of species of flowering plants is counteracted by various mutations in a number of species belonging to families as different as the Rosaceae, Leguminosae, Solanaceae, and Pedaliaceae (*Sesamum*). All of the known mutations, however, alter carpel number in the direction of increase in the number of carpels. This is in striking contrast to the trends toward decrease in carpel number that are regarded by most taxonomists, with good justification, as having been the predominant ones in the evolution of genera and families of flowering plants. The most logical explanation of this apparently paradoxical situation is the hypothesis that the most common pressures of natural selection are unfavorable to the more frequent mutations that increase carpel number, whereas they permit the establishment of the much rarer mutations or gene combinations that reduce the number of carpels. On the other hand, artificial selection by man in cultivated species such as the strawberry (*Fragaria*), tomato (*Lycopersicon*), and sesame (*Sesamum*) has favored the more frequent mutations for increase in carpel number.

This review of mutations in flowering plants that have conspicuous effects on the phenotype suggests that the establishment of such mutations in populations, though by no means impossible, occurs so rarely that it must be regarded as exceptional rather than the usual basis for evolutionary trends. This point of view is supported by numerous analyses of variation in the segregating F_2 and later-generation progeny from hybridizations between natural races and species (Clausen and Hiesey 1958, Stebbins 1950, Grant 1963). These analyses have shown consistently that the morphological differences between such populations are governed by many genes, each of which by itself has a relatively small effect. With few exceptions, therefore, the mutations that have been significant in evolution have been those having relatively slight effects on the phenotype. Fisher's (1930) dictum, that the selective value of mutations is inversely related to the size of their effects, is now supported both by the results of hybridization experiments and by theoretical considerations based upon present knowledge about gene action. As maintained by Simpson (1953) on the basis of an analysis of the fossil record, and by Dobzhansky (1951) and Mayr (1963) on the basis of genetic comparisons between contemporary species of animals, major evolutionary changes come about through the action of gene recombination and natural selection in reorganizing the gene pool, the variability of which is constantly enriched by mutations that individually have small effects on the genotype.

The arguments of Fisher against the belief that "parallel mutations" of a particular kind can establish new characters in a population without the aid of natural selection are equally well supported by our present knowledge of the complexity of gene action. His mathematical calculations show that, without the aid of differential reproduction and isolation, the establishment of even a single mutant gene in all of the individuals of a population requires mutation rates far higher than any that have been recorded. The improbability of directed change by mutation pressure becomes even greater when we realize that any significant change in the gene-controlled reaction system of development requires harmonious adjustments of many different genes having diverse primary effects. Consequently, modern knowledge about mutation rates, sizes of populations, and the complexity of gene interaction in development renders any interpretation of directed evolution on the basis of mutation alone so improbable as to be untenable.

On the other hand, the complexity and integration of the developmental reaction system are such that internal selection pressures, based upon compatibility or incompatibility of new mutations with the existing system, can be a powerful directive force. I cannot follow Cronquist (1968) in making a sharp distinction between characters that are subject chiefly to external control by natural selection and those that are subject to internal control *and therefore* are governed only by mutation pressure. A substantial amount of internal control can be exerted by internal selection for a harmonious course of development. Moreover, the more precisely integrated and determinate is a particular pattern of development, the stronger will be the controlling action of internal selection. Such integrated patterns as the development of a floret in a species of Compositae are much more strongly buffered against the effects of mutation and external selection than is the branching pattern of a plant. One can, therefore, uphold the dictum of Cronquist, that the evolutionary constancy and hence the taxonomic value of a character are directly proportional to the extent to which its development and evolutionary modification is controlled by internal forces. The forces that operate, however, are internal selection rather than directed mutation pressure.

Hybridization as a Means of Expanding Gene Pools

Since the possibilities for new adaptive shifts depend upon the contents of finite gene pools, the importance of hybridization as a means of enriching gene pools must be considered. Of course, not all hybridizations will have this effect, since in many instances the hybrid products are sterile or otherwise less fit under any set of conditions. Nevertheless, many examples are now known of successful hybridizations between members of differently adapted populations, and of the proliferation of the resulting hybrid swarms in newly available habitats (Stebbins 1950, 1959*a*, Grant 1963, 1971). The catalytic effects of such hybridization must be considered with respect not only to the individual populations concerned, but also to their position in the biotic community that they occupy.

Whenever a major disturbance of the environment affects a number of adjacent plant communities and makes new ecological niches available for colonization, many different plant populations in the surrounding communities are potential candidates for occupying them. All of these populations, however, must become genetically modified to some degree. Before they can become adapted to the new conditions, they must undergo adaptive shifts of varying extent. Success will depend upon (*a*) a certain amount of preadaptation, thus minimizing the amount of genetic change that is required, and (*b*) a maximum diversity of the gene pool. Consequently, if two differently adapted populations can combine their gene pools by hybridization, this newly enriched gene pool or hybrid swarm will, *ipso facto*, have an advantage over nonhybrid gene pools. The only restriction on this potentiality is the fact that many hybrids produce progeny that are highly sterile or have conspicuously lowered fitness in other respects. This restriction is more often decisive in animals than in plants. The great success of hybrid swarms of flowering plants in colonizing new habitats has now been demonstrated by a large number of examples, as has already been mentioned.

The Role of Chance in Determining Evolutionary Trends

The extent to which chance determines evolutionary direction, which has long been a subject of violent controversy among evolutionists concerned with micro-evolutionary processes, is now being hotly debated with respect to its influence on major evolutionary trends. On this subject, the spectrum of contemporary opinion runs the entire gamut from complete acceptance to violent rejection of chance as a factor in evolution. The most recent advocates of evolution by "random-walk" or chance establishment of mutations having no adaptive value are either biochemically oriented evolutionists who are impressed with the high degree of "degeneracy" or synonymy of the genetic code (King and Jukes 1969) or population geneticists who work primarily with mathematical models, and who point out that, given the very large number of probably neutral mutations that must occur during the evolutionary history of a given species, some of these must become established by chance alone, even if populations remain relatively large (Kimura 1968, Kimura and Ohta 1972). These biologists have had minimal experience with actual populations living under natural conditions, and this only second or third hand.

At the other end of the spectrum are certain zoologists whose primary concern has been with natural populations of animals. From his experience with small insular populations of Lepidoptera, Ford (1964) has concluded that even repeated reduction of a population to a small size is insufficient to bring about significant changes in gene frequencies, unless the reduction is accompanied by significant changes in the environment. After reviewing several examples of animal populations in which significant effects of chance fixation have been claimed, Mayr (1970:128) concludes that "it would be entirely misleading to say that chance directs the course of evolution."

This book is not an appropriate place to discuss the arguments presented by these supporters of diametrically opposite points of view. For a recent review of them, the reader may consult the symposium volume containing papers delivered at Berkeley in 1971 (LeCam, Neyman, and Scott 1972). I favor an intermediate point of view, corresponding in general to that of Dobzhansky (1970) and to a certain extent to that of Mayr (1970). Mutations having very small effects on the phenotype, particularly those from one synonymous DNA codon to another, have almost certainly been established occasionally in populations by chance alone or in association with other phenomena, such as genetic linkage with adaptive genes or gene complexes. For the most part, however, these chance establishments have constituted a kind of "evolutionary noise" which has had little effect on the general direction of evolutionary trends. On the other hand, during those critical situations when populations are exploiting new environments and several "evolutionary strategies" for acquiring new adaptations are possible, chance events may well play an important role in deciding which particular method of adaptation will become established. As Mayr (1970:128) has suggested, chance may "jar" evolution "at frequent intervals and may occasionally be responsible for a jump to another track."

Selection and the Population-Environment Interaction

Of particular significance with respect to transspecific evolution is the stability or the alteration of the environment. The three phenomena of stabilizing selection, directional selection, and diversifying selection (Dobzhansky 1970), sometimes called "disruptive selection," are now widely recognized as distinctive population–environment interactions. Stabilizing selection is probably the commonest of these. It is the result of environmental constancy with respect to the adaptive needs of the population. This last phrase is particularly significant. Simpson (1953) has pointed out that in animals "bradytelic" species or evolutionary lines, which change very little or not at all over millions of years, have two important characteristics. First, they inhabit biotic communities that are relatively constant, such as the great forest belts, the intertidal zones of the seashore, and the vast expanses of the open ocean. Second, they exploit their environment in such a general way that they are unaffected by many of the changes that it undergoes. The opossum (*Didelphys*), which is omnivorous and relies for defense and survival on the death-feigning reaction and on high reproductivity, has remained virtually constant while occupying during the Tertiary period the same deciduous forests within which mammals having more specialized diets and defense mechanisms, particularly the rodents, have undergone extensive and relatively rapid evolution.

The same situation exists in plants. In the temperate forest belts, genera of wind-pollinated trees such as *Platanus* and various conifers have remained virtually constant for as much as a hundred million years, since the middle of the Cretaceous Period (Axelrod 1960), while herbaceous groups such as Caryophyllaceae, Compositae, Scrophulariaceae, and Gramineae have been actively proliferating new species and genera. The same situation exists in tropical forests when trees belonging to such families as the Magnoliaceae and Lauraceae are compared with undershrubs in genera of Melastomaceae, Rubiaceae, and other families, and particularly with animal-pollinated epiphytes or cliff-inhabiting plants such as Orchidaceae, Bromeliaceae, and Gesneriaceae. Depending upon the nature of their adaptations, the same overall environment can have a stabilizing effect upon some evolutionary lines, and can promote rapid change in others.

The prevalence of stabilizing selection is due in part to the fact that populations rarely if ever evolve radically different and more efficient adaptations to a particular habitat while remaining constantly in that habitat. An evolutionary line is more likely to evolve the optimum adaptation that is possible on the basis of its original gene pool plus mutational changes that can take place with relatively little modification of the overall adaptive mode. In Chapter 9 evidence is presented to suggest that radically different and more efficient ways of becoming adapted to a particular habitat are acquired in an indirect fashion. The evolutionary line first becomes adapted to a different environment, in which the environmental challenge in terms of selection pressures is particularly strong. Under those more stringent conditions it may acquire new adaptive properties. If these are sufficiently general, affecting such widely important properties as seed dispersal, establishment of seedlings, and protection from animal predators, they may later become modified and supplemented so that new adaptations to various environments, including the original one, are evolved.

Consequently, evolutionists who are interested in transspecific evolutionary trends must pay particular attention to *directional selection,* which takes place when populations are responding to a continuously changing environment, and to *diversifying selection,* which accompanies the diversification of a previously homogeneous environment into many new ecological niches. In Chapter 3, examples are presented which suggest that many, and perhaps most, major groups of angiosperms, such as genera and families, have originated in marginal habitats or ecotones, where habitat diversity and consequent diversifying selection are predominant. They have then diversified by means of adaptive radiation, involving directional selection of many kinds, each of which has been characteristic of a particular line. Finally, the end products of each line, living in habitats that are more extreme with respect to either favorable conditions or severity, have evolved more slowly or have become completely stabilized, and through extinction of intermediate lines have become progressively more distinct from one another. The usual history of a particular evolutionary line is, therefore, its origin in association with diversifying selection, followed by a longer or shorter period of directional selection, and ending with stabilizing selection at the maximum level of fitness possible for the gene pool in question. The subsequent fate of the line may be either extinction, long-time persistence or bradytely, or entrance either immediately or eventually into a new cycle of advance through diversifying selection followed by directional selection. Which of these three fates will come to the evolutionary line will depend upon the kinds of environments, whether constant or changing, to which it is exposed.

The Significance of Reproductive Isolation for Transspecific Evolution

As was mentioned some time ago (Stebbins 1950), the principal importance of reproductive isolation with respect to transspecific evolution is that it permits related populations to occupy the same habitat without exchanging genes, and so to exploit that habitat in different ways. This is, necessarily, the first step of divergent evolution. It cannot be achieved unless the isolation is virtually complete and exists with respect to the entire genotypes of the sympatric populations. Failure to recognize this fact has recently caused some authors (Ehrlich and Raven 1969) to minimize unjustifiably the evolutionary significance of reproductive isolation. To be sure, sympatric populations, or even the same population, may become strongly differentiated by natural selection in response to habitat diversity, and may maintain this differentiation in spite of extensive and constant possibility for gene exchange. Good examples are hybrid populations of oaks (*Quercus*) in the mountains of southern California (Benson, Phillips, and Wilder 1967) and British populations of grasses that live on and adjacent to soils having a high content of toxic minerals (McNeilly and Antonovics 1968, Antonovics 1971; Fig. 1-2). Nevertheless, as the diagrams presented by Antonovics clearly show, the differential gene frequencies that are maintained in such situations exist only with respect to those genes that are directly connected with the adaptation in question. With respect to other genes, the ecologically segregated populations or portions of populations may show no significant difference. The complete inhibition of

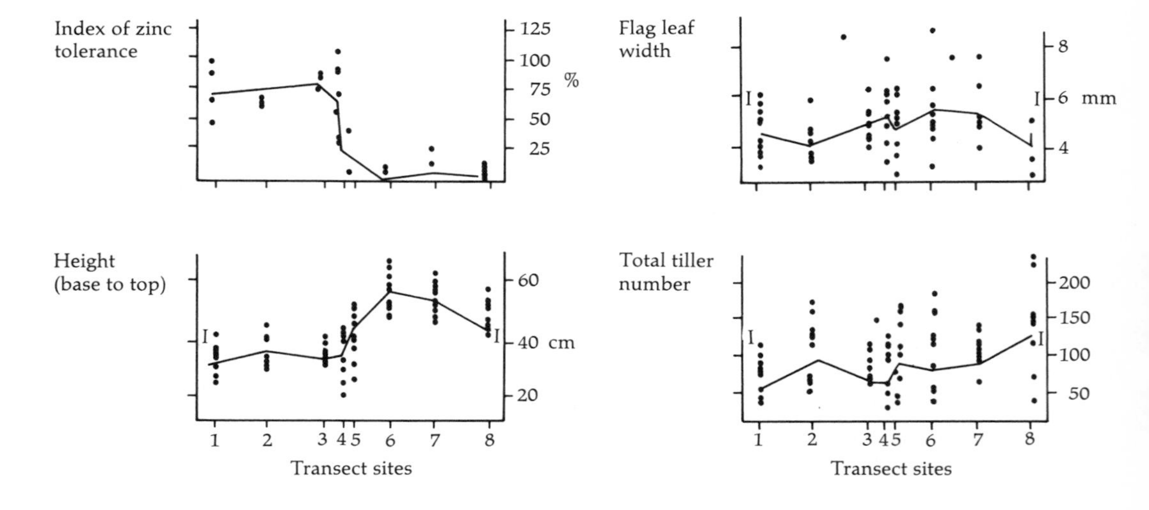

Fig. 1-2. Diagram showing differences with respect to certain phenotypic characters in populations of *Agrostis* found on either side of the boundary of an area containing soil having a high content of copper. Note that the frequency of copper tolerance changes abruptly at the boundary, and that the change in plant height is nearly as abrupt, but that with respect to leaf and tiller characters the transition is either gradual or insignificant. (From Antonovics 1971.)

gene exchange that is necessary for divergent evolution is accomplished only by means of reproductive isolating mechanisms based upon internal divergences in the genotypes themselves, particularly those that lead to hybrid inviability or sterility. The almost universal inviability or sterility of hybrids between species that belong to different genera or even subgenera is ample testimony in favor of this generalization.

The Principle of Genetical Uniformitarianism

In the early and middle part of the 19th century, the interpretation of the sequence of fossils that was being discovered was much aided by the development of the concept or principle of geological uniformitarianism by Hutton, Lyell, and others (Eiseley 1958, Simpson 1970). This principle emphasized an obvious deduction: the processes that give rise to the deposition of geological strata, as well as to the deposition and preservation of fossils, have been the same and have taken place in the same way throughout the history of the earth. This principle enabled geologists to interpret the past in terms of processes that are still taking place.

Now that most biological evolutionists agree that mutation, genetic recombination, and natural selection are the dominant processes governing contemporary evolution at the level of populations and species, and that counterparts of modern populations have existed ever since life arose on the earth, a basis exists for establishing a principle of uniformitarianism similar to that of geologists. It has been called the principle or hypothesis of *genetical uniformitarianism* (Stebbins 1972). It states that the processes of evolution—mutation and genetic recombination as sources of variation, natural selection as a guiding factor, and reproductive isolation as a basis of diversification and canalization—have operated in the past essentially as they do now, even though the genotypes and phenotypes upon which they operated, as well as the environmental conditions that created selection pressures, were different. In other words, processes of evolution are constant; phenotypes, genotypes, and environmental conditions are variable and specific to certain places and certain times in the earth's history.

This hypothesis is implicit in the research of most modern evolutionists, both those whose research is at the level of populations and species, such as Dobzhansky (1970) and Mayr (1963), and those whose major contributions have been toward the understanding of transspecific evolutionary trends, such as Simpson (1953). On the other hand, many botanists have held to a philosophy that places great emphasis upon the morphology and geographic distribution of modern phenotypes that appear to them to be primitive and perhaps even ancestral, and pays little attention to evolutionary processes as understood by experimental evolutionists (Cronquist 1968, Takhtajan 1969, 1970, Smith 1970).

The principle of genetical uniformitarianism has five corollaries:

1. Adaptive radiation, which is the primary basis of diversification among contemporary populations and species, was

also the primary basis of diversification among those populations that, in the remote past, gave rise to the distinct genera and families of modern times.

2. The origin of major categories, such as genera and families, does not require the evolution of distinctive characters that are qualitatively different from those that separate at least some contemporary populations and species.

3. The initial divergence that led to the origin of major categories took place under conditions similar to those that promote maximal diversification of populations and speciation in the modern world.

4. Archaic forms, which among modern groups resemble most closely the putative ancestors of major taxonomic categories, are most likely to be preserved under conditions that are the least favorable for the origin of new, more efficient genotypes that would compete successfully with them and drive them out. Consequently, biologists who wish to interpret evolutionary phylogenies must distinguish clearly between biotic communities that function as evolutionary "cradles," in which new adaptive complexes arise, and those that act as "museums," in which archaic forms are preserved.

This fourth corollary requires further explanation. It does not imply that a particular community must be in its entirety either a "cradle" or a "museum." Many, and probably most, plant communities are "cradles" for some of their species groups and "museums" for others. The tropical rain forest is probably a "museum" for most of the woody plants that it contains, since these plants respond chiefly to selective pressures determined by climate, soil, and competition indiscriminately with a great diversity of other woody species. All of these factors have been operating in very much the same fashion in these habitats for millions of years. On the other hand, the same tropical forest is a "laboratory" for the diversification of species in such groups as epiphytic orchids, in which evolutionary diversification occurs chiefly in response to specific pollinators, such as insects or birds. Similarly, the semiarid chaparral communities of California are "laboratories" for most of the genera, both woody and herbaceous, that occupy them, but are "museums" for certain taxonomically isolated and morphologically stable genera, such as *Crossosoma, Carpenteria, Simmondsia,* and *Lyonothamnus.* The importance of this fourth corollary lies in its bearing upon the fifth corollary, which is as follows:

5. Transspecific evolution is not best studied by focusing primary attention upon relictual modern phenotypes that are believed on morphological grounds to be primitive but certainly cannot be regarded as ancestral to any modern forms. Much more profitable is the analysis of differences between modern forms, regardless of their supposed phylogenetic position, that are analogous to the differences between major taxonomic groups.

The discussions in this book are based largely upon the principle of genetical uniformitarianism and its five corollaries that have just been mentioned. The next two chapters, in particular, are devoted to an elaboration of these corollaries.

Summary

The five basic processes of evolution—mutation, genetic recombination, natural selection, chance fixation of genes, and reproductive isolation—have been briefly reviewed, with emphasis upon those aspects of them that are relevant to an analysis of evolutionary trends in flowering plants. One factor that determines such trends is the content of rich but finite gene pools. Individual mutations having large effects contribute only occasionally to evolutionary trends, but a review of their effects provides an insight into what kinds of changes are most likely to occur at the level of the gene. Since the guidance of evolutionary trends by mutation pressure alone is very unlikely, the adaptive basis of such trends must be sought. Hybridization, as well as the effects of chance fixation, may occasionally, but not regularly, play significant roles. Given an environment that is constant with respect to the adaptive needs of the population, stabilizing selection will bring about evolutionary constancy of a line for millions of years. Evolutionary change is based upon directional and diversifying selection, which takes place in association with a changing environment. The significance of reproductive isolation for transspecific evolution is that it permits related populations to exist in the same environment and to exploit this environment differently without disruption of their integrated gene complexes through gene flow.

As a guide to extrapolation from populations and species to the historical phase of evolution, the principle of genetical uniformitarianism is proposed. It states that the basic processes of evolution are constant; phenotypes, genotypes, environmental conditions, and patterns of geographic distribution are variable. Five corollaries to this principle have served as major guidelines for arguments to be presented in later chapters.

2 / Evolutionary Processes and the Origin of Higher Categories

Our understanding of evolutionary trends on the basis of the processes responsible for them depends primarily on the kind of answer that we give to the following question: Can we account for such trends by postulating the continuation, through long periods of time, of the processes that are responsible for the origin of races and species, or do we need to search for additional, as yet unknown, processes or forces?

Unfortunately, a direct answer to this question cannot be given. The trends involved have taken place over such long periods of time that evidence concerning them must be derived from comparative observations of modern and fossil forms, when the latter are available. The inferences made from such comparisons cannot be tested experimentally, as is often possible when one is dealing with processes at the level of races and species. Indirect evidence must be of three kinds: (1) that which enables the evolutionist to infer the degree of similarity between the actual course of events at transspecific and subspecific levels; (2) that which permits the evolutionist to estimate the probability of the hypothesis that particular transspecific evolutionary trends have been guided and directed by the effects of natural selection; and (3) that which establishes the probability or improbability of alternative explanations, particularly internal direction of mutations so as to produce "orthogenetic" trends, as well as the role of chance. The first seven chapters of this book are devoted to a consideration of these three kinds of evidence.

With respect to the similarity between subspecific and transspecific evolutionary levels, the first question to ask is: Are there any individual character differences, either morphological, physiological, or biochemical, that separate higher categories from one another, but that never vary significantly at the level of species, races, or populations? Many systematists might give a positive answer to this question. If one's attention is confined to a single evolutionary line, one might be very likely to do so. This is because the characters that separate a particular genus or family from its relatives are usually different from those that distinguish species or races within the family concerned. In the family Compositae, for instance, the species of a genus are usually distinguished on the basis of either vegetative characteristics or relatively trivial characters of the flowers and seeds, such as size and hairiness of parts, presence or absence of ray flowers, color of rays, and so on. The distinctions between genera and tribes, on the other hand, are based almost entirely upon certain reproductive characters, such as the pappus, the anthers and style branches, and the distribution within the capitulum of hermaphroditic ("perfect"), staminate, and pistillate florets. Finally, the Compositae are separated from other families on the basis of still different reproductive characters, such as whether the anthers are united or separate, whether the calyx lobes are pappuslike or foliaceous, and the capitate inflorescence. Considering only the Compositae and their relationships, therefore, one might be able to make a case for distinctive familial, generic, and specific characters. This has, in fact, been done by some botanists, such as Anderson (1937) with reference to the Amaryllidaceae.

When more information is acquired about such characters, however, this position becomes untenable. In the first place, the rule that in any one group the diagnostic characters of genera and families are different from those that separate species and subspecies has many exceptions. Occasional genera of Compositae contain species that differ from one another with respect to pappus, anthers, style branches, and sex of florets in essentially the same way as genera differ from one another. Moreover, within the entire class of angiosperms, every diagnostic character that has been or can be used to separate families or orders may occasionally serve to separate different species of the same genus. Table 2-1 illustrates this point. In order to keep this table relatively simple, only one out of numerous possible examples was chosen to represent each category.

One character deserves special comment in this connection. The number of cotyledons in the seedling is the name-giving diagnostic character that separates the two principal subclasses of angiosperms, monocotyledons and dicotyledons, from each other. Nevertheless, variations in cotyledon number occur occasionally among different species of the same genus. For instance, most of the species of *Claytonia* (Portulacaceae) have seedlings with two normal cotyledons; yet *C. virginica* has only one cotyledon (Haccius 1954). Several other examples exist of species having only one cotyledon but belonging to normal dicotyledonous genera (Haskell 1954).

In the other direction, species of the genus *Pittosporum* from Asia (*P. tobira*) or Australia (*P. undulatum*) have embryos possessing two normal cotyledons (unpublished personal observations), but in several species of New Zealand—*P. crassifolium* (Lubbock 1892:200–204), *P. tenuifolium* (unpublished personal observations), *P. rigidum* (Cockayne 1899), *P. anomalum, P. divaricatum, P. crassicaule, P. lineare* (Laing and Gourlay 1935)—three to four cotyledons are always present (Fig. 2-1).

The same independence of taxonomic rank exists with respect to ecological, physiological, and biochemical characteristics. Succulence, as an adaptation to deserts or dry, rocky situations, can vary at the level of species within a genus, as in *Euphorbia, Oxalis, Senecio, Lewisia* (Portulacaceae), and many other genera; it can be a diagnostic character to separate the genera of a family, as in Portulacaceae and Asclepiadaceae; or it can be one of the principle diagnostic characters of a family, as in Cactaceae. Adaptation to life under water was probably normal for the

Table 2-1. Distribution of character differences at various hierarchial levels.

Character difference	Diagnostic at species level	Diagnostic at genus level	Diagnostic at family or order level
Woody vs. herbaceous growth habit	*Mimulus longiflorus* vs. *M. clevelandii*	*Zanthorhiza* vs. *Coptis* (Ranunculaceae)	Myrsinaceae vs. Primulaceae
Compound vs. simple leaves	*Ranunculus repens* vs. *R. cymbalaria*	*Eschscholtzia* vs. *Dendromecon* (Papaveraceae)	Oxalidaceae vs. Linaceae
Capitate vs. umbellate or other kind of inflorescence	*Arenaria congesta* vs. *A. macradenia*	*Trifolium* vs. *Melilotus* (Leguminosae)	Dipsacaceae vs. Valerianaceae; Asterales (Compositae) vs. Campanulales
Bilateral (zygomorphic) vs. radial (actinomorphic)	*Saxifraga sarmentosa* vs. other *Saxifraga* spp.	*Tolmiea* vs. *Heuchera* (Saxifragaceae)	Violaceae vs. Cistaceae
Foliaceous vs. awnlike or pappuslike calyx lobes	*Marrubium Alysson* vs. *M. vulgare*	*Dracocephalum* vs. *Galeopsis* (Labiatae)	Dipsacaceae vs. Caprifoliaceae; Asterales vs. Campanulales
Tetramerous vs. pentamerous perianth	*Rhamnus crocea* vs. *R. californica*	*Ludvigia* vs. *Jussiaea* (Onagraceae)	Cruciferae vs. Moringaceae
Corolla lobes separate vs. united	*Crassula Zeyheriana* vs. *C. glomerata*	*Monotropa* vs. *Pterospora* (Monotropaceae)	Pyrolaceae vs. Ericaceae
Perianth biseriate vs. uniseriate	*Sagina nodosa* vs. *S. decumbens*	*Agrimonia* vs. *Sanguisorba* (Rosaceae)	Portulacaceae vs. Chenopodiaceae
Carpels separate vs. united	*Saxifraga Lyallii* vs. *S. arguta*	*Delphinium* vs. *Nigella*	Dilleniaceae vs. Actinidiaceae
Ovary superior (hypogynous) vs. inferior (epigynous)	*Saxifraga umbrosa* vs. *S. caespitosa*	*Tetraplasandra* vs. other Araliaceae	Loganiaceae vs. Rubiaceae
Placentae axial vs. parietal	*Hypericum perforatum* vs. *H. anagalloides*	*Boykinia* vs. *Heuchera* (Saxifragaceae)	Theaceae vs. Cistaceae
Ovules numerous vs. solitary	*Medicago sativa* vs. *M. lupulina*	*Spiraea* vs. *Holodiscus* (Rosaceae)	Campanulales vs. Asterales

first multicellular plants, and the conquest of the land was a notable series of accomplishments, carried out separately by a number of different evolutionary lines. Nevertheless, readaptation to a completely submerged aquatic existence has taken place several times during the evolutionary history of angiosperms. Its products may be certain species of a predominantly terrestrial genus (*Ranunculus aquatilis*), distinct genera of a terrestrial family (*Subularia aquatica*), or distinct families (Najadaceae). The same is true of adaptations to a saprophytic or a parasitic existence. They may be developed at the level of species or even races (*Pyrola aphylla*), of genera in a family (various Orchidaceae and Scrophulariaceae), or of distinct families (Orobanchaceae, Rafflesiaceae).

A most interesting biochemical characteristic having this kind of distribution is the C-4 dicarboxylic acid pathway of photosynthetic CO_2 fixation (Hatch, Slack, and Johnson 1967, Laetsch 1968). This pathway is found regularly in certain chloroplasts of all investigated species of the tribe Andropogoneae of the Gramineae; it is apparently always present in the family Amaranthaceae; but in the grass genus *Panicum*, and especially *Atriplex* in the Chenopodiaceae (Björkman *et al.* 1971), closely related species exist, some of which have the C-4 dicarboxylic acid pathway, and others have only the usual Calvin pathway that has 3-phosphoglycerate as a major early product.

Still further evidence that the characters usually employed to distinguish between families and orders are not intrinsically distinctive "family" or "order

Fig. 2-1. Seedlings of *Pittosporum crassifolium* and *P. undulatum*, showing differences in the numbers of cotyledons. (From specimens collected under cultivated shrubs in San Francisco and Berkeley, California.)

Pittosporum crassifolium

Pittosporum undulatum

characteristics" comes from observations and experiments that have revealed genetic differences with respect to some of them within individual populations. An example is the research of C. A. Huether (1968, 1969) on the annual species *Linanthus androsaceus* of the Phlox family (Polemoniaceae). In this family, which contains 16 genera and hundreds of species, the flowers are pentamerous, with very few exceptions. Nevertheless, in every population of most of the annual species of *Linanthus* and *Gilia* in California a small proportion of plants, usually ranging from 0.5 to 4 percent, contain one or more flowers that deviate from the pentamerous pattern toward either more (6–7) or fewer (4) corolla lobes. These deviants are not merely due to developmental accidents occurring in otherwise normal genotypes. They are, on the contrary, the rare visible signs of an extensive pool of hidden genetic variability. Huether demonstrated this fact by using abnormal plants as bases for five generations of selection, for either increased or decreased numbers of corolla lobes. In the most favorable "selection-up" line, all of the plants in the fifth generation bore at least some flowers having more than 5 corolla lobes, and in some of these plants the mean number of lobes per flower was between 8 and 9. Selection for decreased numbers of corolla lobes was less successful, but significant results in this direction were also obtained (Fig. 2-2).

Fig. 2-2. Top row: A normal pentamerous flower of *Linanthus androsaceus,* and extreme variants of corolla lobe number found in natural populations. *Bottom four rows:* Extreme variants produced by five generations of selection, plus exposure of some of the genotypes to an abnormal, "stress" environment. (From Huether 1968.)

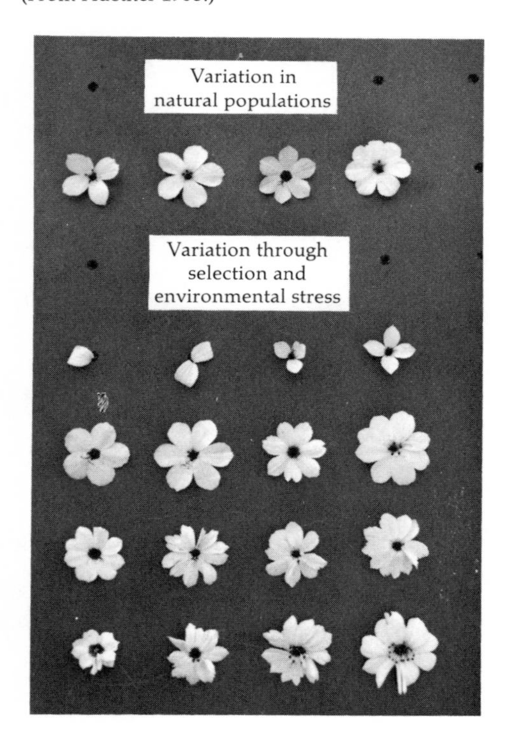

These results show that the normal pentamerous condition in *Linanthus androsaceus* and probably other species of Polemoniaceae does not result from fixed homozygosity with respect to all gene loci that might contribute to variation in corolla form. On the contrary, it represents a genic balance or equilibrium. Its constancy is best explained by the hypothesis that, for some unknown reason, it is adaptive and is maintained by natural selection. It appears to be an example of genetic homeostasis, in the sense of Lerner (1954). The adaptiveness may not, however, depend upon the pentamerous condition *per se,* but upon certain developmental processes that contribute to this condition. This conclusion was reached from the following facts. In the first place, careful field observations of insects (Bombylidae and Syrphidae) pollinating the flowers of *L. androsaceus* failed to reveal any discrimination on their part against flowers having 4 or 6 corolla lobes. Since the only conceivable function of a corolla is attraction of pollinators, the low frequency of these deviations cannot be explained by assuming that tetramerous or hexamerous corollas are disadvantageous by themselves. Secondly, plants of the selection-up line that had the highest mean lobe number were all relatively weak and slow growing, even before they began to flower. Deviation with respect to corolla-lobe number was apparently associated with weaker and slower growth of the plant as a whole. Apparently, the origin of a vigorous and successful population of *L. androsaceus* having more or fewer than 5 corolla lobes would require a more complete reorganization of the gene pool than was possible by means of the relatively simple selection procedure used by Huether. However, there is good reason to believe that such reorganization would be possible, given the large gene pool that exists in the widespread and common species *L. androsaceus* plus the related "species" or subspecies with which it can exchange genes, *L. parviflorus* and *L. bicolor.*

An experiment which shows that character differences of transspecific evolutionary importance can be established and stabilized by introducing modifier genes via hybridization with a different population is that of Stubbe (1963) on polycotyly in *Antirrhinum majus.* Previous to his experiment, a gene had long been known which, when homozygous, gives rise to a low percentage of plants having three cotyledons rather than two. Many attempts had been made to increase the penetrance of this gene, that is, to increase the proportion of tricotylous plants in lines homozygous for it. Neither selection nor environmental manipulation had any appreciable effect until Stubbe crossed tricotylous individuals with the species *A. tortuosum,* which is so closely related to *A. majus* that the hybrid between them is fertile in both the F_1 and F_2 generations. From tricotylous segregants in the F_2 progeny of this cross, Stubbe selected for four generations progenies having increasing proportions of tricotylous plants. He finally obtained a true-breeding line in which 97 percent of the plants were tricotylous.

This experiment shows that drastic reorganizations of the developmental pattern, which produce character differences of fundamental significance, can be produced by strong selection pressures acting upon a large store of genetic variability. The presence of a "switch gene," which by itself has conspicuous effects on the phenotype, probably aids greatly in this process, but it must not be regarded as essential.

These two examples are by no means isolated, unusual situations. Observations on natural populations of various species belonging to the families Polemoniaceae, Boraginaceae, Plantaginaceae, Rubiaceae, Rosaceae, Rhamnaceae, and Ranunculaceae have revealed the presence of occasional deviants from the predominant floral symmetry, and there is every reason to believe that selection in these populations would produce the same results as those of Huether on *Linanthus androsaceus*. Occasional tricotylous individuals have been recognized in garden cultures of species belonging to more than 20 genera in various families (Haskell 1954). In the genus *Brassica*, Holtorp (1944), selecting for increased frequency of tricotyly, obtained a few tetracotylous individuals in the selected lines. Haskell concluded that the persistence of dicotyly in most dicotyledons, in spite of the frequent presence of genes that would make possible a deviation from this condition, is due to the fact that such genes invariably bring about an imbalance of development that renders their bearers less adaptive than are genotypes which are buffered for the normal dicotylous condition.

The population structure suggested by these observations and experiments is one that would make highly probable the origin of the kind of variation pattern that has been observed. Certain characters, such as cotyledon number, floral symmetry, and the structure of the gynoecium, are constant, in spite of hidden genetic variation, because their development is highly canalized, in the sense of Waddington (1962). Successful deviations from the mode are possible, but they are highly improbable, since they require extensive reorganization of the genotype through the establishment of a combination of relatively uncommon genes. From time to time, however, such reorganizations do occur. If they then give rise to only one or a few new populations, which in other respects are closely similar to the parental stock, these products are regarded by systematists and evolutionists as nothing more than unusual species belonging to recognized, familiar genera. If, however, these deviations occurred a long time ago, and then gave rise to successful and diversifying evolutionary lines, a higher rank is accorded to their modern derivatives.

On the basis of these considerations, a tentative answer to the question posed at the beginning of this section is given in the form of the following working hypothesis. The origin of higher categories of flowering plants involved major adaptive shifts, which were based primarily upon the same kinds of processes that can be studied at the level of races and species. Nevertheless, a change in emphasis has been of basic importance. The characters and combinations of characters that have figured most prominently in these shifts are highly canalized because they are produced by coordinated and integrated interactions between many different genes. The greater is this canalization and integration, the more constant is the character or character combination, and the greater the probability that it will serve to distinguish higher categories. To understand prolonged evolutionary trends, therefore, we need to consider the adaptive value of genes or gene combinations with reference to their interaction with both the external environment and other genes that are present in the population. At every level of the taxonomic-evolutionary hierarchy, natural selection has included both an external phase, consisting of the interaction between genotypes and the surrounding environment, and an internal phase, consisting of interactions during development between different genes present in the same individual (Whyte 1965, Stebbins 1969). For the kinds of adaptive shifts that have given rise to higher categories, the internal phase of selection has been more important relative to its external phase than it has in the origin of the great majority of races or species within a genus.

This hypothesis may be summarized and characterized as the hypothesis of *evolutionary canalization and decanalization through internal selective pressures.* It represents the principal shift in emphasis that distinguished the origin of higher categories from the more usual processes of differentiation between populations and the origin of species. Before considering the phenomena of adaptation, therefore, we need to explore more thoroughly the nature and basis of evolutionary canalization. The term "canalization" has been chosen deliberately in order to emphasize the analogy between evolutionary canalization, as characterized in the next section, and developmental canalization, as recognized by Waddington. The canalization of development in a complex animal or plant and the course of evolution in a population that is undergoing a major adaptive shift under the guidance of directional selection have the following properties in common. Both phenomena are characterized by *epigenesis:* the nature of each successive change is strongly determined by changes that have occurred previously. Moreover, the canalization of development or evolution results from complex interactions between different genes, at the level in one case of the development of the individual, and in the other of adaptive shifts affecting populations. Evolutionary canalization can be defined as the tendency for populations to respond adaptively to new environments in ways that are determined by characteristics acquired as a result of previous adaptive radiations.

The Basis of Evolutionary Canalization

Evolutionary canalization depends upon three other principles: *selective inertia, conservation of organization,* and *adaptive modification along the lines of least resistance.* Each of these principles will be discussed in turn.

Selective inertia

This principle states that the intensity of selection which is required to establish a new adaptive gene combination is many times greater than that required to maintain or modify an adaptive mechanism, once it has been acquired. In other words, progressive diversifying selection requires much higher selection pressures than stabilizing selection.

A simple mathematical model will serve to illustrate this point. Let us assume that a particular adaptive state is determined by alleles at each of five independently segregating loci, the alleles interacting epistatically with each other to produce the adaptation. The adaptive alleles are assumed to be dominant with respect to their epistatic interaction, so that only one desirable allele need be present at each locus.

In accordance with the fact that most cross-fertilizing populations contain two or more alleles at most of their gene loci, we shall assume that at each locus a second allele occurs at a low frequency. The pairs of alleles may be designated a_1a_2, b_1b_2, c_1c_2, d_1d_2, and e_1e_2. For convenience of calculation, the frequencies of alleles a_1, . . . , e_1 are assumed to be $p = 0.95$ for each allele, so that a_2, . . . , e_2 each are present at a frequency $q = 0.05$. If these conditions exist, the fraction of individuals that would constitute "breaks" of the adaptive combination, that is, that would be homozygous for a divergent allele at one of the five loci, namely, a_2a_2, b_2b_2, c_2c_2, d_2d_2, or e_2e_2, would be $5q^2$, or 0.0125. Hence elimination by stabilizing selection of between 1 and 2 percent of the population every generation would be enough to maintain the adaptive combination.

A model for the establishment by selection of a particular new adaptive combination might make the assumption that a second adaptive "peak," optimal for a new environment, would be one in which all of the alternative alleles, a_2, . . . , e_2, were present in the heterozygous or homozygous condition. The chances of obtaining such an individual in the original population would be $q^5 = 0.05^5$ or 3.1×10^{-7}. Even if this very remote chance should be occasionally realized, the likelihood that it would lead to a new population similar to it would be very low, because of outcrossing and the breaking up of the favorable combination by genetic processes. Because of this fact, adaptation to a new environment is likely to occur via a smaller number of gene substitutions that modify rather than completely reorganize the original adaptive system, even if such modification fails to give rise to the optimum adaptation to the new conditions.

The conservation of organization

The consequence of selective inertia is the *conservation of organization* which has been characterized as follows (Stebbins 1969:124–125): "Whenever a complex, organized structure or integrated biosynthetic pathway has become an essential adaptive unit of a successful group of organisms, the essential features of this unit are conserved in all of the evolutionary descendants of the group concerned." The evolutionary implication of this principle is twofold. In the first place, it provides an answer to a question that biologists have frequently asked as a challenge to the modern synthetic theory of evolution: How can random fluctuations in the environment interact with random mutations and gene combinations to produce the progressive increase in complexity of organization that is a well-recognized feature of the evolutionary history of organisms? The answer given by the principle of conservation of organization is that, once a population has acquired an adaptive structure or system of organization that is determined by specific interactions between many different genes, mutations or gene combinations which tend to destroy or weaken that particular adaptive system are more likely to be eliminated than are those which modify it, or which alter the number of different organizational units that are present in an individual. The organizational unit, once acquired, can therefore serve as a basis for the origin of further complexity.

Second, the conservation of organization can explain why certain features that are peculiar to a particular family or genus persist with remarkable constancy, becoming modified in many, often bizarre, ways in response to various adaptive challenges. Examples are the petal structure and single carpel of the Leguminosae, subfamily Papilionoideae, the perianth and staminal column of the Orchidaceae, and the fruit structure of the Cruciferae and Umbelliferae. The floral organization of families such as legumes and orchids was probably built up gradually over a long succession of intermediate stages, involving high selective pressures which acted over this entire period. Once acquired, however, these structures can be maintained by much lower pressures of stabilizing selection. Furthermore, given the highly pleiotropic action of most genes in higher organisms (Chapter 6), we may expect to find that some of the gene loci that contribute to these structures will have other adaptive functions, so that the structure as a whole cannot be modified with impunity.

From these considerations we can reach the conclusion that many complex adaptive structures will be retained long after the strong selective pressures that were required to establish them have ceased to exist. This is true, for instance, of the elaborate floral structures in self-fertilizing representatives of Leguminosae, Compositae, and other families. A corollary to this conclusion is that, particularly in ancient families, we can expect to find structures which appear to have little adaptive value in the modern representatives of the family, but which were highly adaptive in its original, now extinct, members. This is particularly likely to be so in view of the well-recognized ability of successful families to become highly diversified ecologically, and to spread into many niches having extremely different environmental conditions (Chapter 3).

Examples of the conservation of organization are particularly instructive if there exist a few exceptions to the conserved structure or organizational system. These exceptions tell us that deviations from the conserved organization are not impossible, but just highly improbable. Such examples can be found at nearly every level of organization in the plant body.

A good example at the level of the cellular organelle is the chloroplast. The shape and fine structure of chloroplasts became standardized in the evolutionary line that led to the Archegoniates, and with a few exceptions these organelles are basically similar in all higher plants. They are ellipsoid, and contain grana which consist of stacked lamellar disks, connected to each other by frets (Weier, Stocking, and Shumway 1966). In a few groups, however, particularly the Gramineae, subfamily Andropogoneae, as well as the Amaranthaceae, a different kind of chloroplast has evolved. These have extensive stroma lamellae and lack grana. Furthermore, they form many large starch grains, which are absent from typical chloroplasts. The exceptional chloroplasts are found in the cellular sheaths which surround the vascular bundles; elsewhere in the leaf chloroplasts having a normal structure exist. They are associated with the C-4 dicarboxylic acid pathway of CO_2 fixation, mentioned earlier in this chapter.

Two examples of the conservation of organization at the level of the cell deserve mention. The first of these is the stomatal apparatus. In all vascular plants the stomatal opening is flanked by two guard cells, which next to the opening have highly thickened, specialized walls. In dicotyledons these guard cells contain well-developed chloroplasts, and the epidermal cells which surround them are only slightly or not at all different from epidermal cells elsewhere in the leaf. The same condition exists in the Liliales and a few other orders of monocotyledons (Stebbins and Khush 1961; Fig. 2-3). In other monocotyledons, however, two types of divergent pattern are found. In both of them, the guard cells themselves are relatively small, and their chloroplasts contain little chlorophyll. They are, moreover, flanked by two or more subsidiary cells which are markedly different in structure from the normal epidermal cells, and originate in a completely different way. In the families Alismataceae, Juncaceae, Cyperaceae, Gramineae, and several others the number of subsidiaries is consistently two, whereas in the Commelinaceae, Araceae, Palmae (Arecaceae), and a few other families four or more subsidiary cells are usually formed.

Fig. 2-3. Stomatal apparatus of mature leaves of monocotyledons, showing variable numbers of subsidiary cells: (*A*) *Northoscordum inodorum* (Liliaceae); (*B*) *Juncus effusus* (Juncaceae); (C) *Commelina communis* (Commelinaceae). (From Stebbins and Khush 1961.)

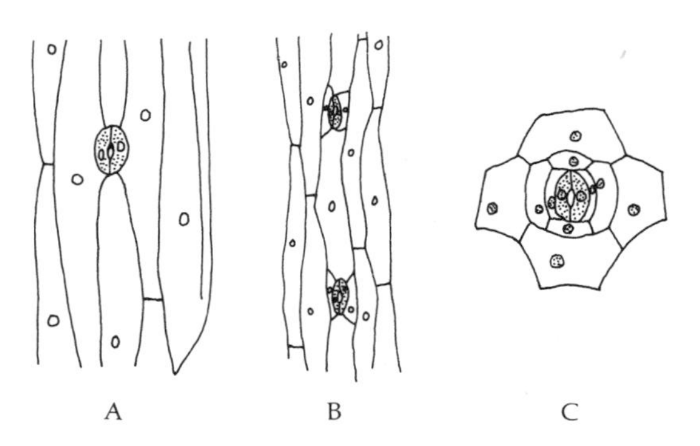

The function and adaptive significance of these different kinds of stomatal apparatus are still obscure. Nevertheless, the distribution of the various patterns, particularly among the different families of monocotyledons, tells us two things. In the first place, a shift from one kind of organization to another is not impossible, at least in certain groups. For instance, the reduction of the number of subsidiaries from several to two occurs in certain specialized, temperate representatives of the Araceae, such as *Symplocarpus* and *Lysichiton.* The presence of two or more subsidiaries in most of the primitive, tropical monocotyledons, including the genus *Astelia* in the Liliaceae, suggests that the absence of subsidiaries in the Liliales and Amaryllidales is a derived condition. Leaves of Liliaceae which are strikingly grasslike in superficial appearance, such as those of *Xerophyllum* and *Asphodelus,* nevertheless agree with others of their family in the absence of subsidiaries. On the other hand, certain species of rushes (*Juncus* sect. *ensifolii*), which live in habitats similar to those occupied by members of the Iridaceae, have evolved leaves which are *Iris*-like in their vertical position and their two-faced equitant structure above a sheathing base. This morphological convergence has not, however, been accompanied by convergence with respect to epidermal structure. The stomatal apparatus of *Juncus ensifolius* and its relatives always contains two subsidiaries, whereas leaves of Iridaceae lack them.

From the strictly functional point of view, this histological difference would appear to be meaningless. If, however, we postulate that in the organizational system of *Juncus* the loss of subsidiaries and the simultaneous acquisition of well-developed chloroplasts in the guard cells is difficult or impossible, whereas the acquisition of subsidiaries is equally difficult in a plant having the developmental organization of *Iris,* the histological difference between these two morphologically similar leaves becomes comprehensible. One can, in fact, erect a reasonable hypothesis to explain the rigidity of the developmental patterns in these two advanced monocotyledons, and their comparative flexibility in families such as aroids and palms. In the latter families, the differentiation of subsidiaries occurs when the stomatal initial or guard-cell mother cell is hardly different in size and cytoplasmic content from the surrounding epidermal cells. On the other hand, in both the Iridaceae and the Juncaceae, as well as in most Liliaceae, Cyperaceae, and Gramineae, the epidermal cells become much elongated immediately after the differentiation of the stomatal initial, and when paired subsidiaries are formed, they are produced by epidermal cells that are already well differentiated and are very different from the guard-cell mother cell. Most probably, the reorganization of stomatal complexes in these advanced groups could take place only if the entire timing sequence of cellular development in the leaf were reorganized, so that the epidermal cells would be in a more undifferentiated condition at the time of subsidiary formation.

The embryo sac of angiosperms is another example at the cellular level of the conservation of organization, with occasional exceptions. The presence of the eight-celled embryo sac in the great bulk of families and genera of both monocotyledons and dicotyledons has often been invoked as an indication of the common origin of these two subclasses of angiosperms (Maheshwari 1950). Nevertheless, exceptions to this pattern are well known: the four-celled embryo sac of the Onagraceae, the extra nuclei and nuclear fusions of the Piperaceae, the extra antipodals of the Gramineae, and the remarkable condition in *Lilium* and its relatives of the Liliaceae. The functional significance, if any, of these deviations, as well as the reasons for the constancy of embryo-sac organization in most angiosperms, are obscure. Nevertheless, until much more is known about the physiological and cytological processes that underlie the development of the embryo sac, as well as the relation of these processes to ovule development in general, the assumption that these deviations are merely accidental vagaries having no functional significance is just as unwarranted as is the erection of hypotheses that might purport to explain them. When the essential facts are still unknown, the wisest course is to maintain an open mind, and to recognize the almost equal probability of either the presence or the absence of functional significance and an origin via the guidance of natural selection.

At the level of the organ, striking examples of the conservation of organization can be found in gynoecia of which the number of carpels per flower or of ovules per carpel has been reduced to one. In the very large family Leguminosae, for instance, the almost completely constant condition of one carpel per flower exists not because reversion toward higher numbers of carpels is genetically impossible, through lack of the right kind of mutations, but because the organizational pattern sequence of floral development makes the occurrence and establishment of functional reversions highly improbable.

Evidence that this is the correct interpretation of the observed constancy exists in the form of flowers which have two carpels, but which are otherwise highly spe-

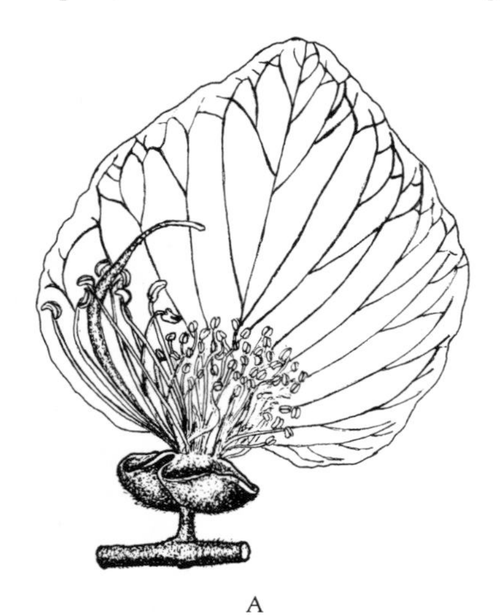
A

B

C

Fig. 2-4. Flowers of three species of *Swartzia* (Leguminosae, Caesalpinoideae): (*A*) *S. Trianae,* showing the reduction of the perianth to a single petal, the increased number and strong dimorphism of the stamens, and the gynoecium consisting of a single carpel, as is typical of the family; (*B*) *S. ingifolia* and (*C*) *S. Littlei,* showing gynoecia having increased numbers of carpels. (From Cowan 1968.)

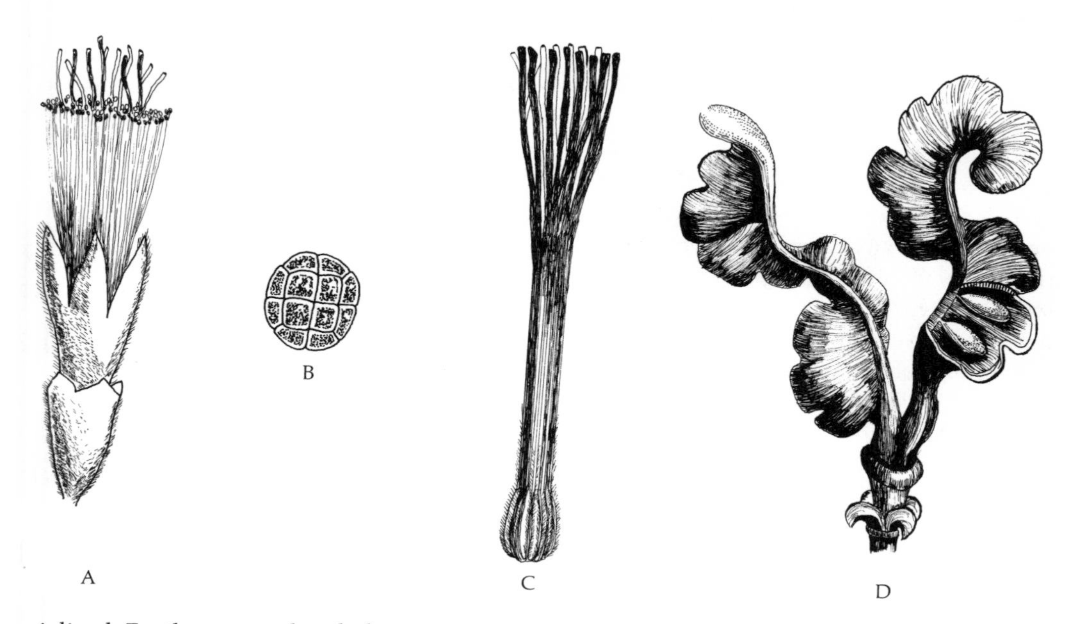

Fig. 2-5. Archidendron Vaillantii F. Muell: (*A*) single flower, showing united sepals and petals; (*B*) single multicellular pollen grain; (*C*) gynoecium at anthesis, showing several carpels; (*D*) mature follicle at dehiscence. (From Taubert 1894.)

cialized. Furthermore, they belong to species which are clearly related to others having less specialized overall floral structure, and have the conventional condition of one carpel per flower. The best examples are *Swartzia ingifolia* and *S. littlei* in the subfamily Caesalpinoideae of the Leguminosae (Fig. 2-4, Cowan 1968). The hypothesis that the bicarpellate condition in these species is a relictual holdover from a similar condition in some unknown ancestor of the Leguminosae would be difficult to maintain. Much more plausible is the assumption that in *S. ingifolia* and *S. littlei* an unusual set of developmental circumstances made possible the origin of flowers having increased seed production via an increase in the amount of gynoecial meristem such that two carpels could be differentiated from it rather than one.

Three other examples are the genera *Archidendron, Affonsea,* and *Hansemannia,* of the subfamily Mimosoideae (Fig. 2-5).

These genera have been regarded as primitive, ancestral prototypes of the family and links to the family Connaraceae (Taubert 1894:99), for apparently no other reason than the presence of 2 to 15 carpels per flower. One would expect, however, that an ancestral prototype of the family would have a generalized floral structure which would provide a connecting link between at least the two primitive subfamilies of Leguminosae, the Mimosoideae and Caesalpinoideae. This, however, is by no means true of the genera in question. In their inflorescence structure, their sympetalous, radially symmetrical perianth, and their many-celled pollen grains, *Archidendron, Affonsea,* and *Hansemannia* are typical, relatively specialized members of the Mimosoideae tribe Ingeae. Furthermore, the legumes of *Archidendron* appear to be specialized in their irregular,

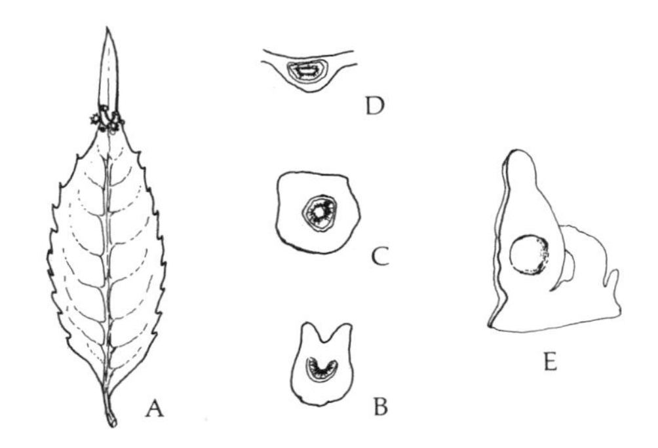

Fig. 2-6. Phyllonoma laticuspis: (*A*) foliage leaf, showing inflorescence emerging from the distal portion of the midrib; (*B*) transverse section of petiole just above its base, showing normal vasculation of an appendage; (*C*) transverse section of the middle of the petiole; (*D*) transverse section of the midvein just below the insertion of the inflorescence; (*E*) young leaf primordium, showing protuberance from which the inflorescence is differentiated. (From C. de Candolle 1891.)

lobed, indehiscent, or irregularly dehiscent character. With respect to the single character that suggests primitiveness, the number of carpels, reversion rather than an actually primitive condition would appear to be the most probable interpretation.

At the level of the architecture of the plant as a whole, an excellent example of the conservation of organization is the almost universal presence of a continuous axis, which produces appendages laterally according to a definite phyllotactic scheme. Branches, lateral inflorescences, or single flowers are produced in the axils of the appendages.

An unusual deviation from this kind of organization is the emergence of inflorescences from the surface of lateral appendages. In wild species, this condition exists in the genera *Phyllonoma* (Saxifragaceae), *Chailletia* (Chailletiaceae), *Helwingia* (Cornaceae), and a few others (de Candolle 1891, Harms 1917). In some of these species, the inflorescence is borne on the adaxial surface of the midvein of the leaf (Fig. 2-6); in others, at the distal end of the midvein; and in a species having compound leaves, in the axils of leaflets. In terms of conventional morphology, this situation is so unusual that some morphologists have argued that there exists an axillary "peduncle" which is so completely adnate to the leaf that its existence cannot be detected even on the basis of anatomical or developmental evidence.

An alternative explanation of this situation is suggested by the development of the hooded genotype in barley (Stebbins and Yagil 1966, Yagil and Stebbins 1969). This dominant gene converts the awn of the lemma into an elaborate structure which usually bears one or two rudimentary florets on its adaxial surface. These florets develop from a "cushion" of meristematic tissue, similar to that shown in Fig. 2-6, *E*. It has a tunica-corpus organization of its cellular layers essentially similar to that of a floral meristem, and apparently has similar developmental potentialities. The "cushion" arises through an increased frequency of mitoses taking place locally in one region of the adaxial surface of the lemma primordium when it is about 300–700 microns long. At no stage of development is there any evidence of adnation of an axial branch to the lemma primordium, which before the appearance of the "cushion" is indistinguishable from the primordium of a normal awned lemma.

The development of the hooded genotype tells us, therefore, that, given an unusual alteration of hormonal balance, the cells of an appendage, when still in a meristematic condition, can develop the potentiality for producing an axis *de novo*. The usual continuity of the axis in vascular plants is not determined by any genetically based inability to produce another kind of organization. It is, rather, the most generally adaptive kind of organization, which is the result of a canalized direction of evolution.

Adaptive modification along the lines of least resistance

In Chapter 1 the fact was emphasized that many possible and alternative pathways exist for adaptation to a particular environmental situation. To a certain extent, the particular pathway that an evolutionary line will take depends upon chance combinations of genes that exist in the initial population of the line. Nevertheless, an equally and perhaps even more important factor in determining the pathway of adaptation is the innate, genetically controlled pattern of development that exists in the population at any stage of its evolution. This is because the direction of adaptation will often be determined according to the principle of adaptive modification along the lines of least resistance (Ganong 1901, Stebbins 1950: 497). I have elsewhere (Stebbins 1967) given an example which shows that, on the basis of this principle, the same selection pressures, acting simultaneously on different adaptive systems, can cause these populations to diverge further from each other.

This example compares the probable response to selection for increased seed production in three different kinds of plants, a tulip, a buttercup (*Ranunculus*), and a sunflower, or any other species of Compositae (Figs. 2-7 and 2-8). In the tulip, the flower is borne at the end of a single shoot that emerges from the bulb, so that increasing seed number by increasing materially the number of flowers per plant would require a drastic reorganization of its shoot structure. Its large flowers are strictly trimerous, and produce a single ovary with three locules, which correspond to three carpels. Increasing the number of locules or carpels per flower would require a complete reorganization of the floral architecture. On the other hand, the numerous ovules per locule are produced by an active intercalary meristem. Increasing the activity of this meristem, and so

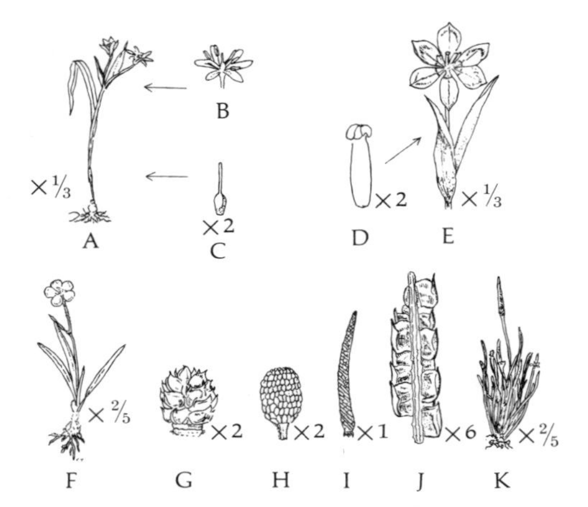

Fig. 2-7. (*A–E*) Differences in flower and capsule size between members of the family Liliaceae, tribe Tulipeae: (*A–C*) *Gagea lutea* Ker-Gawl, (*A*) whole plant, (*B*) individual flower, (*C*) ovary and style at anthesis; (*D, E*) *Tulipa strangulata* Reb., (*D*) ovary style and stigma at anthesis, (*E*) uppermost leaves and flower. With respect to size and shape of flower, as well as ovary style and stigma, *Gagea* more nearly resembles relatively unspecialized members of the family Liliaceae. Increase in gynoecial size of *Tulipa* is associated with increase in both seed size and seed number per ovary locule. (*F–J*) Differences in carpel number per gynoecium in *Ranunculus* and *Myosurus:* (*F*) *R. pyrenaeus* L., habit sketch; (*G–I*) gynoecia at maturity of *R. repens, R. sceleratus,* and *M. minimus;* (*J*) enlargement of a portion of (*I*); (*K*) *M. minimus,* habit sketch. In this species group, differences with respect to seed number per plant are determined largely by differences in number of carpels per gynoecium. (From Fiori 1921.)

producing more ovules per locule, would be adaptive modification along the line of least resistance, since it would involve the smallest amount of reorganization of the architecture of either the plant as a whole or the flowers.

In *Ranunculus* the large flower is usually produced at the end of a long branch, and the production of an inflorescence containing numerous flowers would probably involve considerable reorganization of the architecture of the shoot, as in *Tulipa.* The flower itself, however, is so constructed that in *Ranunculus* adaptive modification along the lines of least resistance would involve an increase in the number of carpels per gynoecium. This is because each carpel is a unit of highly determinate growth, bearing only a single ovule and seed, which at maturity is a highly adaptive structure for seed dispersal. Increase in number of seeds per carpel would require considerable reorganization of this structure. On the other hand, increase in carpel number per flower would involve only an increase in the amount of floral meristem that is produced before the carpel primordia become differentiated from it. This has been, apparently, the course of evolution that has been followed in the evolution of the small genus *Myosurus* from its *Ranunculus*-like ancestor.

In *Helianthus,* the sunflower, the individual flowers or florets are highly determinate structures, each of which produces only a single seed. An increase in number of seeds per floret by any means would involve a drastic reorganization of floral development. On the other hand, increase in number of florets per capitulum involves merely an increase in the amount of undifferentiated floral meristem that is produced before the floret primordia are differentiated. This is, therefore, the way seed production has been increased by plant breeders, who have selected high-yielding cultivated varieties from their lower-yielding wild ancestors.

Evolutionary Canalization and Alternative Pathways of Adaptation

The principal argument presented in the previous sections can be summarized as follows. At every level of the taxonomic hierarchy, differences between categories are related to adaptation. Nevertheless, this relation is indirect, for a number of reasons. Some of this indirectness is based upon the complexity of the relation between genes and characters.

Fig. 2-8. Inflorescences of four species of the Compositae, tribe Anthemideae, to illustrate differences in number of seeds per plant that depend upon size and number of capitulae: (*A*) *Achillea nana* L.; (*B*) *Achillea millefolium* L.; (C) *Chrysanthemum Parthenium* (L) Bernh.; (*D*) *Chrysanthemum segetum* L. (From Fiori 1921.)

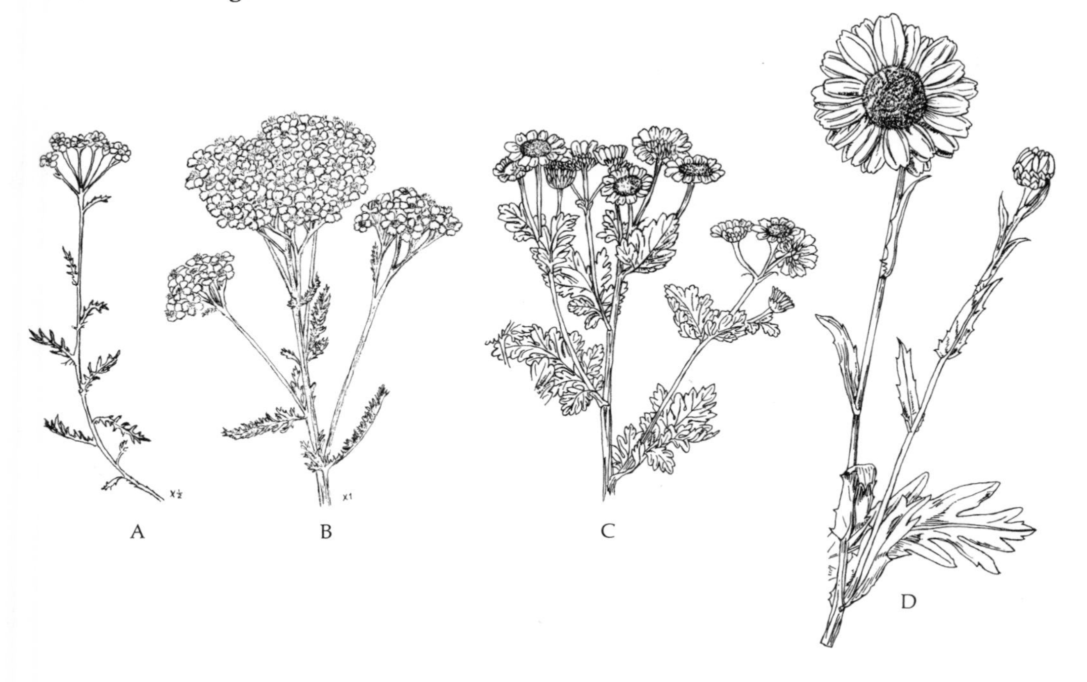

Table 2-2. Representative small genera of angiosperms having wide geographic distributions or local dominance. The assignment to families follows the system of Takhtajan (1959).

Genus	No. species in genus	Family	No. genera in family	Geographic	Ecological
Acorus	2	Araceae	100	N. temperate zone	Swamps, streams
Adenostoma	2	Rosaceae		California	Scrub (chaparral), Mediterranean climate
Adoxa	1	Adoxaceae	1	Temperate Eurasia, North America	Moist woodlands
Ailanthus	3	Simarubaceae	30	E. and SE Asia, East Indies; introduced in N. America	Dry to moist woodlands
Alhagi	3	Fabaceae (Papilionaceae)	350	Mediterranean, S. Asia; introduced in N. America	Deserts
Aruncus	3	Rosaceae		N. temperate zone	Mesic woodlands
Berula	1	Umbelliferae	250	N. temperate zone	Swamps and streams
Barachyelytrum	2	Gramineae	450	Eastern N. America, E. Asia	Mesic woodlands
Brasenia	1	Cabombaceae	2	N. America, Cuba, E. Asia, tropical Africa, Australia	Lakes and slow streams
Calla	1	Araceae	100	N. temperate zone	Swamps and bogs
Calluna	1	Ericaceae	60	Temperate Eurasia	Open country, mesic
Cannabis	1	Cannabiaceae	2	Central Asia, worldwide spread as weed by *natural* means	Open fields and waste places, mesic
Cetunculus	3	Primulaceae	28	Worldwide	Moist open places
Ceratophyllum	1	Ceratophyllaceae	1	Worldwide	Fresh water
Chamaedaphne	1	Ericaceae	60	N. temperate zone	Swamps and bogs
Chelidonium	1	Papaveraceae	25	Temperate Eurasia; introduced in N. Amer.	Fields and woodlands, mesic
Conium	2	Umbelliferae	250	Eurasia, Africa, introduced worldwide	Moist fields, waste places
Decodon	1	Lythraceae	21	Eastern N. America	Swamps
Dulichium	1	Cyperaceae	60	Temperate N. America	Swamps
Empetrum	2	Empetraceae	3	Cool temperate regions, both hemispheres	Moist open places
Glaux	1	Primulaceae	28	N. temperate zone	Salt marshes

Table 2-2. (continued)

Genus	No. species in genus	Family	No. genera in family	Geographic	Ecological
Hippuris	1	Haloragaceae	8	Cool temperate regions, both hemispheres	Swamps and bogs
Hottonia	2	Primulaceae	28	N. temperate zone	Lakes and ponds
Humulus	3	Cannabiaceae	2	N. temperate zone	Open places, mesic
Lilaea	1	Lilaeaceae	1	N. and S. America	Swamps and marshes
Lilaeopsis	1	Umbelliferae	250	Worldwide	River banks, marshes
Linnaea	1	Caprifoliaceae	14	N. temperate zone	Coniferous woods, mesic
Liquidambar	2	Altingiaceae	5	Eastern N. America, Southeast Asia	Moist forests
Menyanthes	1	Menyanthaceae	2	N. temperate zone	Swamps and bogs
Mitchella	2	Rubiaceae	340	Eastern N. America, E. Asia	Woodlands, mesic
Penthorum	1	Penthoraceae	1	Eastern N. America, E. Asia	Swamps, wet places
Phragmites	3	Gramineae	450	Worldwide	Swamps, wet places
Spirodela	3	Lemnaceae	4	Worldwide	Lakes, pools, rivers
Subularia	2	Cruciferae	200	N. temperate zone	Lakes and ponds
Symplocarpus	1	Araceae	100	Eastern N. America, E. Asia	Swamps, bogs
Trapa	3	Trapaceae	1	Eurasia, Africa	Lakes, ponds, streams
Zannichellia	2	Zannichelliaceae	6	Worldwide	Pools and streams

This problem is discussed in Chapter 6. In addition, plants can respond in many different ways to the various kinds of selection pressure to which they are exposed. Any particular adaptive shift, such as increased resistance to drought, cold, or predation by animals, change in pollinating agents, increased seed production, or increased seedling vigor, can be accomplished in any one of several different ways. The particular kind of adaptation that is carried out by any particular evolutionary line may depend to some degree on chance combinations of genes that are present in a population when it is first exposed to a divergent selection pressure. To a much greater degree, however, the nature of this response will depend upon the adaptive complex that already exists, and will be based upon the principles of selective inertia, conservation of organization, and adaptive modification along the lines of least resistance. The operation of these principles will bring about a basic resemblance between related evolutionary lines having very different adaptations, since it will cause certain characteristics to be retained that existed before these adaptations were developed. These resemblances are the foundation upon which higher categories, such as genera, families, and orders, can be based.

Evolutionary Success and Evolutionary Opportunity

A taxonomist who is interested in the nature of higher categories is likely to regard a genus or family as successful if it has evolved a large number of species. The families of flowering plants that are usually mentioned as having reached pinnacles of success are the pea family (Leguminosae or Fabaceae), the grass family (Gramineae), the orchid family (Orchidaceae), and the sunflower family (Compositae or Asteraceae). This designation is certainly justified. Nevertheless, we should not forget that many very small families or genera, which contain only one or a few species, are nevertheless common and widespread over a large number of habitats, climatic zones, or both, and have probably existed in these habitats for tens of millions of years. A representative list of such genera and families, taken from temperate floras with which I am acquainted, is presented in Table 2-2. To call these genera and families unsuccessful would certainly be a mistake. Their existence tells us clearly that evolutionary success is not necessarily associated with speciation. On the animal side, the species that exemplifies to the greatest extent the achievement of evolutionary success in terms of widespread dominance over the environment, unaccompanied by speciation, is *Homo sapiens.*

On the other hand, there are very few genera of angiosperms that contain many species of which none is widespread and abundant. The only examples that approach this condition known to me from temperate North America are *Malacothamnus* (Malvaceae), *Chorizanthe* (Polygonaceae), and *Hemizonia* (Compositae). Even in these genera one or two species are abundant and may even be weedy in a localized area. This comparison suggests that, although evolutionary success can be attained without subsequent speciation, any group that is capable of extensive speciation will give rise occasionally to one or more highly successful species.

The Significance of Extinctions and Alterations of Distribution Patterns

During the long periods of time over which evolutionary trends extend, environmental changes are certain to cause many species and genera to become extinct. These extinctions are not at random with respect to the trends themselves. As was pointed out by Darwin in the *Origin of Species,* and is amply documented by the fossil record of mammals and conifers, the imperfectly adapted forerunners of a particular evolutionary line are particularly likely to be eliminated by competition with their more efficient successors. This elimination has two effects. In the first place, it is largely responsible for the wide gaps that exist between most successful families, such as Leguminosae, Compositae, Orchidaceae, and Gramineae, and their nearest living relatives.

Secondly, the consistent elimination of these generalized forerunners makes virtually impossible the persistence in the modern flora of the direct ancestors of any major group. This fact is far too often neglected when botanists attempt to construct phylogenetic trees on the basis of contemporary forms.

In addition to widespread extinction, extensive shifts in geographic distribution have taken place during angiosperm evolution (Stebbins 1950, Axelrod 1960, and many other references). Because of this fact, speculations about the place and conditions of origin of a group made on the basis of modern distribution patterns are always hazardous, and can be completely misleading. This topic is discussed extensively in later chapters.

Summary

Although in a particular family the characters that separate species usually differ from those that distinguish genera within the family, and still other characters separate the family from its nearest relatives, nevertheless a review of angiosperms as a whole shows that every morphological character that is used to distinguish families and orders can in some groups vary at the level of genera and species. In some cases this kind of variation can be detected within a single population, for example, *Linanthus androsaceus* and *Antirrhinum majus*. An important principle that is valid for long-continued evolutionary trends but is not a significant factor at the level of contemporary populations is evolutionary canalization. This is the epigenetic nature of successive adaptations. The way in which a population will respond adaptively to a changing environment depends to a large degree upon the adaptations that it has already acquired as a result of previous adaptive radiations.

Evolutionary canalization depends upon the principles of selective inertia, conservation of organization, and adaptive modification along the lines of least resistance. Evolutionary success may consist either in the long-continued persistence and wide ecological distribution of one or a few forms, or in the diversification of a group to give rise to a large number of genera and species. In considering the action of evolutionary processes over long periods of time, both extinction and extensive alterations of geographic and ecological distribution patterns must be recognized.

3 / Adaptations for Survival

Both evolutionary success and the potentiality for further diversification depend upon the success with which the evolutionary line solves three very different problems: (1) survival of the growing and the adult plant under a variety of different environmental conditions; (2) cross pollination; and (3) development and dispersal of seeds and establishment of seedlings. The selective pressures associated with these three kinds of adaptation affect the plant at different times of its life cycle, and differ from one another so greatly that adaptations to them must usually be accomplished by modifications of different gene systems. Consequently, one must make a clear distinction between selection for maximum ability to survive and selection for optimum reproductive fitness. The distinction between these two kinds of selection with reference to trends of evolution is much more significant in plants than it is in animals.

The reason for this difference lies in the great difference between animals and plants with respect to functional and developmental integration. In animals, a high degree of integration is required for both survival and reproduction. Motility, whether walking, flying, or swimming, sensory awareness, the capture, eating, and digestion of food, and the excretion of waste are all essential for the survival of animals, and require a high degree of integration and cooperation between different parts of the body. Furthermore, the more complex and specialized is the animal, the more complex are both the integration of the adult body and the pattern of its developmental sequences. Consequently, developmental patterns associated with survival become so complex and highly integrated that selection along the lines of least resistance usually produces variations or modifications of the same basic pattern. Only rarely can new patterns of organization evolve which represent novel adaptations for survival. Animal taxonomists, therefore, can establish orders and families on the basis of such character complexes as are associated with survival. Differences between orders of mammals, for instance, depend upon similarities between related animals with respect to diet (herbivores, carnivores), method of locomotion (running, leaping, flying), habitat (aquatic, terrestrial), and similar properties. Differences with respect to reproduction also represent different highly integrated adaptive systems, and are often valuable diagnostic characteristics of major groups, such as the classes of terrestrial vertebrates and the subclasses of mammals. They are not, however, of overriding importance.

In plants, survival depends to a much lesser degree upon integrated activities than in animals. This is because plants usually lack motility, sensory awareness, and digestion. To be sure, leaves cannot perform photosynthesis efficiently unless their supply of water and carbon dioxide is well regulated, and they must be efficiently placed with relation to the source of light. Furthermore, the efficient functioning of the nonphotosynthetic tissues depends upon well-integrated mechanisms of food transport. The integration required for these systems is far less, however, than that needed for animal activity. Consequently, we find that, with the exception of wood structures, the mechanisms for performing these functions are very similar in all groups of vascular plants. Interestingly enough, the structure of the stem, which has developed greater efficiency in higher classes of vascular plants, provides botanists with almost the only diagnostic characters of the vegetative system that can be used for distinguishing higher categories.

On the other hand, the reproductive functions of plants require integration of organization and of developmental pattern quite comparable to that found in animals. For cross-pollination by animals with specialized habits, such as bees, birds, and bats, the flower must become a highly integrated structure, with all of its parts precisely adjusted relative to one another. In many plants, the maintenance of a large gene pool by means of obligate cross-fertilization is accomplished by elaborate immunochemical systems which result in self-incompatibility. In addition, the integration between the receptivity and nonreceptivity of the stigma and the accomplishment of pollination and fertilization is essential for seed development.

With respect to seeds, even more integration between widely separated parts and functions is required. The development of the ovule into the seed must be coordinated with that of the ovary into the fruit, and both of these processes must be adjusted to the favorable growing season of the plant. In addition, precise mechanisms have often been evolved for seed dispersal, by wind, water, or animals, which require adjustment of buoyancy to seed size, attractiveness to animal dispersers, adherence in various ways to animals that serve as involuntary dispersing agents, or retention of viability after passing through the animal's digestive system. Finally, embryo size and structure, stored food, seed coats, and other auxiliary structures must be adjusted both morphologically and chemically to insure germination and seedling establishment when conditions are favorable, and to prevent their occurrence under unfavorable conditions. This integration often confers the greatest adaptive value on systems that involve various kinds of compromise. For instance, increased seed size often confers an advantage because it makes possible either more vigorous growth of seedlings or the continuation of seedling growth for longer times on the basis of stored food before photosynthesis is begun. Large seeds, however, are less easily dispersed than small ones, and they usually require a longer period of development. Furthermore, if the resources of stored food plus the ability to photosynthesize are limiting for a plant, increased seed production may be acquired only at the cost of reduction in seed size.

Obviously, the number of possible compromises between these conflicting demands is very large. Furthermore, when a particular integrated system has evolved, adaptive modification along the lines of least resistance is more likely to give rise to modifications of this system than to the origin of entirely new adaptive complexes. Consequently, we might expect that adaptive radiation of integrated character complexes concerned with seed development and dispersal, as well as with the establishment of seedlings, would often give rise to recognizable orders, families, and genera of flowering plants. In Chapter 5 the hypothesis is advanced that this kind of evolution has, in fact, been one of the most significant bases, and perhaps the most significant single basis, of major trends of evolution in flowering plants.

The Role of Vegetative Adaptations in the Origin of Higher Categories

Adaptive modifications of the vegetative organs—roots, leaves, and stems—are highly important factors in the differentiation of races or ecotypes within species. They are the principal basis of the ecotype concept (Clausen, Keck, and Hiesey 1940, Stebbins 1950, 1970*a*). On the other hand, such changes have played a subordinate role in the origin of higher categories. To be sure, many families have restricted or specialized ecological distributions. For example, Dipterocarpaceae are largely confined to tropical rain forests; Najadaceae, Vallisneriaceae, and Nymphaeaceae have mainly aquatic habitats; Cactaceae, with a few notable exceptions, are xerophytic; Loranthaceae, Rafflesiaceae, and Orobanchaceae are either parasitic or saprophytic. Such ecologically specialized families and orders are, however, side branches rather than major lines of angiosperm evolution.

Amplitude of climatic adaptations within genera and families

Two other facts are much more striking and significant. In the first place, unrelated evolutionary lines have often become adapted to similar environments, either via morphological and physiological convergence or by evolving different ways of exploiting the same environment. Second, adaptations to widely divergent ecological conditions have evolved within the confines of individual plant genera. Extreme examples are *Euphorbia* and *Senecio,* both of which contain about 1000 species. In both genera, there are woody shrubs, perennial herbs, annuals, and stem succulents. Their habitats range from forests through savannas, scrub lands, deserts, and (in *Senecio*) high alpine slopes. Hardly less diverse in their vegetative adaptations are other large genera such as *Cassia, Lupinus, Astragalus, Solanum,* and *Salvia.*

Even at the level of species and complexes of closely related species a great diversity of ecological adaptations can be found in certain instances. The example of *Potentilla glandulosa* is now classic, and such species complexes as those of *Achillea millefolium, Phacelia magellanica,* and *Festuca ovina* equal or exceed it in the diversity of their ecological adaptations. Most of the larger plant families contain among their genera and species a wide spectrum of adaptations to various climatic and edaphic conditions. This fact is evident from the summary of the ranges of climatic distribution of 350 families presented in Table 3-1, which is taken largely from the information provided by Takhta-

Table 3-1. Synopsis of the geographic distribution of angiosperm families.

Size	Tropical	Temperate	Tropical, temperate	Temperate, arctic-alpine	Tropical, temperate, arctic-alpine	Total
Small (less than 50 species)	66	46	35	3	0	150
Intermediate (50–1500 species)	41	13	94	4	10	162
Large (more than 1500 species and 30 genera)	0	1	21	6	10	38
Total	107	60	152	14	17	350

Large families

Tropical, temperate		Temperate, arctic-alpine	Tropical, temperate, arctic-alpine	
			Large	*Medium*
Chenopodiaceae	Solanaceae	Ranunculaceae	Leguminosae	Papaveraceae
Annonaceae	Convolvulaceae	Caryophyllaceae	Boraginaceae	Polygonaceae
Lauraceae	Verbenaceae	Cruciferae	Scrophulariaceae	Plumbaginaceae
Moraceae	Labiatae	Ericaceae	Campanulaceae	Violaceae
Cactaceae	Gesneriaceae	Rosaceae	Compositae	Salicaceae
Malvaceae	Acanthaceae	Umbelliferae	Cyperaceae	Onagraceae
Myrtaceae	Rubiaceae		Gramineae	Gentianaceae
Melastomataceae	Bromeliaceae		Liliaceae	Polemoniaceae
Euphorbiaceae	Palmae		Orchidaceae	Plantaginaceae
Rutaceae	Araceae		Iridaceae	Potamogetonaceae
Apocynaceae				
Asclepiadaceae				

jan (1966). Contrary to a popular impression, that most plant families are strictly tropical in distribution, the strictly tropical families (66) represent only 21 percent of the total number. Furthermore, most of these families are small, so that the total number of species contained in them is only about 7–8 percent of the known species of angiosperms. Several of the larger families that are often regarded as exclusively tropical, such as Annonaceae, Melastomataceae, Flacourtiaceae, Gesneriaceae, and Bromeliaceae, contain a few genera that are chiefly temperate in distribution (though in many instances only in the south temperate zone). These temperate genera, within their families, are by no means always specialized derivatives. In flowering plants, adaptations to new climates can be accomplished by relatively slight reorganizations of their gene systems.

Specific Adaptations of Roots, Leaves, and Shoots

Even though modifications of the vegetative parts are less important for major evolutionary trends than are alterations of reproductive structures, nevertheless they are significant enough to deserve special consideration. They will not be discussed in their entirety, but only insofar as their modifications relate to trends of evolution.

Adaptations of roots to soil conditions

Among the most striking illustrations of the parallelism between discontinuity of the external environment and that of populations is the presence of distinctive races of plants inhabiting localized areas having distinctive characteristics of the soil. Among them are the serpentine-inhabiting races found in various species of the California flora (Kruckeberg 1951), the races of *Festuca ovina* in England which have radically different calcium requirements (Snaydon and Bradshaw 1961), and most particularly the races of *Agrostis tenuis* and other species that tolerate high concentrations of heavy metals and have already been mentioned in Chapter 1.

The work of Bradshaw and his associates on these grasses has shown that many gene pools of certain species are capable of forming gene combinations for heavy-metal tolerance. Given the available habitats, such adaptive properties can be acquired or lost with relative ease. The same is true of the adaptation of certain common and widespread species, such as that of *Festuca ovina* and *Trifolium repens* (Snaydon 1962) to calcareous or acid soils.

Nevertheless, these experiments have by no means solved the problem of genetic adaptation to particular types of soils. To do this, one must not only answer the question: "Why have certain species evolved adaptations to unusual soil conditions?" but also the complementary question: "Why are certain species, and even large groups of related species, restricted to certain types of soils?" Such restrictions are very familiar to both field botanists and horticulturists. The strongly calciphilic species and subgenera such as *Saxifraga* and *Primula* are well known. Equally familiar is the restriction of such genera as *Sarracenia, Lechea,* and *Hudsonia* (Cistaceae), *Sabatia* (Gentianaceae), and various genera of Ericaceae to acid soils. One suspects that the genetic mechanisms which adapt these species to their particular kinds of soil are more complex and deep-seated than are those which have evolved during recent times in the heavy-metal–tolerant grasses. Exploration of such mechanisms would be both highly rewarding and very difficult. It would require collaboration between ecologists, physiologists, biochemists, and geneticists.

The principle of selective inertia can be invoked to explain the tendency of many plant families to contain a great preponderance of species adapted to a particular kind of soil. Examples are the preferences for acid soil exhibited by most species of Ericaceae, which is associated with and may depend upon the symbiotic association of their roots with mycorrhiza, the tendency of most Caryophyllaceae to grow on mineral-rich soils, and that of Chenopodiaceae to grow on saline or subsaline soils. The constancy of these soil preferences suggests that they are based upon highly integrated genetic systems that are difficult to alter. Certain features of the cellular physiology of these plants probably exhibit conservation of organization.

Such constancy may be used to suggest the kind of ecological conditions under which the families concerned originated. Selection pressures favoring the establishment and diversification of this kind of gene system would be strongest in regions where the soils to which it is adapted are particularly well developed and widespread. Thus the Ericaceae probably originated on some land mass underlaid by acid igneous or other crystalline rocks, such as the greater part of the Laurentian Shield of North America. The Caryophyllaceae were probably differentiated in regions underlaid by calcareous sediments, and the Chenopodiaceae very likely arose in brackish or saline habitats, either along the seashore or in alkaline depressions of desert regions. Once these adaptations had become firmly established, the evolutionary lines containing them could diversify more easily in areas having the right kind of soil, but could also occasionally modify their adaptive systems to become adjusted to soils of somewhat different kinds.

Adaptive modifications of leaves

The leaf is probably the most plastic single organ that the plant body possesses. This plasticity is evident both from the amount of variability that can exist between different leaves of the same plant and from the extensive differences in leaf shape that often occur between different races of the same species. Phenotypic plasticity is well exemplified by seedling heterophylly (Stebbins 1959*b*). Most of the individual characteristics by which leaves may differ from one another can, in certain instances, be produced by differential action of genes belonging to the same genotype. Furthermore, these phenotypic modifications are reversible. In many species of woody plants, stump sprouts produced from the woody base of an old individual produce leaves that resemble those of seedlings. Another kind of heterophylly exhibits these reversals to an even more striking degree. This is the ability of certain aquatic and subaquatic species to produce leaves that may be either strongly dissected or nearly entire, depending upon external stimuli provided by variations in either photoperiod, light intensity, or the level of the water.

Cook (1968) has shown that in *Ranunculus flammula* phenotypic plasticity can vary adaptively from one population to another. Populations inhabiting environments that are immature ecologically and are subject to unpredictable alterations in water level consist of genotypes that respond strongly to environmental changes and can easily be induced to produce leaves of different shapes. Populations inhabiting more strictly terrestrial habitats, on the other hand, contain genotypes having less plasticity with respect to heterophylly. Nevertheless, the genetic variation between individuals, which is revealed when they are grown in a constant environment, is greater in the less heterophyllous populations inhabiting the more stable habitats than it is in the strongly heterophyllous populations found in immature habitats. There is, therefore, an inverse correlation between mean phenotypic plasticity and the amount of genetic variation in the population.

These observations support the generalization made by Bradshaw (1965) that populations become adjusted best to short-term fluctuations in the environment by means of phenotypic plasticity, but that adjustment to variations over longer periods of time is best achieved by genetic variability and heterozygosity.

Association between leaf form and edaphic factors

Recent physiological studies of photosynthesis and transpiration have established an adaptive basis for differences in leaf form between ecotypes of the same species. In a careful comparison of a species that is widespread in northern Europe, ecotypes within *Geranium sanguineum*, Lewis (1969) showed that highly dissected leaves with narrow lobes are found in ecotypes that inhabit steppe communities, where they are subject to large annual variations in temperature and to strong light in summer. On the other hand, leaves belonging to coastal ecotypes, compared with those of ecotypes when both were growing under the same cultural conditions in the same garden, are relatively entire and have broad lobes. The same difference has been found between coastal and interior ecotypes of *Nigella arvensis* in Israel (Waisel 1959) and of *Layia gaillardioides* in California (Clausen 1951). A tendency in this direction has been observed in the complex of *Clarkia unguiculata*, a species having entire leaves (Vasek 1964).

This difference is adaptive in two ways. In the first place, the chief limiting factor for photosynthesis of plants growing under conditions of high solar radiation is the availability of CO_2. Its availability to the cells of the leaf is limited at least to some extent by the resistance of mesophyll cells to CO_2 diffusion (Slatyer 1967, Holmgren 1968). Consequently, a high ratio of leaf surface area to volume decreases the distance of chloroplast-bearing cells from the nearest stomatal opening through which CO_2 enters the leaf, and so increases its overall availability. When leaves are growing in low light, on the other hand, light becomes the limiting factor, so that thin leaves having a maximum surface exposure to the light are the most efficient for photosynthetic activity (Holmgren, Jarvis, and Jarvis 1965).

A second factor that determines the adaptiveness of such differences in leaf form is the convection of heat. A very small leaf, or a compound leaf having many leaflets, has a very thin boundary layer, in which a small difference in temperature is enough to bring about an effective convection (Raschke 1960). Consequently, small leaves as well as compound leaves can remain near the ambient temperature, or even below it, in hot weather, whereas leaves having a large flat surface exposed to the sun may become much warmer. In the deserts of western North America, the temperature of small leaves remains within 3°C of air temperature, whereas large flat surfaces, like the stems of *Opuntia* species, may rise to temperatures that are 10° to 16° above ambient temperature (Gates, Alderfer, and Taylor 1968). In hot weather, such temperatures would be damaging to plant organs that are not equipped with a strongly protective cuticle. Consequently, two alternative strategies exist for survival in desert regions. On the one hand, a plant may have relatively delicate and finely dissected leaves that photosynthesize actively when moisture is available and either fold and become inactive or drop off under drought conditions. Alternatively, it may have leaves or phyllodes having large surfaces that can carry out photosynthesis even in hot weather, and are protected from the heat by thick cuticles, dense coverings of matted trichomes, or both.

In some climates having equable temperatures, low rainfall, and abundant dew or fog, leaves having a large, entire surface are an advantage to the plant from the standpoint of water relations. Under such conditions, large flat leaves are more likely at night to cool below the dew point of the water vapor in the air, and so to become covered with moisture (Raschke 1960). This effect may not succeed in actually adding to the moisture content of the leaves above the amount that can enter through the roots and the vascular system, but it can conserve moisture by greatly reducing the effects of transpiration.

Convergence and Divergence of Adaptations for Survival

Convergence is characteristic of all kinds of adaptations, but is particularly evident in adaptations to climatic conditions. It is most easily recognized in extreme environments, such as deserts, arctic-alpine habitats, and aquatic habitats. Most perennial plants inhabiting deserts conform to a relatively small number of highly differentiated growth forms: sclerophyllous shrubs, deciduous shrubs with evanescent leaves, stem succulents, and bulb-forming geophytes. These similar forms are often attained by representatives of very different families. The scleropyllous Zygophyllaceae (*Larrea, Bulnesia*) of the American deserts resemble Rutaceae and Proteaceae from Australia; the deciduous, dissected-leaved Compositae-Ambrosieae (*Hymenoclea, Oxytenia*) are parallelled by species of *Artemisia,* belonging to an entirely different tribe of Compositae; New World cacti are hardly distinguishable in the vegetative condition from African species of *Euphorbia.* These convergences are commonplace, and serve to emphasize further the ability of individual plant families or even genera to evolve adaptations to a very wide range of different climatic and edaphic situations. The complexity of the convergent adaptations to the Mediterranean climate that are displayed by various evergreen shrubs has been well described by Mooney and Dunn (1970).

Less obvious, but of equal importance, is the fact that, given different adaptations to similar but less extreme environmental conditions, further adaptation to an extreme environment can cause evolutionary divergence in response to the same selective pressure. This point is well illustrated by the life forms of plants inhabiting the high mountains of tropical Africa, as described by Hedberg (1968). He recognizes five of these (Fig. 3-1): giant rosette plants, sclerophyllous shrubs, cushion plants, acaulescent rosette plants, and tussock grasses. Except for the grasses, which probably had acquired their tussock form before their arrival on the African mountains, species having each of these extreme life forms are related to other African species which inhabit less extreme environments and have more conventional life forms. Thus the giant Senecios are related to montane forest species which have large leaves but normal trunks and some branches; the rosette-forming *Dianthoseris* is related to conventional species of *Launeaea* and *Sonchus;* the sclerophyllous species of *Alchemilla* have shrubby relatives with larger leaves; and the cushion-forming *Sagina afroalpina* is related to herbaceous species of the genus with normally branched stems.

Fig. 3-1. Diagrams showing five extreme life forms that are adaptations to the severe environmental conditions found in the alpine zone of the equatorial mountains of Africa: (*A*) giant rosette plants; (*B*) sclerophyllous shrubs; (*C*) cushion plants; (*D*) acaulescent rosette plants; (*E*) tussock grasses. (From Hedberg 1968.)

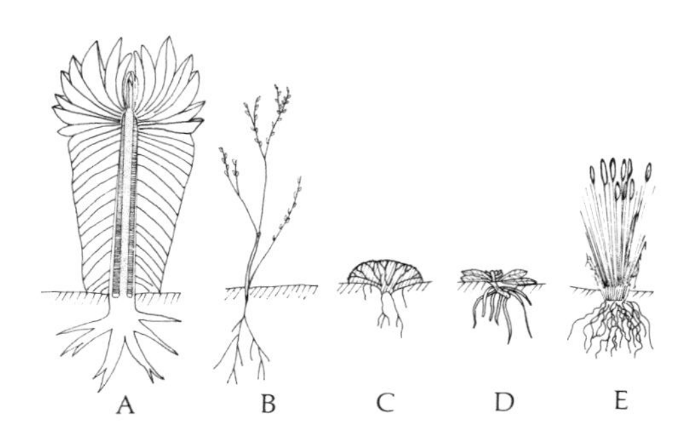

Extreme climatic conditions favor extreme modifications of life form, but in every instance several different ways of becoming adapted to these conditions are possible. The examples of warm deserts and high tropical mountains could be matched by the flora of sand dunes along the seashore, rock crevices and faces of cliffs, peat bogs, alkali sinks, arctic tundra, and many others. The particular pathway that an evolutionary line will take in order to become adapted to any one of these extreme environments will be determined by its mode of adaptation to a less extreme environment. One can imagine, for instance, two related species growing in the same montane forest and having the same stature and area of leaf surface. One of them, however, has relatively few branches and large leaves, whereas the other has numerous branches and smaller leaves. The principle of "modification along the lines of least resistance" would lead one to expect that in becoming adapted to the tropical high-mountain conditions the first of these would be most likely to evolve into a giant rosette plant and the second into a sclerophyllous shrub.

The Adaptive Significance of Secondary Chemical Substances

The taxonomic value of secondary chemical substances has been increasingly recognized during the past two decades and the literature on the subject is now extensive. This taxonomic and phylogenetic aspect is reviewed in Chapter 7. Along with this heightened interest, botanists have speculated many times on what might be the value of these substances to the plant. Why should most conifers of the temperate zone be richly endowed with terpenes; why should alkaloids be widespread in Ranunculaceae, tannins in Fagaceae, essential oils in Labiatae, and so on? Why should highly successful families like Euphorbiaceae, Apocynaceae, and Asclepiadaceae have evolved not only the specialized and highly complex "liquid tissue" known as latex, but also a special and intricate conducting system that ensures the presence of abundant latex in all organs of the plant? Why should one tribe of Compositae, the Cichorieae, have evolved a similar latex system, whereas other tribes have not, but are often equipped with aromatic resins?

One answer to this question has been that these substances are inevitable products of cellular metabolism, which are difficult to excrete and must therefore be disposed of somehow. This answer is unsatisfactory in two ways. In the first place, it does not explain why certain plants should have evolved conducting systems for these substances whose elaborate developmental patterns (Mahlberg 1959, Mahlberg and Sabharwal 1967) must be controlled by a large number of genes. The possibility that these developmental patterns could have evolved without positive selection pressure in their favor is negligibly small. Second, the process of artificial selection in the origin of many cultivated derivatives of these wild species, such as garden lettuce, has greatly reduced their content of these secondary substances, but has not reduced their metabolic efficiency or their vigor of growth. Clearly, a more positive explanation of the presence of these substances is required.

The explanation that is now gaining favor (Whittaker and Feeny 1971) was first proposed by Stahl (1888) in a study that was at the same time a brilliant series of experiments on natural selection and an interpretation that was so far ahead of its time that it was almost completely ignored, and was forgotten until resurrected by Fraenkel (1959). Stahl suspected that these secondary substances, which are bitter, toxic, or both, were evolved by plants as defenses against herbivorous animals, chiefly insects and land snails. He reasoned that this adaptiveness would be difficult for naturalists to detect if all plants had acquired similar defenses, since at least some kinds of animals would have evolved means of dealing with one or more of these substances. He therefore exposed snails in cages to equal amounts of the plants that they normally eat in nature and to cultivated lettuce, which has been selected for a low content of secondary substances. He found that in every experiment the snails showed a strong preference for lettuce over the plants that they normally eat. He then extracted the secondary substances from these plants, and found that this greatly increased their acceptability to snails.

Stahl's conclusion, that the ability to form secondary substances such as phenylpropanes, acetogenins (including flavones and anthocyanidins), terpenoids, and alkaloids has evolved in response to natural selection for defense against such herbivores as insects and snails, is now supported by a large body of evidence (Whittaker and Feeny 1971). Two kinds of defense exist. Compounds that remain in the plant serve as direct defenses against destructive herbivores. In addition, many plants produce allelopathic substances that prevent the growth of competing species in the immediate vicinity of an adult plant, and so increase the likelihood that seedlings of its own species can become established. This method of defense is found most commonly in plant communities like the California chaparral, which are dominated by one or a few species (McPherson and Muller 1969).

The evolution of secondary substances as defense mechanisms appears to be an excellent example of the interaction of natural selection, chance, and the principle of adaptive modification along the lines of least resistance. Without some kind of mechanism of defense for their vegetative organs, plants could never survive in nature, since they would be selectively attacked and destroyed by herbivores. This fact has been clearly shown by Janzen's experiments with *Acacia,* described in the next paragraph. On the other hand, the relative adaptive value of one protective substance as compared with that of another would be very difficult to demonstrate and very likely does not exist. Which substances would evolve in a particular evolutionary line, therefore, could very well depend upon the chance occurrence and establishment of mutations responsible for a biosynthetic pathway leading to the precursor of one particular substance. On the other hand, we must assume that all of the ancestors of any modern group must have had some kind of defense mechanism, so that diversification may well have been largely a matter of evolving new substances as protections against predators that had evolved immunity to the preexisting ones. Such modifications would be most likely to occur by altering as little as possible preexisting biosynthetic pathways, that is, by operation of the principle of modification along the lines of least resistance.

Janzen (1966, 1967) has obtained evidence for a still more elaborate system of defense against insect predators, involving a complex symbiotic relation with other insects that are beneficial to the plant. In the "bull-horn" acacias of Central America (*Acacia cornigera*) the centers of the hollow thorns are occupied by colonies of ants (*Pseudomyrmex ferruginea*). These ants feed on specialized oil- or fat-containing structures, known as Beltian bodies, that arise at the ends of the leaflets of the acacia. In addition, whenever potential predators appear on the plants, they are promptly destroyed by the ants. Finally, when new seedlings of *A. cornigera* become established, ants move to them and establish new colonies. Janzen showed that, if seedlings of *A. cornigera* are sprayed or otherwise treated so that ants of the species *P. ferruginea* do not establish colonies on them, they are destroyed or weakened by other insects to such an extent that they can never reach reproductive maturity. *Acacia cornigera* is exceptional in its genus in that it does not produce alkaloids. Other species of *Acacia* growing in the same part of Central America (*A. chiapensis, A. macracantha*) produce alkaloids, grow more slowly, and live in drier areas than *A. cornigera.* Apparently, most species of *Acacia* pay the price for alkaloid production in terms of a slower growth rate, which enables them to compete only in the drier regions where growth of all plants is slower and less vigorous. The ability of *A. cornigera* to enter the more mesic, lusher plant communities is apparently associated with its evolution of a new defense system that dispenses with alkaloids and depends upon the symbiotic relation with the ant species *P. ferruginea.*

Summary

In plants, in contrast to animals, adaptations for survival, for cross-pollination and fertilization, and for seed dispersal and seedling establishment constitute separate problems and are accomplished by different gene complexes. Short-term climatic adaptations may be accomplished via environmental modifications of the phenotypes, but for most adaptive changes alterations of genotypes are required. A distinctive feature of many genera of angiosperms and most of the larger families is the very wide range of ecological distributions encompassed by their species. This is associated with the relatively low significance of alterations of vegetative characteristics in the origin of major categories. Among the individual plant organs, stem structure is more likely to serve as a diagnostic character for higher categories than are the shape and structure of leaves. This is because the anatomical and developmental patterns of stems are more highly integrated than are those of leaves. With respect to adaptations of all vegetative characteristics, parallelisms and convergences between distantly related families are common. Adaptations for protection against insects and other predators, consisting mostly of repellent chemical substances but including also symbiotic relations with animals, are adaptations for survival that have only recently been generally recognized.

4 / Adaptations for Cross-Pollination*

An important difference between adaptations for survival in higher plants and adaptations for their various reproductive functions is that the latter require much higher levels of integration between different organs, in fact, levels that are quite comparable to those found in animals. For adaptation to cross-pollination by animals with specialized habits, such as bees, butterflies, moths, birds, and bats, the flower must become a highly integrated structure, with all of its parts precisely adjusted to one another. Consequently, in considering the possible origin via adaptation and natural selection of the differences between the flowers of related taxa, one cannot think in terms of separate characteristics but only of syndromes of characteristics that are functionally interrelated. In the present chapter, some of the most striking of these syndromes will be considered, together with examples of adaptive radiation by which one kind of syndrome has evolved from another, different one.

Methods of Pollination

Since the time of Darwin (1877, 1880), botanists have marveled at the variety of devices that have evolved in plants in order to ensure cross-pollination. Even more remarkable is the fact that none of these devices are necessary for survival, or even for the successful perpetuation of a species over thousands or even millions of years. The existence of many species that are largely or entirely self-fertilizing

*The material in this chapter is largely reprinted, by permission of Annual Reviews, Inc., from Stebbins (1970*a*).

has long been recognized (Fryxell 1957, Stebbins 1957*a*). Moreover, such species can maintain stores of genetic variability that are nearly or quite as great as those found in related species which are obligate outcrossers (Allard 1965, 1969, Jain and Marshall 1967). Nevertheless, self-fertilizing species are usually the ends of evolutionary lines, and rarely if ever contribute to major evolutionary trends. Apparently, the genetic structure that is most characteristic of obligate outcrossers, particularly the presence of heterozygosity at a large proportion of gene loci, is a *sine qua non* for the continuance of major evolutionary trends.

Nevertheless, we cannot regard the reduced potentiality for evolutionary change as the chief basis for the persistence of obligate outcrossing in most species. Natural selection has no foresight; only immediate advantages or disadvantages are significant. If, however, a species has for a long time existed as an obligate outcrosser, its populations usually harbor so many recessive alleles that great reduction in vigor results from a sudden shift to self-fertilization. This phenomenon of inbreeding depression has long been known, and has been reaffirmed by recent authors (Müntzing 1961). It is probably responsible for the fact that in many instances evolutionary lines respond to new habitats by evolving new mechanisms for cross-fertilization.

Evolution of Self-fertilizing Species from Obligate Outcrossers

The evolutionary pathway from obligate outcrossing based upon self-incompatibility to predominant self-fertilization has probably been followed by more different lines of evolution in flowering plants than has any other. As I mentioned in an earlier publication (Stebbins 1957*a*), numerous examples of this trend can be found in families of herbs, such as Cruciferae, Leguminosae, Onagraceae, Compositae, and Gramineae. More recently described examples are those of Grant (1964) in *Gilia,* Lewis and Szweykowski (1964) in *Gayophytum,* Raven (1969) in *Camissonia,* Khoshoo and Sachdeva (1961) in *Convolvulus arvensis,* Martin (1967) in *Melochia* (Sterculiaceae), and Ornduff (1969, Ornduff and Crovello 1968) in the Limnanthaceae. In all of these examples, self-compatibility has probably been acquired through loss mutations of genes at the self-incompatibility locus (Lewis 1955). The origin of self-fertilization from cross-fertilizing heterostylous groups, through persistence of one of the monomorphic types that can arise from "illegimate" fertilizations, is well known through the research of Crosby (1940, 1959) on *Primula,* of Baker (1948, 1953*a,b,* 1966) on the Plumbaginaceae, and of Mulcahy (1964) on *Oxalis.*

In most of these examples, the morphological alterations of the flower that are associated with this shift are minimal. In most genera of Gramineae, for instance, cleistogamous flowers evolve from chasmogamous ones merely by the reduction in size of anthers and lodicules (Harlan 1945*a,b*). The self-fertilizing *Eupatorium microstemon* is morphologically almost indistinguishable from its cross-fertilizing relative, *E. Sinclairii* (Baker 1967). In *Lycopersicum,* the self-fertilizing species related to *L. esculentum* and *L. pimpinellifolium* differ chiefly with respect to minor adjustments in the structure and position of anthers and stigma (Rick 1950). The homostylous descendants of heterostylous species contain a favorable combination of characteristics found in the two different flower types present in their ancestors, usually long styles combined with anthers situated high on the corolla tube. This condition can originate through rare crossing over between closely linked genes (Dowrick 1956). In all self-fertilizers, flower size becomes reduced below that found in their cross-fertilizing ancestors.

More profound changes have accompanied the shift from cross- to self-fertilization in the small North American family Limnanthaceae (Ornduff and Crovello 1968). In most species of the genus *Limnanthes,* which are cross-fertilizing, the flowers are large and consistently pentamerous with respect to all of their parts. A few species, however, are self-fertilizing and have flowers of reduced size, though without reduction in the number of parts. Finally, the monotypic genus *Floerkea* resembles *Limnanthes* in all respects except that its flowers are still smaller and are trimerous. *Floerkea* apparently evolved from *Limnanthes* by reduction following the acquisition of self-fertilization. In the grass species *Bothriochloa decipiens,* self-fertilization via cleistogamy has been aided by the acquisition of a distinctive morphological characteristic—a pit on the glumes (Heslop-Harrison 1961).

The shift from obligate crossing to self-fertilization occurs chiefly and perhaps exclusively in species that occupy temporary, pioneer habitats. Such species can persist only through the repeated colonization of new habitats, usually by one or a few individuals. Under these conditions, self-fertilization may be the only way in which the new colonizers can produce offspring. Often, these pioneer habitats are on the margins of the area of distribution for the species and its genus, and may be subject to adverse conditions for cross-pollination, such as excessive moisture in northern and oceanic regions (Hagerup 1950, 1951) or drought and absence of suitable pollen vectors in the deserts. Baker (1953*b*, 1955, 1961*a*) has documented well the association of the shift from cross-fertilization to selfing with transoceanic migration of a species. In order to be successful the change must be gradual, via an intermediate stage of facultative selfing. In this way the adverse effects are minimized, and in a favorable new location, free from competition against vigorous outcrossing ancestors, the new colonist may adjust its genetic system to that of a facultative selfer. Although many genera exist in which the self-incompatible species are predominantly perennials and the self-fertilizers are annuals, the shift from cross- to self-fertilization does not usually accompany the acquisition of the annual habit of growth. In many genera of grasses, such as *Bromus* (Harlan 1945*a,b*), *Elymus* (Snyder 1950), *Festuca* and *Poa* (unpublished personal observations), self-fertilizing perennials are common, and the shift to selfing has apparently occurred at the perennial level. In *Lolium,* on the other hand (Jenkin 1954), both outcrossing and self-fertilizing annuals exist, and the shift to selfing has apparently occurred at the annual level.

The same is true of previously mentioned examples, such as *Clarkia, Camissonia, Gayophytum, Gilia,* and *Limnanthes,* as well as *Amsinckia* (Ray and Chisaki 1957*a,b*) and many other genera. The frequent correlation in a genus of the perennial habit with outcrossing and the annual habit with selfing is probably due to the fact that annuals, because of their general adaptation for frequent colonization of new habitats, are more likely to succeed and spread as selfers than are perennials.

The shift from one mechanism that favors outcrossing to another kind of adaptation having the same effect is best documented with respect to the origin of heterostyly and of dioecism (Vuilleumier 1967). Dioecism can arise directly from heterostyly, as in *Nymphoides* (Ornduff 1966), but it more often evolves in other ways, particularly from facultative selfers having bisexual, monomorphic flowers, as in *Silene* (*Melandrium*) *dioica* and other Caryophyllaceae (Baker 1959). The same is probably true of the origin of heterostyly, as suggested by Vuilleumier (1967) after reviewing a large number of examples.

Shifts Between Insect and Wind Pollination

The three principal ways in which pollen can be borne from one plant to another are by insects, by wind, and, in aquatic plants, by water. The last method has been little studied, and no good examples of adaptive radiation to it are known to me. Adaptive shifts from insect to wind pollination and vice versa are, however, well known. They are accompanied by characteristic alterations of floral morphology (Faegri and van der Pijl 1966). Wind-pollinated flowers differ from those of related species having insect pollination in their tendency to be clustered in dense inflorescences, and in their small size, reduced, inconspicuous perianth, lack of nectar, anthers having abundant pollen and opening with explosive dehiscence, expanded surface of the stigma, and usually the separation of the sexes, either on different inflorescences (monoecism) or different plants (dioecism). Although the majority of wind-pollinated flowering plants occur in distinctive families or genera, the relationships of which are often in doubt, several examples exist in which the direction of evolution can be postulated with reasonable assurance. The following are representative.

Ranunculaceae. The only genus of this family that contains wind-pollinated species is *Thalictrum* (Kaplan and Mulcahy 1971). All of the species of this large genus are apetalous and have flowers in dense panicles or racemes. Some have bisexual flowers, others are dioecious, and still others have various intermediate conditions. The nearest relative of *Thalictrum* is the small genus *Anemonella,* which consists of low-growing perennials having solitary, more conspicuous white flowers, which are probably pollinated by small insects, although good documentation is lacking.

Rosaceae. The only exceptions to insect pollination in this family are genera belonging to the tribe Poterieae, such as *Poterium, Sanguisorba,* and *Acaena.* These genera, found chiefly in the Mediterranean region and South America, with several species of *Acaena* in New Zealand and one in California, possess the syndrome of characteristics mentioned above: apetalous, inconspicuous, unisexual flowers in dense inflorescences. Most of them are vigorous, thick-stemmed herbs or small shrubs adapted to rocky, semiarid sites. In *Sanguisorba,* one group of species (*S. officinalis, S. canadensis, S. microcephala, S. tenuifolia*) consists of tall herbs that inhabit swamps in northern Eurasia and North America. In spite of having apetalous flowers, they produce nectar and are pollinated by insects. They may represent a secondary reversion from wind to insect pollination.

Compositae. In this, the largest family of flowering plants, two groups of genera have diverged from the entomophily that is characteristic of the family and have acquired wind pollination. One of them is the subtribe Ambrosiinae or ragweeds, of the tribe Heliantheae. Its largest genus, *Ambrosia* (including *Franseria*), consists of shrubs, herbs, and weedy annuals, of which the center of distribution is the arid southwestern portion of the United States and adjacent Mexico (Payne 1964). The genus *Iva,* found in the same region, forms a partial link between this subtribe and the subtribe Melampodiinae, particularly the genus *Parthenium.* Other small genera of Ambrosiinae, *Oxytaenia, Dicoria, Hymenoclea,* and *Xanthium,* are radiants from this complex that have retained wind pollination and have acquired various specializations of their fruits. Of these, only *Xanthium* has spread beyond the dispersal center of the subtribe, and it has become worldwide, exceeding in geographical area even *Ambrosia* itself.

A second, independent origin of wind pollination in the Compositae is the genus *Artemisia* of the tribe Anthemidae. This large genus, of which the number of species is estimated at "over 100" (Munz and Keck 1959) or "more than four hundred" (Polyakov 1961), has its main center of distribution in Central Asia, where 174 species are recognized. This is also a region of great diversity for other genera of the tribe, there being 18 genera endemic only to this area. Most of these genera are insect pollinated. The origin of *Artemisia,* therefore, was probably in the cold, arid steppes of Central Asia, where it acquired all of the morphological characteristics of wind-pollinated groups. Its present distribution is in temperate regions throughout the world, including those having moist, equable climates.

Cyperaceae. This family consists almost entirely of wind-pollinated species. Along with the related Juncaceae, it has a long evolutionary history of wind pollination, the origin of which is obscure. Nevertheless, the genus *Dichromena,* which is widespread in moist tropical regions, has reverted to insect pollination (Leppik 1955). In association with this shift, it has evolved a conspicuous group of white leaves surrounding its inflorescence.

Gramineae. Like the Cyperaceae and Juncaceae, the family Gramineae exhibits extreme adaptations for wind pollination, including abundant, light, and easily blown pollen, large, feathery stigmas, and a modification of the perianth, the lodicules, that causes the florets to open when weather conditions are favorable for wind dispersal of pollen. Nevertheless, insect pollination has been recorded in two neotropical genera of grasses, *Olyra* and *Pariana* (Soderstrom and Calderón 1971). In view of the facts that these genera retain many of the morphological features that in other grasses are associated with wind pollination, that both of them are quite specialized relative to many other grass genera, and that they are not obviously related to each other, the hypothesis of Soderstrom and Calderón, that they represent primitive, insect-pollinated ancestors of the wind-pollinated members of the family, appears to me to be improbable. More plausible is the hypothesis that, like *Dichromena* in the Cyperaceae, they represent reversions to insect pollination in a climate where winds are rare and insects are abundant.

Moraceae. The mulberry family probably represents the best example in flowering plants of extensive reversion from wind to insect pollination, even though information on its floral biology is very scanty. It belongs to an order, Urticales, in which all of the other families (Ulmaceae, Cannabaceae, Urticaceae) are wind pollinated. Within the Moraceae, wind pollination is recorded for *Morus* and *Broussonetia* (Knuth 1895–1905). Nevertheless, the largest genus of the family, *Ficus,* which contains about 600 species, is one of the most remarkable examples of entomophily in the plant kingdom. Almost all of the species investigated are pollinated by one or a few species of chalcid wasp, distinctive for that species of *Ficus* (Baker 1961*c,* van der Pijl 1960, 1961, Wiebes 1963). Since *Ficus* has many species in tropical regions of both hemispheres, this condition must be very old.

The Moraceae apparently are an example of a family that evolved from wind-pollinated ancestors adapted to a relatively dry tropical climate. After invading the moist tropics, they evolved at least one secondary adaptation for insect pollination, on the basis of which extensive adaptive radiation took place.

The facts summarized above lead to the conclusion that the shift from insect to wind pollination is most likely to occur in open, often arid habitats, which may be either cold (*Thalictrum, Artemisia*), warm temperate (Poterieae, Ambrosiinae), or tropical (Moraceae). Since the groups involved may be herbaceous, shrubby, or arboreal, this shift is not favored by any particular growth habit. If derivatives of wind-pollinated groups reinvade regions having a moist climate, they usually either retain wind pollination and occupy only relatively open habitats that have free circulation of air (*Ambrosia, Xanthium, Artemisia*) or revert to facultative self-pollination, which enables them to enter closed, forest habitats. Much more rarely, they may evolve secondary adaptations for insect pollination (*Dichromena, Ficus, Olyra, Pariana*).

Adaptative Radiation for Different Pollen Vectors

This subject has been extensively studied ever since the time of Charles Darwin. Until recently, however, most of the observations have been made by naturalists working in a relatively restricted region, who did not, as a rule, apply their observations to the evolutionary relationships between the groups under study. In a paper that to some extent marks a turning point in our understanding of comparative floral morphology in relation to evolution, Grant (1949) pointed out that in genera of plants adapted to pollination by specialized animal vectors, a high proportion of the morphological characters upon which species distinctions are based are concerned with those parts of the flower that adapt it for efficient pollination by these vectors. On the other hand, in genera adapted to unspecialized vectors or to wind or water pollination, the distinctive characteristics of the species affect other parts of the reproductive or vegetative system of the plant.

Upon the basis of this foundation, Grant made a careful comparative study of adaptations to pollen vectors in all genera of the family Polemoniaceae, the results of which will form the central focus of the present section (Grant and Grant 1965). Adaptive radiation for different pollinators was found at all levels of the taxonomic hierarchy, from subspecies of the same species (*Gilia splendens*) to related species of the same genus (*Polemonium, Gilia, Linanthus, Ipomopsis, Cobaea, Loeselia*) and finally to the origin of a genus in association with adaptation to a particular pollinator (Lepidopteran pollination in *Phlox*).

As a result of their investigations, the Grants were able to construct a probable phylogeny of pathways of adaptive radiation in the family (Fig. 4-1). In addition to frequent reversions toward autogamy, six different lines of radiation could be detected. Most of these were followed independently by several different parallel lines. Thus the shift from bee to hummingbird pollination has occurred seven times, that from bees to long-tongued flies ten times, to butterflies (day-flying Lepidoptera) three times, to noctuid moths four times, to beetles twice, and to scavenger flies once.

The second important monograph on adaptive radiation with respect to floral biology is that of van der Pijl and Dodson (1966) on Orchidaceae. These authors have presented a systematic and analytical review of the complex relations that have long been recognized to exist between these remarkable flowers and their pollinators, and have added many observations and illustrations of their own, particularly from the flora of tropical America. As in the Polemoniaceae, the most primitive condition in Orchidaceae is adaptation to hymenopteran vectors. Adaptive radiations to hummingbird, scavenger-fly, beefly (Bombylidae), butterfly, moth, and possibly beetle pollination occur, and there are many instances of reversion to autogamy.

An important feature of adaptive radiation in orchids, which is poorly developed in Polemoniaceae, is adaptation to pollination by distinct groups of Hymenoptera. The more primitive terrestrial genera, including *Cypripedium*, are pollinated chiefly by solitary bees, as are the more primitive Polemoniaceae. A few temperate genera (*Listera, Coeloglossum*) are pollinated by Ichneumonid wasps. Among tropical epiphytic orchids, several genera (*Vanda, Phajus, Oncidium, Barkeria*) have large flowers of firm texture that are adapted to pollination by large carpenter bees (*Xylocopa* spp.). On the other hand, social bees, which are the most common pollinators of specialized flowers belonging to other

Fig. 4-1. Adaptive radiation for pollination by different pollen vectors in the Phlox family (Polemoniaceae). The following species of pollinators and flowers are illustrated as representative of each group: bees, *Polemonium reptans* and *Bombus americanorum;* bats, *Cobaea scandens* and *Leptonycteris nivalis;* cyrtid flies, *Linanthus androsaceus croceus* and *Eulonchus smaragdinus;* bombylid flies, *Gilia tenuiflora* and *Bombylius lancifer;* beetles, *Linanthus Parryae* and *Trichochrous* sp. (Melyridae); noctuid moths, *Phlox caespitosa* and *Euxoa messoria;* hawk moths (Sphingidae), *Ipomopsis tenuituba* and *Celerio lineata;* diurnal lepidoptera, *Leptodactylon californicum* and *Papilio philenor;* birds, *Ipomopsis aggregata* and *Stellula calliope;* self (autogamous), *Polemonium micranthum* (*bottom*), *Gilia splendens,* desert form (*left*), *Phlox gracilis* (*lower right*). (Redrawn from Grant and Grant 1965.)

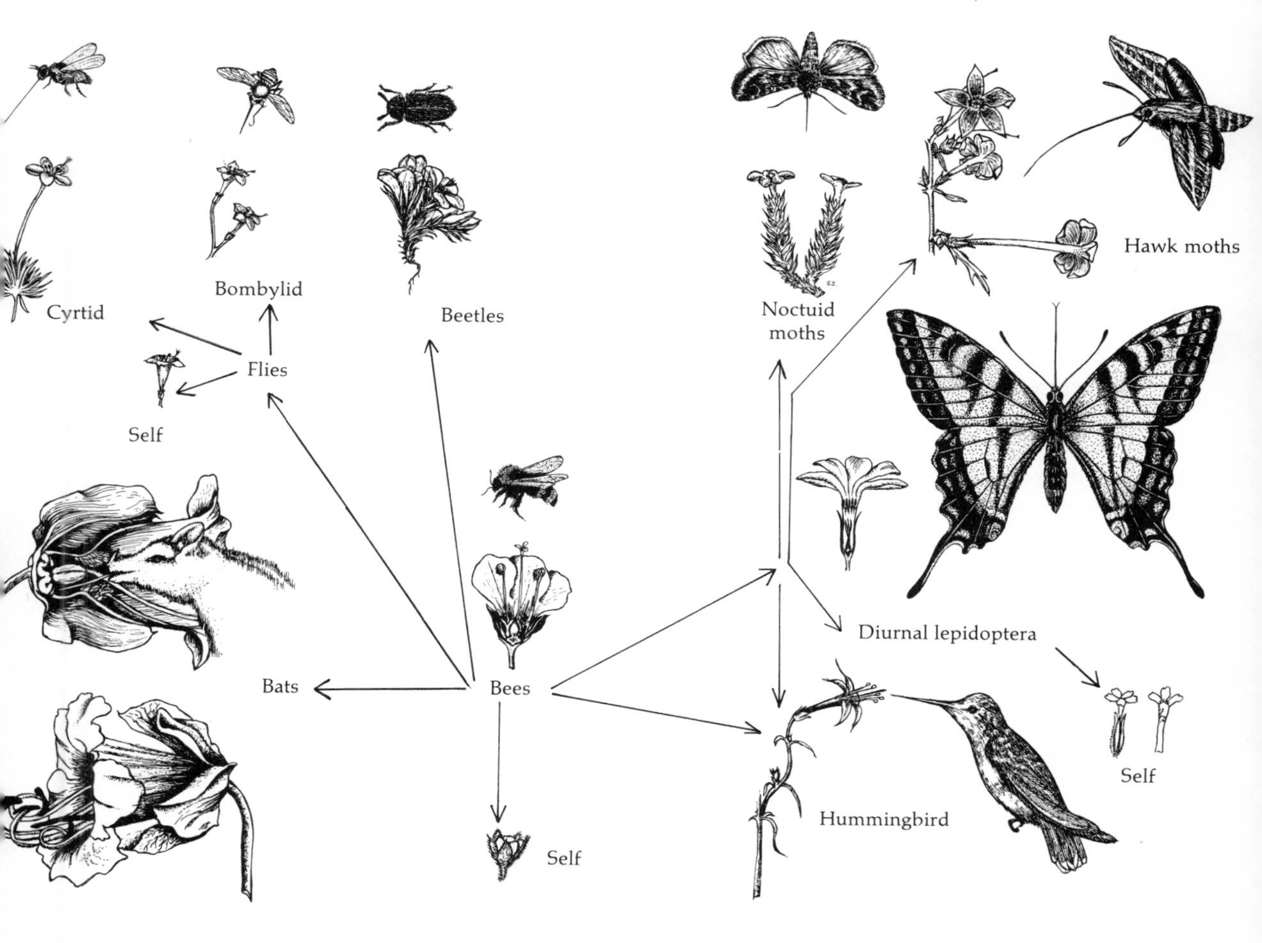

families, have only rarely become regularly associated with species of orchids. The most remarkable adaptive radiations in orchids are to bees of the genera *Euglossa* and *Eulaema* in tropical America, and to males of various groups of Hymenoptera in the Mediterranean genus *Ophrys,* as well as to male wasps in the Australian genera *Cryptostylis* and *Calochilus.* In these two latter examples, pollination is effected by the well-known behavior of pseudocopulation. The relation between male euglossine bees and species of orchids is much more complex. The orchids produce a variety of fragrant substances similar to 1,8-cineole, which attract the male bees (Dodson *et al.* 1969, Dressler 1968). The bees then gather the substances and store them in their hind tibia, where they serve either as sex attractants or to attract other males of their species into clusters about which the females gather. Because of the specificity of the relation between bee and flower, it has brought about extensive speciation of both the bees and the orchids.

Less complete data from other genera support these conclusions about adaptive radiation at the species level in response to different pollinators. The elaborate corollas in species of *Pedicularis* are mostly pollinated by bumblebees (*Bombus;* Macior 1966, 1968), but at least one species, *P. densiflora* of California, is adapted to hummingbirds (Sprague 1962). Asiatic species of *Pedicularis* have evolved very long corolla tubes, similar to those that in other genera are associated with pollination by Lepidoptera or long-tongued Diptera (Li 1951). This shift has occurred independently in several different species complexes. Adaptation of the flowers of *Lantana* (Verbenaceae) to particular species of Lepidoptera has been noted by Dronramju (1958, 1960, Dronramju, Rao, and Spurway 1960) and may be one basis of speciation in this complex genus. In the genus *Penstemon,* different species groups are adapted to various genera and families of Hymenoptera, as well as to hummingbirds (Crosswhite and Crosswhite 1966, Straw 1956*a,b*). A similar situation exists in *Salvia* (Grant and Grant 1964, 1967). In the California flora, at least 45 separate adaptations to hummingbird pollination, in 18 different families, are confirmed by direct observation (Grant 1952, Grant and Grant 1965, 1967). Specificity of visits (oligolecty) is probably responsible at least in part for the diversity of species of *Oenothera* and its relatives in western North America (Linsley, MacSwain, and Haven 1963, 1964), as well as for the similar diversity of Myrtaceae in Australia (Michener 1965). In tropical floras, several genera of various families have evolved large, thick-textured flowers that are pollinated principally by carpenter bees (Xylocopa; van der Pijl 1954).

Two kinds of adaptations to pollinators have not been discussed in the review presented above: primitive beetle pollination and bat pollination. The opinion originally expressed by Diels (1910), that the earliest angiosperms were pollinated by beetles (Coleoptera), has been made increasingly plausible by more recent observations on beetle pollination of various primitive genera, such as *Calycanthus* (Grant 1950*a*), *Eupomatia* (Hotchkiss 1958),

Magnolia (Heiser 1962), and the Annonaceae (van der Pijl 1953, 1960). Many species of Ranunculaceae attract beetles along with other pollinators (Leppik 1964), but few if any members of this family are predominantly or exclusively pollinated by these insects. Leppik (1960) has made the interesting suggestion that beetles and other primitive insects first became pollen gatherers in association with the Jurassic Bennettitales, and later transferred their adaptations to the evolving angiosperms.

Although the most primitive flowers may well have been pollinated by beetles, one cannot conclude from this fact that beetle-pollinated plants in the modern flora are ipso facto primitive. Secondary adaptation to coleopteran pollination has been well documented by Grant and Grant (1965) in *Linanthus* and *Ipomopsis*. The pollination by Coleoptera of the white-flowered Japanese species of *Trillium* (Fukuda 1961) is probably also secondary, since this genus is a relatively specialized member of the Liliales. In view of this fact, the suggestion by van der Pijl (1961) that the Compositae, the most advanced of flowering plants, have retained primitive coleopteran pollination must be regarded with some skepticism.

Adaptations to pollination by bats (Chiroptera) were the last to be recognized by naturalists. This is not surprising, since this kind of adaptation exists only in the tropics, and observations on it must be made at night. The earlier observations on these pollinators by Porsch (1934–35), have been verified and amplified by van der Pijl (1956, 1961), Baker and Harris (1959), Harris and Baker (1958, 1960), Vogel (1968, 1969), and others (Faegri and van der Pijl 1966). The plant genera involved belong to various families. The best-documented of them are *Parkia* (Leguminoseae, Mimosoideae), *Ceiba* (Bombacaceae), *Kigelia* (Bignoniaceae), *Caryocar* (Caryocaraceae), and *Couepia* (Chrysobalanaceae). The bat pollination of *Cobaea* (Polemoniaceae) has been reviewed by Grant and Grant (1965) and verified by Vogel (1969).

Bat pollination is one of the most specialized and probably most recent adaptations of flowers for pollen vectors. As Baker (1961*b*) and Vogel (1968, 1969) have pointed out, bat-pollinated flowers must be large, and are produced on robust flower stalks that extend outward from the foliage or on long, pendent peduncles. Their parts must also be thick and tough, to survive the bitings and scratchings to which they are subjected by their rude pollinators. Furthermore, they must have powerful odors and excrete large quantities of nectar. In all probability, therefore, bat pollination has in every case evolved secondarily from adaptation to other large pollen vectors, such as birds, sphingid moths, and carpenter bees. Close relationships between species of the same genus, some pollinated by sphingid moths and others by bats, have been recorded by Vogel (1968) for South American species of Cactaceae subfamily Cereoideae, and for *Capparis*, *Inga*, and *Nicotiana*. Similar relationships between bat-pollinated and bird (hummingbird)-pollinated species were noted in various genera of Leguminosae, in *Vriesia* (Bromeliaceae), and particularly in *Burmeistera, Centropogon, Siphocampylus,* and their relatives in the Lobeliaceae.

Differential Adaptations to the Same Pollinators

In addition to adaptive radiation for pollination by different vectors, many genera have evolved distinct species adapted in different ways to pollination by the same vector. One of the most clear-cut examples consists of two species of *Pedicularis, P. attollens* and *P. groenlandica,* in the Sierra Nevada of California. These two species occur sympatrically over a wide area; they are very similar in growth habit but have different ecological preferences. They resemble each other in flower color and structure. The chief difference between their flowers is in the length of the beak of the corolla, which encloses the style and stigma. Sprague (1962) has observed pollination of both species by the same species of *Bombus.* Nevertheless in *P. attollens,* the bee receives and transfers pollen on its head, while the longer beak of *P. groenlandica* is such that pollen transfer is on the bee's venter. The pollination of *P. groenlandica* has also been observed by Macior (1968) in the Rocky Mountains, where *P. attollens* does not occur. Although Macior's observations disagree with those of Sprague with respect to the behavior pattern of the bee, the two agree that the pollen is deposited on the bee's venter.

Much more extensive differentiation of species in adaptation to different methods of pollination by the same vectors has taken place in the Orchidaceae, and is well documented by van der Pijl and Dodson (1966). Another family in which this kind of differentiation may have been responsible for the principal morphological differences between species is the Asclepiadaceae (milkweed). Unfortunately, no recent studies of this family are available, the latest known to me being Woodson's (1954) monograph of North American species of *Asclepias.* There he reviews the observations of earlier naturalists, who showed that all of the species may be visited by a large number of different kinds of insects belonging to various families and orders, and the same species of insect may visit on the same day several different species of milkweeds. Nevertheless, hybridization between species is exceptionally rare. This is due partly to the inviability of hybrid embryos, but also because of the peculiar and specific morphology of their pollen sacs, or pollinia. These are borne in pairs that are connected by clips, in such a way that an insect visiting the flower for nectar picks up pollen clips on its hind legs. On a visit to another flower, it may deposit a clip with its pollinia in the stigmatic sac and so effect pollination, but only if the pollinium belongs to the species on which the insect has landed. The fit between pollinium and stigmatic chamber is so precise that only pollination between different individuals of the same species can be carried out, even though the vector may be carrying on its legs pollinia derived from several species of *Asclepias.*

This condition almost certainly is the climax of a long course of adaptive radiation for specificity of pollination. Perhaps noteworthy in this connection are the observations of Holm (1950) on the related genus *Sarcostemma* of Mexico. The species of this genus are apparently visited by oligolectic insects, which confine their visits to a single species of flower. In all probability, this kind of specificity preceded the present situation in *Asclepias,* since otherwise it is difficult to see how the morphological differentiation was brought about.

Two other examples of adaptive radiation for different methods of pollination by the same vectors are in the families Aristolochiaceae and Araceae. In both instances, the flowers act as "fly traps," capturing small flies and imprisoning them until pollination is effected. The European genera of Araceae are described by Knuth (1905), Kugler (1955), and Werth (1956). The same volumes also have accounts of the Aristolochiaceae, including the European species of both *Asarum* and *Aristolochia.* Information on the "fly trap" mechanism of pollination in other species of *Aristolochia* is given by Cammerloher (1923) and Lindner (1928). The only more recent observations known to me are my own and those of others on *Aristolochia californica* (Fig. 4-2). This latter species, along with several others, differs from *A. clematitis* of Europe, which is often figured in textbooks, in not possessing the stiff, downward-pointing hairs that prevent the flies from climbing out of the trap until pollination has been effected, when they shrivel up. Instead, the traps of *A. californica, A. sipho,* and other species have light "windows" arranged in such a way that the flies, because of their phototropism, beat against the walls of the perianth rather than flying out through its opening

Fig. 4-2. Floral structure of two species of *Aristolochia,* illustrating two kinds of "fly-trap" flowers. (*A–C*) *A. clematitis:* (*A*) cluster of flowers; (*B*) single unpollinated flower in cross section, showing downward-pointing hairs that allow the flies to enter but prevent them from emerging, also stigma lobes spreading downward over the undehisced anthers; (*C*) older, pollinated flower, in which the hairs have dried up so that the flies can escape, stigma lobes pointing upward, revealing the dehisced anthers and pollen. (From Knuth 1894.) (*D–I*) *A. californica,* a light-trap flower that functions because of the phototropic responses of the pollinating flies: (*D*) single pendent flower; (*E*) flower viewed from its orifice, as it would appear to the entering fly; (*F*) upper half of the flower, showing (*below*) the opaque upper lip, which excludes light from this direction and prevents the trapped fly from emerging, and (*above*) the stamens and stigma surrounded by a halo of light formed by the thin, colorless wall of the proximal end of the flower; (*G*) longitudinal section of young flower, opaque upper lip and stigma lobes covering the undehisced anthers; (*H*) longitudinal section of older flower, showing stigma lobes turned upward and anthers exposed, also enlarged, paler upper lip that now enables the positively phototropic fly to emerge; (*I*) a fly (family Mycetophilidae) captured inside the flower, and bearing pollen grains (pale dots) on its body. (Original.)

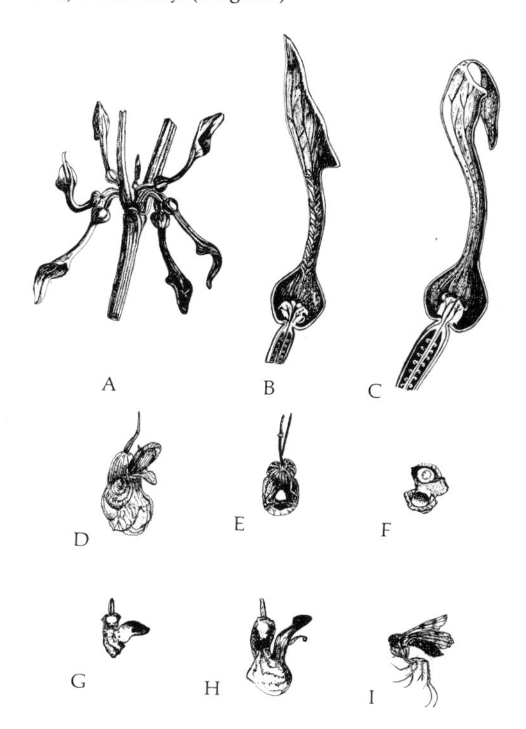

(Fig. 4-2, *F*). Escape becomes possible after pollination, owing to the expansion of the perianth. The approximately 180 species of *Aristolochia*, plus the 20 of *Asarum* and a few that belong to other, smaller genera (M. P. Gregory 1956), present a good example of a family in which differentiation of both species and genera has been based largely upon diversification of adaptations for pollination by one kind of vector. More observations on its species are badly needed.

Principles Involved in Adaptive Radiation for Pollination

If adaptive radiation for characters of floral biology is to be understood as a series of examples of the action of natural selection, observations and interpretations must be made with this factor definitely in mind. The following principles are essential for understanding the process.

1. *The most-effective-pollinator principle.* Since selection is a quantitative process, the characteristics of the flower will be molded by those pollinators that visit it most frequently and effectively in the region where it is evolving. Hence when an evolutionist speaks of a "bee flower," or a "hummingbird flower," he refers to its relation with the predominant and most effective vector. He does not mean that the flower is pollinated exclusively by this vector. Secondary vectors, such as *Bombus* on *Aquilegia canadensis* (Macior 1966), may retard the process of evolutionary modification, but they are not likely either to stop it or to destroy the genetic integration of a floral type once it has evolved. Failure to recognize this principle has resulted in some unwarranted skepticism as to the action of selection by pollinators in the origin of particular forms of flowers.

2. *The significance of character syndromes.* The more specialized vectors are attracted to flowers by a variety of stimuli, of which scent may be even more important than either shape or color (Bateman 1951, Dodson *et al.* 1969, Kugler 1955, Kullenberg 1956*a,b*, 1961, Manning 1956). With respect to color, we must recognize the fact that both the visual sense of the vector and its reaction to color may be very different from our own. For instance, the greater sensitivity of bees to ultraviolet light renders highly significant the much greater contrast of pattern than can be seen in photographs of flowers taken with an ultraviolet-sensitive camera, as compared with the naked eye (Kugler 1963, Lutz 1933). K. Grant (1966) has shown that hummingbirds in a back yard do not prefer red feeders to those of other colors, provided that the food is the same. Nevertheless, the great sensitivity of the bird's eye to red colors must mean that a red flower is more easily detected from a long distance than one of another color. In view of the rapidity of the hummingbird's flight, this factor is probably of considerable importance. Leppik (1953, 1956) has emphasized the ability of bees and other specialized insects to recognize patterns based on definite numbers of parts, for example, tetramerous or pentamerous flowers, as the basis of the constancy of these patterns. This hypothesis needs experimental confirmation.

Leppik (1957, 1964) has attempted to classify and place in an evolutionary sequence the different forms of flowers that are related to specific modes of pollination. This attempt has met with only partial success, since the types are very difficult to classify, and sometimes a particular form is not associated with the expected pollinator. For instance, the cup-like flower of *Trollius* has a form resembling that of most coleopteran-pollinated flowers, but actually it is pollinated by a fly (Leppik 1964). A somewhat more satisfactory classification is presented by Faegri and van der Pijl (1966). These attempts suggest that, at least in the present state of our knowledge, we shall learn more from making direct studies of the functional relations of particular kinds of flowers to clearly identified pollinators than from attempting new classifications or evolutionary generalizations. The important facts, which have already been reviewed, are the numerous correlations that exist in most flowers between scent, color, form, texture, and the recognized behavior patterns of the predominant pollinators.

Perhaps the most important feature of character syndromes is that they favor correlations between characters. One of the most conspicuous of these is between zygomorphic or bilaterally symmetrical flowers and racemose or spicate inflorescences (Stebbins 1951). Pollination by insects that land on a flower, such as bees, is most efficient if the flower possesses a "landing platform" upon which the insect can alight before entering it to seek nectar. If the inflorescence is a flat-topped or rounded cyme or corymb, as in *Ranunculus, Gentiana,* or the Compositae, the flower is borne on an erect stalk, and its horizontally spreading petals or rays can provide the platform. If, however, the inflorescence is a raceme or spike, the flower stalk projects horizontally, so that the petals or rays of a radially symmetrical flower would be vertically oriented, and so would provide a poor platform. Under these conditions, genetic changes that would modify the flower into a zygomorphic shape, providing the platform by modifying the lower petals or corolla lobes so that they assume a horizontal position, would have a high adaptive value. On the other hand, in flowers pollinated chiefly by vectors that hover rather than land on the flower, such as Lepidoptera and hummingbirds, radial symmetry would be compatible with a racemose or spicate inflorescence. This condition is found in *Ipomopsis aggregata* (Polemoniaceae), the genus *Nicotiana,* and elsewhere.

In my opinion, failure to recognize the significance of character syndromes as the principal basis of character correlations has led to many mistaken conclusions about trends of evolution in flowering plants, particularly those of Sporne (1948, 1956, 1959, 1960).

3. *Selection along the lines of least resistance.* The flower-structure–pollen-vector relation is one that offers many different evolutionary pathways involving both adaptation to different vectors and different ways of adaptation to the same vector. When this situation exists, the particular pathway that a given line will take will depend upon the principle of selection along the lines of least resistance (Stebbins 1950:497, 1967; see also Chapter 2 of this volume). The significant factors in the flower-vector relation are (1) frequency of

a particular vector, (2) limitations imposed by the existing structure of the flower, and (3) limitations imposed by the external environment. The evolutionary shift from one vector to another is probably triggered by the entrance of the plant into a habitat where the original vector is scarce and the new vector is abundant. For instance, many of the hummingbird-pollinated flowers of western North America, which for the most part are derived from bee-pollinated flowers, occur most abundantly and probably originated in forests, where bees are uncommon, or flower at times when bees are less than normally active (Grant and Grant 1964, 1967). The shift to pseudocopulation on the part of *Ophrys* species in the Mediterranean region is associated with their adaptation to relatively dry sites, in which early seed maturation is essential, so that dormancy can begin when the dry season comes. Since male hymenoptera usually emerge earlier than workers (Kullenberg 1961), adaptation to these vectors may have originated in response to selection pressure for early flowering.

Limitations imposed by the existing structure of the flower are responsible for many of the radiating pathways shown by Grant and Grant (1965). The long corolla tube characteristic of most flowers pollinated by birds, long-tongued flies and Lepidoptera can evolve only from corollas that are already somewhat tubular, as adaptations to hymenopteran pollination. On the other hand, the relatively broad chambers characteristic of fly-trap flowers are more likely to evolve from coleopteran-pollinated flowers, which have a similar shape.

The environment can be limiting chiefly in its influence on the time available for flower development and the conditions that prevail at the time of pollination. Primitive beetle-pollinated as well as bat-pollinated flowers are most common in the tropics because their massive structure can develop best in climates in which the favorable period for flower and seed development is very long (Vogel 1969, Gottsberger 1970). Adaptation to moth pollination, involving the shift from diurnal to vespertine or nocturnal flowering, is most likely to take place in hot, dry climates, where the high temperature and dry atmosphere that prevail during the day are conditions unfavorable for flowering and successful pollination. At the other end of the climatic spectrum, adaptation to indiscriminate pollination by nonspecific flies and other small insects is most likely to evolve in cool, shady regions, such as temperate forests. Finally, the shift from insect to wind pollination is most likely to take place in regions where, because of either drought or cold, insects are scarce and the openness of the plant community promotes exposure to strong winds.

4. *Transfer of function via an intermediate stage of double function.* The shift of adaptation of flowers from one specialized pollinator to another must be based upon the same kind of change in selection pressures as that which accompanies the change of function in a particular organ of an animal. This shift is generally recognized by zoologists to require an intermediate stage in which either the same organ performs two functions or the same function is performed by two different organs

(Corner 1958, Mayr 1960). Since in either animals or plants the function of reproduction is carried out by only one set of organs, only the intermediate stage of double function is possible for these organs. In terms of flower pollination, this means that the shift from one specialized pollinator to another can be carried out only via an intermediate stage during which both vectors are capable of pollinating the flower. Situations of this sort are described in many of the references already given. They emphasize even more the need for avoiding a classification of flowers into sharply delimited types, and the implication that flowers belonging to one particular type are pollinated exclusively by one particular vector.

5. *Reversals of evolutionary trends.* If we view phyletic evolution as the outcome of a succession of adaptive radiations, we must recognize the probability that reversals of general trends can occur frequently, whenever a reversal of environmental conditions elicits them. This principle is widely recognized by zoologists, and is well documented by the fossil record (Chapter 7). When these reversals occur, however, they do not retrace the original evolutionary pathway with respect to details of structure, but only with respect to general adaptation. In terms of pollinating mechanisms, this means that we can expect reversals with respect to the adaptation of flowers to particular vectors, but these reversals will not restore the primitive adaptation to the original vector. Shifts from wind to insect pollination and vice versa provide numerous examples of this principle. Most systematists who have studied the origin of the flowering plants believe that the original ancestors of the angiosperms were wind pollinated, but that the shift from wind to insect pollination took place along with the origin of the angiospermous condition itself (Cronquist 1968, Takhtajan 1969). According to this theory, all of the numerous groups of wind-pollinated angiosperms must be regarded as the products of secondary reversal. Such secondary reversal is well documented in some instances, such as the genera *Thalictrum* and *Artemisia,* as well as the subtribe Ambrosiinae of the Compositae, mentioned above. Documentation for woody families of temperate regions, such as the Betulaceae, Fagaceae, and other families formerly grouped together as "Amentiferae," is less clear because of their great age, but nevertheless indicates that most if not all of them are the products of secondary reversal to wind pollination.

If this is so, then genera such as *Ficus,* mentioned above, as well as *Salix* and other examples, represent tertiary reversals: primitive ancestors (wind) → earliest angiosperms (insects) → primitive Moraceae (wind) → *Ficus* (insects). In the case of the last reversal, the structural readaptations to insect pollination are entirely different from those that existed in the primitive insect-pollinated ancestors. In many instances, readaptation of wind-pollinated angiosperms to secondary insect pollination involves the appearance of conspicuous colors in the stamen filaments, as in *Salix,* and probably the insect-pollinated species of *Thalictrum.* Here the principle of selection along the lines of least resistance is operating, since the acquisition of colored pigments in stamen filaments involves less reorganization of the developmental pattern than the restoration of petals that have been lost.

Similar reversals from one vector to another and back to the same kind of vector have undoubtedly been frequent, but they are much more difficult to recognize. Possible examples are the two shifts from hymenopteran to coleopteran pollination in the Polemoniaceae, already mentioned (Grant and Grant 1965). If, as seems probable, the earliest angiosperms were pollinated by Coleoptera, the shift from coleopteran to hymenopteran pollinations must have taken place at some point along the line of evolution that led from primitive angiosperms to the relatively advanced Polemoniaceae. The two species that are revertants with respect to the pollen vector, *Linanthus parryae* and *Ipomopsis congesta,* bear hardly even a superficial resemblance to primitive angiosperms in their floral structure. We can conclude, therefore, that the shift from one vector to another, including reversals to the original vector, involves very generalized alterations of selective pressure that can elicit any one of a number of structural modifications of the flower. Under these conditions, the principle of adaptive modification along the lines of least resistance can be expected to appear in its most extreme form.

Relations Between Adaptations for Pollination and Levels of Overall Specialization

Several attempts have been made to relate the kind of adaptation for pollination to the degree of specialization or phylogenetic position of the group concerned. For the most part, these efforts have failed because they did not recognize the degree of differentiation between species of the same genus that can occur in relatively primitive as well as in highly advanced groups. In addition, reversals of the general trends, accompanied by different morphological specializations, are all too common. For these reasons, the type of pollination mechanism that is predominant within a group is by no means a reliable guide to its phylogenetic position. Nevertheless, a general series of parallel trends can be recognized, which has been well characterized by Vogel (1963) as follows:

Lowest rank: Primitive anemophily, coleopteran pollination;

Next rank: Secondary anemophily, hydrophily, generalized insect pollination;

Highest rank: Pollination by specialized vectors, such as social Hymenoptera, carrion flies, sphingid moths, birds, and bats.

The advance through these levels or grades of specialization may take place within the confines of a single family, as in Annonaceae and probably Dilleniaceae; it may be more characteristic of orders that contain several families, as in Rosales and Fabales; or it may progress through a series of different orders, as in most groups of monocotyledons. The correlations between adaptations for different methods of pollination and phylogenetic trends involving the high categories of the taxonomic hierarchy are by no means well marked.

Summary

The principal conclusions that emerge from this review are the following: The diverse floral structures and pollination mechanisms found in angiosperms represent a series of adaptive radiations to different pollen vectors and different ways of becoming adapted to the same vector. Although most of them owe their adaptive success to their promotion of cross-fertilization, adaptations for self-fertilization have arisen repeatedly. They have an adaptive advantage in association with the colonization of new habitats, particularly those that are remote from the main area of a species' distribution and that have unfavorable conditions for pollination. Although the earliest angiosperms were probably insect pollinated, wind pollination arose several times in relatively primitive groups, as well as in advanced ones such as the Compositae. Adaptive radiation toward wind pollination was favored in groups inhabiting dry or cold climates, where pollinators were scarce, the plant formations were relatively open, and winds were strong. Secondary reversion to insect pollination probably took place in some derivatives of wind-pollinated angiosperms, such as *Ficus*, that became readapted to moister, more equable climates.

Adaptive radiation for pollination by animal vectors has been recognized in two families that have been intensively studied, the Polemoniaceae and the Orchidaceae. In these families, adaptation to solitary aculeate Hymenoptera and other relatively unspecialized insects is the primitive condition. Several lines of radiation have led to pollination by more specialized Hymenoptera (*Bombus, Xylocopa*), as well as by long-tongued flies (Bombylidae, Syrphidae), day- and night-flying Lepidoptera, Coleoptera, and hummingbirds (Trochilidae). Each of these adaptations involves characteristic modifications of the color, form, and odor of the flower. In the orchids, two adaptations for attracting male Hymenoptera as pollen vectors have evolved: resemblance to females, which induces pseudocopulation, and chemical sex attractants. Some flowers of more primitive families are adapted to pollination by Coleoptera, and this may have been the earliest form of insect pollination. Pollination by carrion flies, involving the modification of flowers into flytraps, probably represents secondary derivation from coleopteran pollination.

Diversification of species in some genera (*Asclepias, Pedicularis*) is associated with the evolution of different structural adaptations for pollination by the same vector.

Adaptive radiation for pollen vectors involves the following evolutionary principles: predominant influence of the most effective pollinator, adaptive syndromes of characters, selection along the lines of least resistance, transfer of function via an intermediate stage of double function, and reversals of adaptive trends. Each of these phenomena contributes to the complexity of adaptive radiation in higher plants.

5 / Adaptations for Seed Development and Dispersal and for Seedling Establishment*

The point emphasized in the last chapter, that adaptations of the reproductive organs for cross-pollination require higher levels of integration between different organs than do adaptations for survival, applies with even greater force to adaptations for seed development and dispersal and for seedling establishment. Moreover, the adaptive values of seed size, seed number, and the nature of the embryo and of the stored food that is associated with it may conflict with one another to such an extent that reproductive efficiency in a particular habitat depends upon a compromise between these conflicting demands. The number and diversity of such compromises account to a great degree for the diversity of angiosperms with respect to those characteristics that are ordinarily regarded as diagnostic for their major categories. The present chapter is intended as an exploratory review of these characteristics, and of their relation to genotype–environment interactions that take place between the termination of flowering and the successful establishment of the next generation of seedlings.

This review must be at an exploratory level rather than a final exposition because of the paucity of our knowledge concerning seed and seedling ecology. In contrast to flowers and their pollinators, which have been the focus of a greatly revived interest in natural history during the past few years, seeds and seedlings have received relatively little attention, and hardly any comparative systematic

*The material of this chapter is largely reprinted, by permission of Annual Reviews, Inc., from Stebbins (1971*b*).

ecological studies have been carried out on them. One reason for this is the greater difficulty of comparative studies of seeds and seedlings relative to studies of pollination. Flower pollination is a single event that takes place over a relatively short period of time and can be understood on the basis of one set of observations. Seed reproduction, however, includes three different kinds of processes that may be widely separated from one another in both space and time: seed development, seed dispersal, and the establishment of seedlings. Since successful reproduction depends upon compromises between the often conflicting demands of these three processes, all of them must be studied in relation to one another before valid conclusions can be reached about their entire adaptive significance. Moreover, whereas pollination mechanisms depend largely upon coordinations between visible morphological structures (Grant 1949), seed and seedling ecology depends upon both morphological structures and physiological processes, such as fruit ripening, seed dormancy, and seedling growth. For successful research in the latter field, therefore, one must become at the same time a comparative morphologist, a comparative physiologist, and an ecologist.

Significant Characteristics of Seeds and Seedlings

Harper, Lovell, and Moore (1970) have presented a table that illustrates the enormous range in seed size that exists between different species of angiosperms, from seeds weighing 0.000002 g (the orchid, *Goodyera repens*) to those weighing 27,000 g (the palm, *Lodoicea maldivica*). On the other hand, among seeds belonging to the same plant, and usually among the individuals of a population, seed size is one of the most constant characteristics. This constancy is probably associated with the adaptive value of a narrow range in seed size for most species. The table presented by Harper, Lovell, and Moore shows that for many kinds of ecological habitats a particular range of seed sizes is characteristic. Epiphytes (orchids), plants that occupy similar habitats on rocks in tropical forests (Gesneriaceae, Begoniaceae), saprophytes, and parasites have the smallest seeds. In temperate mesic regions, such as Britain, herbs of open ground tend to have smaller seeds than forest herbs, shrubs have larger seeds, and trees have the largest seeds of all (Salisbury 1942). In the moist tropics (Corner 1954*b*), the situation is somewhat similar: a megaspermous (large-seeded) group of plants can be recognized that includes most of the trees forming the upper canopy of the forest (Bombacaceae, Burseraceae, Dipterocarpaceae, Fagaceae, Lauraceae, Lecythidiaceae) as well as many shrubs and vines (Annonaceae, Connaraceae, Myristicaceae), whereas nearly all of the species in the microspermous (small-seeded) group (Compositae, Ericaceae, Hypericaceae, Melastomaceae, Urticaceae) are either herbs, shrubs, or small trees. On the other hand, the situation in arid and semiarid regions is quite different. There, species of herbs inhabiting open country may differ greatly from one another with respect to seed size, as in grass genera such as *Bromus* and *Stipa* (Stebbins 1956*a*), and the larger-seeded herbs tend to have larger seeds than those possessed by any of the shrubs. In these drier regions, the degree of aeration and the moisture-holding capacity of the soil are probably the most important environmental determinants of optimum seed size, but comparative data on this point are still largely unavailable.

Variations in Seed Number

The range of variation with respect to seed number per plant is as great as or greater than that with respect to seed size. A single seed capsule of an orchid (*Cychnoches chlorochilon*) may contain as many as 3,770,000 seeds (Correll 1950). Moreover, a large tree of *Betula, Populus,* or *Salix* probably produces thousands of seeds each year if good conditions for pollination have prevailed, although exact figures on this number do not appear to be available. In trees such as *Ceiba, Adansonia,* or other Bombacaceae, which produce large numbers of capsules each containing hundreds or thousands of seeds, the total number per plant must be even greater. On the other hand, many plants, particularly herbs that bear large seeds, produce relatively few. In the annual flora of regions having a dry season, plants belonging to such species as *Bromus rigidus* or *Erodium botrys* may produce only 10 or 20 seeds during their entire life cycle, particularly if they are growing in an unfavorable habitat.

In contrast to seed size, seed number is subject to very great phenotypic modification, depending upon the environment. In annual plants, seed number is regulated to a large extent by the overall vigor of the plant and the number of flowers that it can produce. In perennial herbs, particularly those having complex flowers like orchids and Asclepiadaceae, successful pollination is an important factor. Finally, in trees and many shrubs, particularly those that have large seeds, the physiological and environmental conditions that prevail during seed development may be of paramount importance. In forest trees such as oaks (*Quercus*) and hickories (*Carya*), seed production may vary greatly from one season to another. In most plants, reproductive success depends upon strong buffering and canalization of those processes that contribute to seed size and shape, with corresponding phenotypic flexibility with respect to those processes that contribute to fecundity or seed number. This may well be due to the high adaptive value of a precise adjustment of the seed and its resulting seedling to the environment that is most favorable for seedling establishment in the species concerned.

Differences with respect to the number of seeds produced by an individual plant can be resolved into differences in the following factors: (1) the number of seeds per ovary locule or "carpel"; (2) the number of ovary locules or "carpels" per flower; (3) the number of flowers per inflorescence; and (4) the number of inflorescences per plant. The operation of these four factors makes possible divergent reactions of different evolutionary lines to particular kinds of selective pressure, such as that for greater fecundity or seed production.

Adaptations for Protection of Developing Seeds

After pollination and fertilization, the next critical stage in the reproductive cycle is the development of the ovule into the seed. During this period two kinds of protection are essential: (1) protection of the tender, growing tissues of embryo and endosperm from drying and other kinds of external damage and (2) protection of these tissues from predation, particularly by insects. Depending upon the family or genus of plants, the first kind of protection is provided by the ovular integuments, the ovary wall, the calyx, the bracts or phyllaries surrounding the flower, or a combination of these structures. In many evolutionary lines of angiosperms, a regular succession of transference of function has occurred, in which the protective function has been taken over successively by structures further and further removed from the seed itself. These successions are discussed below. The point is made that successive transference of function is found in groups having medium- or large-sized seeds, and is virtually absent from groups that have been continuously characterized by small seeds. This restriction can be explained if one assumes that the adaptive value of protection during early development is positively correlated with the length of the developmental period, which will, in general, be longer for large seeds than for small ones.

Protection of developing seeds from insects is sometimes achieved by special developments of the various protective coverings. In the family Compositae, the strategy of many capitula, each containing only a few florets, as compared to a few many-flowered capitula, confers a distinct advantage in this respect (Burtt 1960). In addition, protection from insect predation is in many groups achieved by the presence of toxic substances or other chemical deterrents in the embryo, the endosperm, or both. This subject has been reviewed by both Janzen (1969) and Harper, Lovell, and Moore (1970). The relation between the development of these substances and both seed size and seed number is discussed under "interaction of factors."

Adaptations for Seed Dispersal

The only phase of seed and seedling ecology that has been extensively studied by many authors is seed dispersal. Since reviews and descriptive accounts of the various mechanisms are available elsewhere (van der Pijl 1969, Ridley 1930), the present account will be limited to a listing of the various methods and a commentary on their adaptive significance.

Mechanical dispersal of seeds

Devices for ejaculating seeds forcibly from their capsules are well known and have been described in many genera. They are, however, relatively unimportant with respect to both the overall dispersal of species and genera and the guidance of evolutionary trends. Since the greatest distances to which seeds are ejaculated are only a few meters (15 m in *Bauhinia purpurea* and 14 m in *Hura crepitans*), they serve chiefly to prevent the accumulation of large numbers of seeds within the shadow and root circumference of the adult plant. In order to migrate for long distances, as species of such genera as *Bauhinia* and *Impatiens* must certainly have done, seeds must be adapted for transport by other means.

Dispersal by water. As might be expected, mechanisms for the dispersal of seeds by water have been developed both in plants that live in or near fresh water and in those adapted to seashore habitats. Nevertheless, the kinds of adaptations that these two groups of plants have evolved are of necessity entirely different. Since fresh water does not damage living tissues, mechanisms for aquatic dispersal of freshwater species can be relatively simple. In many instances, seeds or small fruits fall on the surface of the water, are transported for some distance, and finally sink. In the case of species that inhabit stream banks, devices that help to anchor the seeds and keep them from being carried away by the current are likely to have an adaptive value. Ridley (1930) maintains that the ridges, papillae, and other kinds of emergences that often exist on the seed coats of aquatic species are anchoring devices. This interesting conclusion needs to be verified by additional observations and experiments. A number of freshwater species have bladder-inflated seeds or fruits that can float on the water for considerable periods of time. In the large genus *Carex,* the differentiation of certain of its most advanced sections (Pseudo-Cypereae, Paludosae, Vesicariae), as well as of species within these groups, is associated with a greater or lesser development of this kind of dispersal mechanism.

Maritime dispersal of seeds and fruits is possible only if insulating structures have evolved that protect the living structures from the damaging effects of seawater. In the case of certain palms, such as *Cocos* and *Nipa,* as well as the genus *Hernandia,* protection is afforded by greatly thickened seed coats and ovary walls. In the genus *Gossypium,* Stephens (1958) has shown that the dense mat of hairs that surrounds the seeds can protect them for days or even weeks from the damaging effects of seawater. Fruits and seeds such as these are admirably adapted for dispersal by ocean currents from one island or continent to another.

Nevertheless, adaptation for transport by water cannot be regarded as a major factor controlling either distribution or evolutionary trends in the great majority of flowering plants. Since bodies of fresh water are, in general, of relatively limited extent, species living near them can attain wide distributions only if they have, in addition to mechanisms for water dispersal, the ability to be carried at least occasionally over land.

In the case of maritime species, the naturalists who have studied this problem all agree that the proportion of species even on oceanic islands that are adapted for this means of dispersal is relatively small (Carlquist 1966*a,* Ridley 1930). Moreover, these species are remarkably constant, and they are so highly specialized that they have rarely given rise to additional species in the regions where they have become established. They form a ubiquitous, constant, and monotonous maritime flora, composed of elements that belong to widely divergent and for the most part highly specialized genera and families.

Dispersal by wind. Structural modifications for dispersal by wind are either membranous wings, hairs, or plumes or the modification of the entire plant or large portions of it into "tumbleweeds" that can be blown bodily for long distances. The various appendages may be developed on seeds, fruits, or aggregates of fruits.

Compared with modifications for dispersal by animals, those that favor dispersal by wind are less effective over long distances, and are adaptive in a more restricted range of habitats. If the fruit and seed are very small, as in Betulaceae and Salicaceae, long distances can be covered, but large fruits, such as those of Dipterocarpaceae, are blown for only short distances from the parent plant. Massive scattering of seeds by wind, as in tumbleweeds and the pappose seeds of Compositae, is highly effective only in open plains and on mountain slopes. Furthermore, most of the modifications for wind dispersal affect only the superficial features of the plant. Epidermal hairs and winglike outgrowths can be formed on various structures without any profound modifications of their architecture or developmental pattern. Although this is true of many modifications for dispersal by animals, it is by no means true of all of them, as is indicated below.

Dispersal by animals. Seed dispersal by animals is probably the most effective method for the largest number of different kinds of seed plants. Moreover, both the ecological factors involved and the structural modifications that may result are more numerous and often more profound than those associated with other methods of dispersal. Three kinds of animal dispersal can be recognized: active transport, ingestion of edible seeds and fruits, and passive transport through adhesion of seeds and fruits to fur, feathers, and feet.

Active transport may be carried out by small mammals or birds that store edible seeds either in their nests or in special caches. Though most of these seeds are eventually eaten, a large number may escape, either because the animal gathers more seeds than it can use, or because it dies or is killed before the supply of seeds is used up. Observations by Schuster on jays (*Garrulus*) indicate that a single bird, on the average, transports 4600 acorns, and may fly with them for a distance of 4 km (van der Pijl 1969). This is probably the principal method of dispersal for trees and shrubs having large, edible nuts, such as genera of Fagaceae, Juglandaceae, and Betulaceae (*Corylus*). In such species, selective pressures will favor those species having a sufficiently hard seed coat that it cannot be cracked easily and the animal is likely to leave undisturbed some of the fruits that it has gathered, but which can be opened easily enough that it is attractive to the animal in question. This kind of selective pressure can bring about diversification with respect to the size of the nut, as well as the thickness and character of its wall.

One of the most interesting forms of active transport is that accomplished by ants (myrmechochory). The older observations on this method of transport by Sernander have been well reviewed by Ridley. More recently, intensive and careful observations have been made by Nordhagen (1932, 1959) and Berg (1954, 1958, 1959, 1966). The most interesting feature of this mode of transport is that it results in the development of an entire syndrome of structural modifications. In comparison to their relatives, which usually have seed dispersed by wind or by adhesion to animals, ant-dispersed species have the following characteristics. Their flower stalks are relatively low, and the peduncle becomes recurved when the seed is ripe, so that the mature capsule is close to the ground (Fig. 5-1). The capsule, when present, does not dehisce regularly by valves or pores, but irregularly and over a relatively long period of time. This makes possible repeated visits by ants to the same plant. The seeds themselves can have several modifications. Sometimes they have a special oily seed coat, which the ants remove after they have carried the seeds to their nests, after which they discard the stripped seeds. More often, each seed has a particular fat-bearing ap-

Fig. 5-1. Adaptations for seed dispersal by ants in the genus *Trillium*. (*A–C*) seed of *T. ovatum* Pursh.: (*A*) surface view and (*B*) cross section, showing elaiosome or fat-storage body (*el*), consisting of loose spongy tissue with thin-walled cells; (*C*) cells of elaiosome in cross section showing oil globules. (*D–J*) modifications in gross structure that accompany adaptation to ant dispersal: (*D*) hypothetical ancestral form having erect, pedunculate flowers and fleshy berries, adapted to internal dispersal by animals; (*E*) *T. erectum* L., flower initially erect, becoming declined in fruit, berry fleshy, adapted to either internal dispersal by animals or ant dispersal; (*F*) *T. ovatum* Pursh., similar to (*E*), but fruit is a capsule lacking fleshy tissue; full adaptation to ant dispersal; (*G*) *T. cernuum* L., similar to (*E*), but flower initially declined; (*H*) *T. rivale* Wats, similar to (*F*), but leaf petioles and flower peduncles elongated; (*I*) *T. chloropetalum* (Torr.) Howell flower sessile, main stem becoming prostrate in fruit; (*J*) *T. petiolatum* Pursh., leaves long petioled, flowers sessile, main stem shortened so that flower is at ground surface. (From Berg 1958.)

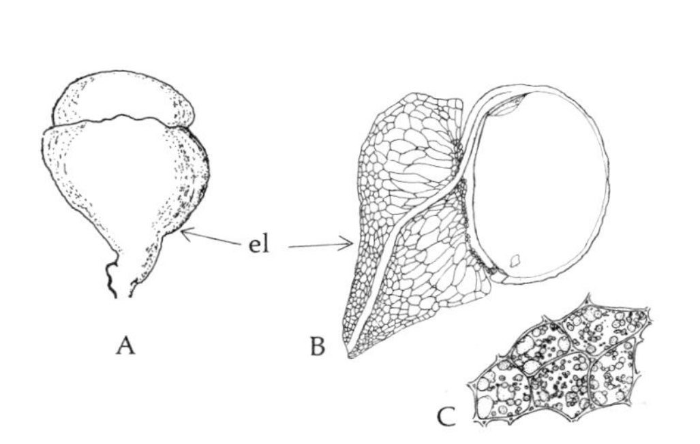

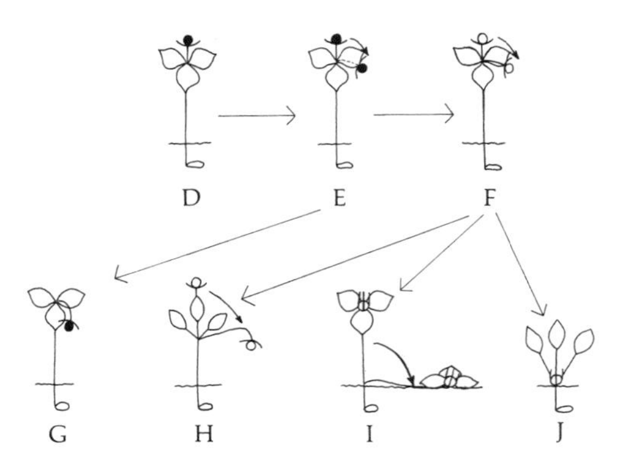

pendage, or elaiosome, which the ant clips off with its mandibles and carries underground, after it has transported the seeds to the vicinity of its nest. Among the best-known species having these characteristics are in Europe *Primula vulgaris* and *Pedicularis sylvatica,* and in western North America species of *Trillium, Nemophila* (Hydrophyllaceae), and *Vancouveria* (Berberidaceae). In the grass genus *Melica,* the elaiosome is formed by the club-shaped lemmas of sterile florets. Ant dispersal is one of the most effective means for herbs living in dense forests.

Transport by ingestion. Adaptations of fruits and seeds for ingestion by birds, and to a lesser extent by mammals and reptiles, are among the most widespread and significant adaptive mechanisms for seed dispersal in flowering plants. The evidence that large numbers of different kinds of seeds can pass unharmed through the digestive tracts of these animals is now extensive, and has been reviewed by Ridley (1930), van der Pijl (1969), Rick and Bowman (1961) and others (DeVlaming and Proctor 1968). In many instances, germination is improved by such passage, since the hard seed coat is partly digested.

The complexity of the relations between plant species and the animals that eat their seeds but at the same time are effective in the dispersal of those seeds that are not eaten has been very well described in an extensive review by Janzen (1971*b*). The commonest kind of adaptation for this purpose is the development of an indehiscent fruit with a fleshy wall, which is eaten whole. Some families, however, such as the Bombacaceae, Cucurbitaceae, and some tropical groups of Leguminosae, have developed large fruits which the animals eat without removing them from the plant, ingesting seeds along with the flesh of the fruit.

In a smaller but highly significant number of plant groups, the seeds themselves have developed adaptations for attracting birds. The outer integument of the ovule itself may become fleshy, bright colored, and conspicuous, forming a sarcotesta, or there may be developed various kinds of special outer coverings, known as arils or arilloids. The complex problem of the homology of these coverings has been discussed by van der Pijl (1969). Families having dehiscent fruits with conspicuous seeds are mainly tropical, such as the Annonaceae, Myristicaceae, Dilleniaceae, Bombacaceae, Sterculiaceae, and arboreal Leguminosae. The temperate groups that have such adaptations are relatively primitive, such as the genera *Magnolia, Paeonia,* and *Celastrus.*

Efficient mechanisms for transport by ingestion necessarily involve a complex succession of developmental processes that affect profoundly both the morphology and the physiology of the ovary and fruit. Moreover, a precise timing sequence of these processes is required, so that a carefully balanced and timed sequence of hormonal interactions must be evolved. While the seeds are developing, they must be protected from premature attacks by animals. In dehiscent fruits of which the seeds are taken, this protection is secured by the development of hard or tough ovary walls and a dehiscence mechanism that is timed to coincide with seed maturation. Fleshy, indehiscent fruits are protected when unripe by the firm texture of the outer covering at this stage, and usually by the presence of various substances of a bitter, astringent, or poisonous nature. In either case, a succession of chemical processes must take place synchronously with the completion of seed growth and the onset of dormancy. These include alterations in the texture of the outer covering, disappearance of the repellent substances, and their replacement by attractive substances, which usually have a high content of sugar, various oils, or both. At the same time, various pigments are synthesized, which give the fruit or seed a red, yellow, or other conspicuous color. From the genetic point of view, the evolution of the complex sequences of physiological processes and biochemical pathways that make possible these orderly changes probably required just as complex a sequence of mutations and gene recombinations as that required for the evolution of complex flowers. Consequently, one need not be surprised at the fact that, once such mechanisms have become highly developed and efficient, they persist for indefinite periods of time, and become modified in various ways as new species and genera evolve.

Systems like these always require compromises between adaptations that attract the potential vector and those that ward off unwanted predators or that reduce the amount of predation on the part of the vectors themselves. In addition, the number of potential vectors as well as harmful predators is always large, and is constantly changing as the animals themselves evolve. This situation might be expected to lead in many instances to complex patterns of evolution, resulting in a great variety of morphological and chemical differences between species and genera. As indicated by Janzen's review (1971*b*:473), the literature of entomology and to a lesser extent that of vertebrate ecology is replete with examples to show that fine details of fruit morphology and chemistry, such as husk hardness, rates of husk expansion, hairiness, constrictions separating seeds, liquid resins, and direct intoxicants like gossypol, may be highly adaptive.

Transport by adhesion. Many fruits and seeds are transported passively by adhering to various parts of animals. Small seeds, particularly those of aquatic plants, are transported in the mud that adheres to the plumage and feet of waterfowl, as well as to the hair and feet of grazing mammals. Darwin, Guppy, Ridley, and many other authors have reported examples that illustrate the number of different kinds of germinable seeds that can be taken from such mud samples.

In the case of larger seeds and of fruits, various devices for more efficient adhesion have evolved. Mucilaginous or gelatinous seed coverings are not uncommon, as in Plantaginaceae and Polemoniaceae. These coverings may also aid in germination, by protecting the germinating seed from desiccation. In other instances, seed hairs or other processes serve for adhesion to animals as well as for wind transport. Much more frequent, however, are devices for the bodily transport of indehiscent fruits by means of special structures on their surfaces or coverings. The burr-like pods of species of *Medicago,* the pod segments of *Desmodium,* and the spine- or hooklike appendages on the nutlets of various Boraginaceae are familiar examples. More spectacular are the elongate, hornlike appendages on the capsules of the Martyniaceae. In some genera, such as *Geum* and its relatives in the Rosaceae, the style may become the adhesive structure.

Frequently, the calyx is modified in various ways for transport by adhesion. The acute spreading calyx lobes of species of *Trifolium,* the stiff, barbed pappus of *Bidens* and other genera of Compositae, and the spiny calyces of *Agrimonia, Acaena,* and other Rosaceae are good examples. In still other instances, an involucre containing one of more fruits develops adhesive spines, hooks, glands, or other processes, as in the genus *Chorizanthe* (Polygonaceae), as well as *Ambrosia, Xanthium,* and various genera of Inuleae, subtr. Filagineae (Compositae). In the Gramineae, the enveloping bract or lemma may have adhesive properties, via either its own or floccose hairs at its base, as in the genus *Poa.* More elaborate adhesive mechanisms may be evolved through modification of sterile florets, as in *Hordeum,* or of clusters of sterile branchlets, as in *Setaria, Pennisetum,* and *Cenchrus.*

This summary of adhesive dispersal mechanisms shows us that, particularly with respect to indehiscent fruits having medium-sized or large seeds, the development and modification of adhesive dispersal mechanisms can be responsible for a large amount of adaptive radiation and structural modification at the level of species and genera.

Comparative Effectiveness of Wind and Animal Dispersal

Dispersal by animals has two conspicuous advantages over wind dispersal for carrying seeds over long distances. In the first place, the incidence of strong winds is erratic and unpredictable, so that they may or may not occur when seeds are ripe and ready to be transported. Birds and mammals, however, migrate at regular intervals, and so could exert a selective pressure on the time of ripening, favoring those plants that ripened their seeds synchronously with the animals' migratory habits. Furthermore, the direction of winds is variable, and they tend to scatter seeds over a large territory, including both favorable and unfavorable habitats. Animals, on the other hand, tend to move from one place that is attractive to them to another similar location. They are, therefore, likely to transport seeds selectively to habitats resembling those occupied by the parental plant. After an extensive review of the pertinent literature, Janzen (1971*b*) has concluded that, even allowing for the destruction of the great bulk of seeds by the vectors, there is probably a much greater chance that a large, heavy seed will reach a safe site by such means than could be achieved through inanimate dispersal.

Data that might indicate how important this advantage might be for the spread and establishment of species and genera of flowering plants are still very scanty. Ridley (1930) has presented an instructive comparison of the floras of three islands that are shown to have been colonized relatively recently by migrants from the East Indies. These are Krakatau, Christmas Island, and Cocos-Keeling Island. His data are presented in Table 5-1. They show that, as expected, the seaborne maritime plants have the greatest chance of reaching both the nearer and the more remote islands. These plants, however, can colonize only a few restricted habitats once they have reached the islands, and they are unlikely to contribute evolutionary descendants that play an important role in colonizing the nonmaritime habitats. Furthermore, their role in the spread of plants over continental areas is negligible, so that the other categories are much more relevant to the general problem of dispersal in relation to evolutionary opportunity.

With respect to the three other means of dispersal, the distance of transport has a conspicuous effect on the kinds of seeds that arrive and become established on each island. For transport over the relatively short distance to Krakatau, wind is almost as effective as animal transport. With respect to the much longer distance to Christmas Island, animal transport is much more effective than wind, and includes both those seeds that are ingested and those that are carried externally by adhesion. Finally, the still longer distance to Cocos-Keeling was traversed successfully only by the maritime group and a few species having adhesive seeds.

These data are far too scanty, but they probably indicate a general trend. That this trend exists also with respect to dispersion over continents is suggested by the distributional range of genera having various means of dispersal in several of the major plant families. Three families—Ericaceae, Myrtaceae, and Rhamnaceae—include one series of genera having capsular fruits and seeds that are not regularly dispersed by animals and another series of genera having fleshy fruits that are, in general, adaptations for seed dispersal by birds.

In the Ericaceae, the two largest genera having capsular fruits are *Erica,* which is confined to Europe and Africa, and *Rhododendron,* which is distributed through Asia and North America, but of which the great majority of the species occur in south, central, and eastern Asia. Among

Table 5-1. Number of species with seeds having various means of dispersal in the floras of three islands situated at various distances from Java. Data from Ridley (1930).

Island	Distance from Java (miles)	Means of dispersal					
		Sea	Wind	Berry or drupe	Adhesion	Mud on birds	Doubtful
Krakatau	25	60	34	34	9	3	4
Christmas	140	44	9	36	15	0	7
Cocos-Keeling	700	17	0	0	5	0	0

the genera having fleshy fruits, the two largest are *Vaccinium* and *Gaultheria.* Including *Pernettya,* a close relative of *Gaultheria,* both of these groups are distributed almost worldwide, the only conspicuous absence being that of *Vaccinium* from Australia and New Zealand. In the Myrtaceae, the genera having capsular fruits (Leptospermoideae) are largely confined to Australia and neighboring regions, only *Metrosideros* being widespread (Andrews 1913). The fleshy-fruited Myrtoideae, on the other hand, include four genera—*Eugenia, Jambosa, Myrtus,* and *Syzygium*—that are widespread on several continents, and several more that are well represented in two continental areas. In the Rhamnaceae, the two largest genera having capsular fruits are *Phylica,* which is confined to Africa and Madagascar, and *Ceanothus,* which is confined to North America. The largest of the genera containing species that have fleshy fruits—*Rhamnus, Berchemia,* and *Zizyphus*—on the other hand, are relatively widespread.

Transference of Function from Protection to Dispersal

In preparing this chapter, my attention was attracted to the existence in certain evolutionary lines of repeated cycles based upon transfers of the function of a particular structure from protecting the developing ovules to aiding the dispersal of mature seeds (Stebbins 1970*b*). Each cycle is completed when protection of the dispersal unit during its early stages of development is taken over by an external structure that previously either was poorly developed or served a different function. The purpose of this section is to review these cycles and to discuss their probable significance. Although this discussion is somewhat hypothetical and speculative, I believe that it will be justified if it stimulates other botanists to look for similar examples, and to investigate them in greater detail.

The first cycle begins with forms having dehiscent follicles, in which the seed is the dispersal unit and the function of the ovary wall is entirely protective. In many instances, such as *Magnolia,* the ovular integuments develop into seed coats that are fleshy and bright colored, and attract animal vectors. In other instances, the seed coats develop adhesive trichomes or become covered with mucilaginous substances that aid in seed dispersal. In still other instances, the ovular integuments produce winged or plumose appendages when the seeds are ripe (Fig. 5-2, *1*). The first step in the cycle itself is either the conversion of a dehiscent follicle into an indehiscent achene or a septicidally dehiscent capsule into an indehiscent fleshy berry. In these examples, the function of dispersal has been transferred from the ovule either to the ovary wall and the tissue surrounding the ovules, as in fleshy fruits, to the ovary wall by itself, as in the samara of *Acer* and the pods of some species of Leguminosae (Fig. 5-2, *2*), or to the modified style, as in the achenes of such genera as *Ranunculus, Anemone, Clematis,* and *Geum.* Depending upon the way in which the structures of the ovary have become modified, the adaptation may be to dispersal by wind, by adhesion to animals, or by animal ingestion.

Fig. 5-2. Examples that illustrate stages in cycles of transference of function for seed protection and transport: (*1*) milkweed (*Asclepias*), protection provided by ovary wall, seed transport effective via its hairy coma; (*2*) tick trefoil (*Desmodium*), ovary wall provides both protection and a transport mechanism, by adhesive surface hairs; (*3*) sour cherry (*Prunus cerasus*), protection provided by modified calyx, transport by fleshy ovary wall; (*A*) section of flower (*B*) section of fruit; (*4*) agrimony (*Agrimonia*), both protection and transport provided by ovary wall; (*5*) blow wives (*Achyrachaena*), a typical composite, protection provided by involucral phyllaries or bracts, transport by modified calyx lobes (pappus); (*A*) entire capitulum, (*B*) single floret; (*6*) California cudweed (*Micropus californicus*), both protection and transport through adhesion provided by modified involucral phyllaries, calyx and ovary wall have both become essentially functionless; (*A*) inflorescence, (*B*) phyllary containing achene, (*C*) achene with style. (From Gajewski 1964.)

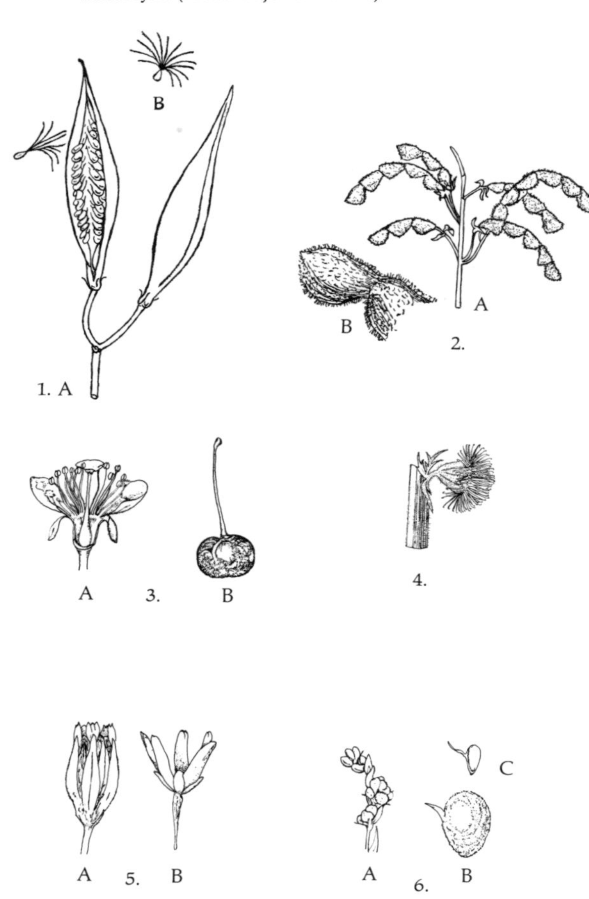

In some examples, such as *Ranunculus*, having dispersed achenes, and various genera of Lauraceae, having fleshy fruits, the ovary wall or a part of it has the double function of protecting the ovules during early stages of development and of dispersal, so that no extra protective coverings are present. In a much larger number of examples, however, the function of protection has been transferred to an outer structure. In some species of *Anemone* and *Clematis*, it is performed by the thick bases of the enlarged sepals. In families such as the Boraginaceae and Labiatae, the calyx tube, formed by "union" or "fusion" of sepals, performs the protective function. In families having perigynous or epigynous flowers, the protective function is taken over by calyx tube, receptacular tissue, or tissues that may be derived from both calyx and receptacle (Fig. 5-2, *3*). This completes the first cycle of transfer. The ovular integuments, which in the original angiosperms formed the coverings of the dispersal unit, have now become much reduced and their function is minimized. The function of aiding in dispersal has been transferred from the integuments to the ovary wall or the style; and the function of protection has been taken over by the calyx, the receptacle, or both.

The second cycle of transfer begins with the modification of the calyx into a dispersal mechanism. This modification is found occasionally in plants having a superior ovary, as in Labiatae (*Marrubium*) and Leguminosae (*Trifolium*). It is more frequent in association with an inferior ovary, as in *Agrimonia* (Rosaceae) (Fig. 5-2, *4*), various Rubiaceae, Valerianaceae, Dipsacaceae, and particularly the Compositae. In the first-mentioned groups, the calyx (or modified receptacle, if one accepts the hypothesis that epigyny involves its adnation to the ovary) serves the dual purpose of protection and dispersal. On the other hand, in Dipsacaceae and Compositae, the bracts or phyllaries surrounding the flower have taken over the function of protection, and the entire floret, except for the dehiscent corolla and stamens, serves as the unit of dispersal (Fig. 5-2, *5*).

Within the family Compositae, certain evolutionary lines have gone through a third cycle of transfer. In genera such as *Madia, Hemizonia,* and various genera of the tribe Inuleae, subtribe Filagineae (*Micropus, Evax, Psilocarphus*), the protective bract that surrounds some or all of the florets has developed glandular hairs or appendages having adhesive properties, so that the floret plus this bract has become the unit of dispersal (Fig. 5-2, *6*). In some of these species, particularly the genera *Evax* and *Psilocarphus,* the outer involucral phyllaries, the upper leaves, or both have taken over the protective function. In other genera of Compositae (*Ambrosia, Xanthium, Arctium*), the involucral phyllaries have acquired the dual function of protection and dispersal, so that the entire capitulum has become the dispersal unit.

In the monocotyledons, the first cycle of the transfer began in those representatives of the Liliales and Commelinales in which the capsule became modified into an indehiscent nutlet or berry. In monocotyledons, however, the perianth usually became reduced along with the ovary, so that it rarely has taken over the function of protection. In both the Cyperaceae and the Gramineae, the protective function is first acquired by the bract subtending the flower. In the Cyperaceae, a second cycle of transfer has taken place only in the genus *Carex.* In the ancestors of this genus, the number of florets in the original spikelets has been reduced to one, and the bract subtending this floret, by "fusion" of its margins, has been modified into an indehiscent, saclike structure, the perigynium. The perigynium containing an entire floret has consequently become the unit of dispersal. The protective function has been taken over by the bract that subtended the original spikelet.

In the Gramineae, the initiation of the second cycle is accompanied not by a drastic reduction in number of florets per spikelet, but by the suppression of floret formation in its lower bracts, giving rise to the "sterile lemmas" or glumes. This is the condition that exists in the Bambuseae, Festuceae, Eragrosteae, and most Hordeae and Chloridae. In these tribes the fertile lemma, that is, the bract that envelops the floret, often acquires adhesive awns or hairs, and so serves both the protective and the dispersive functions. In other tribes, particularly the Arundineae (*Danthonia*), Stipeae, Aveneae, Phalaridae, Agrostideae, and the subfamily Panicoideae, the sterile bracts, either glumes or lemmas, have become larger than the fertile lemma, and have taken over the protective function, thus completing the second cycle of transfer.

Similar cycles of transfer from the protective to the dispersive function, accompanied by modification for the protective function on the part of an external structure, have taken place in various other families, such as the Chenopodiaceae (*Chenopodium, Atriplex*), Labiatae (*Monardella*), Nyctaginaceae (*Mirabilis*), and Leguminosae (*Trifolium*). These cycles have several features in common. In the first place, they always involve reduction to a single ovule per gynoecium (or, in the Labiatae, per gynoecial unit). Second, the seeds in the groups concerned are moderate to large in size, and are usually larger than the seeds of related groups that have not passed through these cycles. Finally, all the groups involved are adapted to climates having alternation between seasons favorable for growth and those that are markedly unfavorable, owing to either cold or drought.

An intriguing aspect of these cycles is that they can probably be extrapolated backward through the evolutionary line that led to the angiosperms from the earliest vascular plants. The hypothesis that the angiosperm ovule is homologous to the anatropous cupules of the Caytoniales (including both the Corystospermaceae and the Caytoniaceae), and that its outer integument is homologous to the cupule wall, has been advanced by Gaussen (1946) and is supported by Walton (1953). Further evidence in favor of this hypothesis is presented in Chapter 10. According to this hypothesis, the ancestors of the angiosperms passed through a stage in which a unitegmentary ovule, homologous to that of other gymnosperms, was the dispersal unit, and the cupule wall, the future outer integument, had a protective function. This stage may have lasted a very long time, since the Caytoniales existed in the Triassic and Jurassic Periods, whereas the earliest cupule-bearing seed plants occur in the lower half of the Carboniferous Period.

Preceding this stage, there existed forms (*Genomosperma*) in which the primitive seed was not included in a cupule, but apparently was nevertheless detached and dispersed as a unit, so that the single integument served both for protection and, presumably, as an aid to dispersal (H. N. Andrews 1963, Long 1966). The gap between these most primitive seed plants and the spore-bearing vascular plants is large. Nevertheless, we can be reasonably sure that the evolutionary line that led to seed plants included forms in which the spore was the unit of dispersal and the naked sporangium wall served the protective function. On the basis of the fossil evidence now available, we would have to postulate that the sporangium wall did not undergo a transference of function from protection to aid in dispersal, since no fossil forms are known in which the sporangium or the nucellus became detached from the surrounding integument and dispersed as a unit, enclosing the megaspore. The first postulated cycle, therefore, is different from the others, in that the original protective structures become surrounded by a second protective layer, without ever assuming the function of dispersal.

Table 5-2 summarizes the five cycles of transference that are postulated in the line leading from the earliest spore-bearing plants to the most advanced Compositae. In the monocotyledons, the cycles that terminate in *Carex* and the more advanced Gramineae are similar, except that the two stages in which the calyx or receptacle have protective or dispersive functions are omitted.

Review of the Phenomenon of Transference of Function

The idea that new adaptations can arise through a change or transference of the function performed by existing organs originated with Darwin, who devoted considerable attention to it in the sixth edition of the *Origin of Species.* Several zoologists, particularly Anton Dohrn, have discussed the problem, citing numerous examples. Mayr (1960) has presented an excellent review of zoological examples. He points out that, in all examples known to him of such transfer, there is an intermediate stage in which both structures perform the same function simultaneously. Usually, also, a structure that is becoming adapted to a new function performs both the old and the new functions simultaneously. The latter condition is characteristic of the examples in plants.

The first botanist to realize the importance of transference of function was Cor-

Table 5-2. Summary of a series of cycles of transference of function in a major evolutionary line of vascular plants. From Stebbins 1970*b*.

Cycle	Functions		Examples
	Protection	Dispersal	
Fifth	Outer bract or foliage leaf	Phyllary, including floret	Advanced Filagineae (*Evax, Psilocarphus*)
	Phyllary or involucre		Certain Compositae (Madineae, Ambrosieae, Filagineae)
Fourth	Bract or phyllary	Floret, including calyx	Compositae, Dipsacaceae
	Calyx or receptacle		Rosaceae, tribe Poterieae, Umbelliferae, Rubiaceae, Caprifoliaceae, and many others
Third	Calyx or receptacle	Carpel or ovary wall	*Anemone, Prunus,* Boraginaceae, most Verbenaceae and Labiatae, and many others
	Carpel or ovary wall		*Ranunculus,* Lauraceae, Berberidaceae, and other angiosperms bearing hypogynous flowers and indehiscent fruits
Second	Carpel or ovary wall	Outer integument (= cupule wall)	Primitive angiosperms that bear dehiscent follicles or capsules
	Cupule wall		Corystospermaceae
	Cupule wall	Integument	Cupuliferous Pteridosperms, Caytoniales
	Integument		Noncupuliferous Pteridosperms (*Eurystoma*)
First	Integument plus		
	Sporangium wall	Spore	Early Pteridosperms (*Genomosperma*)
	Sporangium wall	Spore	Early spore-bearing plants (*Archeopteris*)

ner (1958). He describes several examples, involving shoots, leaves, inflorescences, flowers, fruits, and seeds. His discussion is, however, somewhat confusing, since he uses the term "function" in two very different ways, and the term "transference" is used to cover examples of restriction of function or division of labor as well as the adoption of a new function by a structure that previously had a different one. Most of Corner's examples from fruits are similar to those in the present chapter. He notes, for instance, that in the Cruciferae and Sterculiaceae certain evolutionary lines progress from dehiscent fruits, in which the seed is the unit of dispersal, to indehiscent fruits, in which the ovary wall has the functions of both protection and dispersal. He does not mention examples in which the transference of function has become complete, in that the old function (protection) has been taken over by a different structure.

Corner's "floral examples" are, however, a mixed lot. In *Parkia,* section *Paryphosphaera,* the capitulum has acquired a function similar to that of a single flower. Its outer flowers have become sterile, so that the function of the ancestral flower, which was both attraction and reproduction, has in these flowers become restricted so that only the attractive function is performed. This is more appropriately designated division of labor than transference of function.

Another example that Corner presents is the evolution of the syncarpous ovary, in which the position of the placenta has been transferred from the separate carpels to various positions in the part of the ovary that results from intercalary growth, which he characterizes as a "new intercalation." Apart from the problems that this interpretation raises regarding homologies, this example is inappropriate because it does not involve transference of function at all, according to the ecological meaning in which the concept was originally used by Darwin, as well as by Dohrn, Mayr, and many others, including myself. It illustrates, rather, a reorganization of the developmental pattern, so that ovules that retain their original function acquire a new position of origin. Corner's statement that "most of the properties of the apocarpous ovary become transferred to the new intercalation" is particularly inappropriate. The position at which ovules appear is not a property or characteristic of the carpel or the ovary, but of the developmental pattern that produces it. If we are to understand better the complex interrelations between structures, function, and development that exist in higher plants, we must define our concepts more clearly than Corner has done.

Returning to transference of function in the ecological sense, the zoologists who have dealt with the problem, particularly Mayr, have emphasized the fact that these radical changes have always accompanied drastic alterations in the animal's environment. These have sometimes involved changes of external factors, but more often they have accompanied the animal's entrance into a new ecological niche. Does the same situation exist with respect to plants?

This question is much more difficult for botanists to answer than for zoologists. Differences between animals with respect to locomotion, food getting, use of sense organs, digestion, and the like are relatively easy to recognize. On the other hand, the only kinds of selective pressures that could have affected the structure of gynoecia and seeds, those associated with seed development, seed dispersal, and the establishment of seedlings, are much harder to recognize. In addition, the great plasticity of higher plants, particularly angiosperms, permits a genus that has evolved a particularly efficient reproductive mechanism to evolve species adapted to a great variety of habitats. When this has happened, one can do little more than guess at the nature of the species in which the original reproductive mechanism arose, and at the kind of environment that this species occupied. When dealing with genera such as *Ranunculus, Anemone,* or *Clematis,* the botanist is at a loss to decide which ones among the scores or hundreds of contemporary species are most like the ones in which the distinctive gynoecium of separate, indehiscent achenes first appeared.

The best that we can do in such situations is to compare the most common or modal ecological requirements of those genera that have progressed along one of the cycles of transference of function in the gynoecium with their relatives that have retained the original condition. If the results of such comparisons indicate a consistent trend, we can hope that careful comparisons of seed and seedling ecology between the species concerned will be made, and will test the hypotheses that are suggested by the overall trends.

Five significant comparisons can be made in this connection. The first one is in the Ranunculaceae. The largest genus having achenes, in which the carpel has the double function of protection and dispersal, is *Ranunculus.* Its nearest relatives having dehiscent follicles are *Caltha* and *Trollius.* The habitat of the latter two genera is exclusively mesic: *Caltha* grows in swamps or on wet stream banks and *Trollius* grows in moist forests. Among the numerous species of *Ranunculus* there are some that occupy similar habitats, but there are also many species that are adapted to drier, open situations, particularly fields and mountainsides where the soil is often disturbed. The two other large genera of achene-bearing Ranunculaceae, *Anemone* and *Clematis,* include species with a range of habitats similar to those found in *Ranunculus.* Species of *Anemone,* particularly of the subgenus *Pulsatilla,* are well adapted for the colonization of dry habitats having disturbed soil.

A similar comparison can be made in the Rosaceae. The most widespread genera in the tribe Spiraeae, in which the follicles are dehiscent, are *Physocarpus, Spiraea,* and *Aruncus,* all of them adapted to moist habitats. Species belonging to the numerous genera that form indehiscent fruits or berries occupy a great variety of habitats, but those adapted to dry habitats having disturbed soil are frequent in various genera belonging to every one of the tribes Potentilleae, Poterieae, Roseae, and Pruneae. The species that have advanced the furthest along cycles of transference belong to the tribe Poterieae. The largest genera of this tribe (*Alchemilla, Acaena*), as well as one of its woody representatives, *Poterium spinosum,* are adapted chiefly to montane or rocky habitats.

In the Caryophyllales and their relatives, the Polygonales, similar ecological relations exist between species that have progressed along cycles of transference and those that have not. In the Portulacaceae and most Caryophyllaceae, the unit of dispersal is the seed. The ovary wall, sometimes supplemented by the synsepalous calyx, retains an exclusively protective function. In the tribe Illecebreae of the Caryophyllaceae, however, the fruit has become an indehiscent utricle or achene, and the calyx has taken over the protective function. In some genera of this tribe, such as *Pentacaena,* the calyx has developed spiny lobes that adhere to the fur and feathers of animals, and so have provided the calyx with the double function of protection and dispersal. The Illecebreae are ecologically distinctive in that most species occupy dry, rocky habitats in montane areas, or the well-drained soils of sand dunes and similar formations.

In the Polygonaceae, the ovary is always indehiscent, so that in most species it is the unit of dispersal. This condition supports the belief of Hutchinson (1959) that the Polygonaceae have been derived from plants similar to the Illecebreae, and would suggest that Polygonaceae which in growth habit and ecological preference resemble the Illecebreae, that is, the perennial, subshrubby species of *Polygonum* subg. *Avicularia,* are the most primitive modern representatives of the family.

Various genera of Polygonaceae have progressed along new cycles of transference. These are most highly developed in the genus *Eriogonum* and its relatives, which inhabit dry montane areas of western America. In several species, the papery calyx becomes detached from the plant with the achene, and aids in wind dispersal. In all of them, the function of protection has been taken over by the involucre, which consists of "fused" floral bracts. In the most advanced genus, *Chorizanthe,* the involucre, via its spiny lobes, has assumed the function of dispersal also. The species of this genus are particularly noteworthy for their adaptation to dry, pioneer areas on montane slopes, or to sandy soils in less arid areas.

In another family of Caryophyllales, the Nyctaginaceae, the cycles of transference have progressed to the point at which the ovary plus the calyx is the unit of dispersal, and an accessory involucre has taken over the protective function. This family, also, is most characteristic of dry montane areas.

In the fourth example, the Compositae, the most primitive species have already completed the cycle that ends with the floret as the unit of dispersal and the bract or phyllary as the primary protective structure. The majority of the species of this family are adapted to dry areas. The chief areas for diversification of Compositae genera—the Cordillera of western North and South America, the mountains of central and eastern Asia, and South Africa—are notable for their rugged mountain topography. The few groups of Compositae that have gone through the final cycle, in which the functions both of dispersal and of protection have been taken over by involucral phyllaries or leaves, are noteworthy even within their family for their adaptation to dry situations. The center of distribution for the tribe Ambrosieae is in the arid and semiarid portions of the southwestern United

States. That of the Madiinae is in the Mediterranean climatic zone of California, and of the Filaginae in the Mediterranean region itself. Moreover, the Madiinae are conspicuous among the Compositae of California in their ability to mature their achenes during the hot, dry summer months.

The final example is that of four families of monocotyledons, the Juncaceae, Cyperaceae, Gramineae, and Restionaceae, which have in common wind pollination and great reduction of the perianth. The only one of these families that has not progressed at all along cycles of transference is the Juncaceae, in which the capsule is dehiscent and the unit of dispersal is the seed. The species of this family are exclusively mesophytic or hydrophytic, and the small seeds found in most of them are adapted to germination in heavy, saturated soils. This is also true of most species of Cyperaceae, although in this family adaptations to relatively dry, sandy soils are more common than they are in the Juncaceae. The final cycle of transference, which culminated in the genus *Carex*, is clearly associated with montane habitats. The genera that resemble the probable ancestors of *Carex*, *Kobresia*, and *Uncinia*, as well as the most primitive sections of *Carex* itself, are all adapted to montane or alpine habitats. Furthermore, the completion of the final cycle within the genus *Carex* itself has been accompanied by a great increase in the diversity of habitats that it occupies.

In the Gramineae, adaptations to semiarid conditions are predominant and probably characterized the original members of the family (Stebbins 1956*a*). The final cycle of transference is, moreover, best developed in genera such as *Avena*, *Andropogon* and its derivative *Heteropogon*, and *Cenchrus*. In each of these genera, the final cycle of transference has progressed independently of the others. They are all characteristic of well-drained, sandy, and often semiarid soils.

Within the family Restionaceae, many species are adapted to swamps or other moist habitats. On the other hand, the numerous species of the genus *Restio* found in South Africa include many that are adapted to relatively dry mountainsides and rocky places.

These examples all support the hypothesis that cycles of transference progress in groups that are primarily adapted to montane habitats having loose, sandy or gravelly soils, and usually in semiarid or arid climates. The reason why this should be so probably lies in the fact that two opposing selective pressures exist in such regions. In well-drained soils of semiarid regions, germinating seeds and young seedlings are often subjected to periodic drought. The best way of overcoming this hazard is the development of vigorous seedlings having deep root systems. Seedling vigor is closely correlated with seed size, and large seeds contain much stored food, which enables the seedling to produce a large root system before it develops an extensive leaf surface, with its associated hazard of excessive transpiration. One kind of strong selective pressure to which these plants are exposed is, therefore, selection for increased seed size.

Opposed to this pressure are selective pressures for rapid seed development, which is likely to be particularly strong in regions characterized by seasonal drought or cold, and for ease of seed dispersal. The pioneer habitats that these species occupy are temporary, so that the ability to migrate occasionally over long distances to new favorable sites has a high adaptive value. Since large seeds generally require longer times to develop than small ones, plants that are adapted to climates having short seasons favorable for growth, and that also have large seeds, are most successful if they possess special protective devices that enable the seeds to complete their development under relatively unfavorable conditions. Large seeds also are more difficult to transport over long distances than are small ones, unless they are equipped with special mechanisms for dispersal. Furthermore, one of the commonest ways by which small seeds are transported for long distances, being enclosed in bits of mud that adhere to the feet of birds (Ridley 1930), is less likely to happen in the case of plants that are adapted to well-drained, sandy soils. Consequently, species adapted to well-drained soils in semiarid habitats and having large seeds are particularly subject to selection pressures favoring special mechanisms for seed dispersal.

A final fact to consider is that mountainous regions have a diversity of habitats, and over the long periods of time that are necessary for the differentiation of genera and families many changes in the nature and location of these habitats can be expected. Cycles of transference of function are, therefore, probably produced by alternating selective pressures under conditions that promote a succession of altered compromises between the conflicting selective pressures favoring large seed size, rapid seed development, and efficient seed transport. The distribution of these cycles among the various orders and families of angiosperms is in accord with this hypothesis.

If this hypothesis is correct and generally applicable, then the possibility that a cycle of transference of function was responsible for the evolution of the angiosperm carpel becomes particularly significant. It would suggest that angiosperms originated in mountainous regions, as Axelrod (1952) and Takhtajan (1969) have maintained, but would not support their belief that these regions were equable rain or cloud forests. A much more likely habitat would be semiarid subtropical or tropical mountainsides, in loose, disturbed soil of pioneer situations of the ecological succession. Further evidence is presented in Chapter 10 to support my belief that these were, in fact, the conditions under which the earliest angiosperms evolved.

Factors Affecting the Establishment of Seedlings

Along with the transport of seeds into the area, the establishment of seedlings is the weakest link in the colonization of new territory by plant species. Many seedlings, being small and weak, are easily crowded out by other plants, devoured by animals, or killed by extremes of drought or cold. My experience in many attempts to establish perennial species of grasses in natural areas even slightly different from those that prevailed in their native home indicates that nearly all of the mortality suffered by these plants occurs during the first year of their life. Because of this critical condition of young seedlings, the provision made for them by the parent plant, in terms of stored food or the size and potential vigor of the embryo itself, is a major factor determining their survival.

The importance of seed size in relation to survival of seedlings in various kinds of environments was first realized by Salisbury (1942). His studies have been amplified and made more precise by later workers, particularly Harper and his collaborators (Harper 1965, Harper and Chancellor 1959, Harper and Clatworthy 1963, Harper and Obeid 1967, Harper, Williams, and Sagar 1965, Ludwig and Harper 1958, Sagar and Harper 1961). These studies have revealed with striking clarity the precision and subtlety of the relations between seedling and environment that determine the difference between mortality and successful establishment.

This heterogeneity of the microhabitat is beautifully described in a paragraph from an unpublished manuscript by John Harper, which he has given me permission to quote: "At the scale of the size of a seed the physical environment is exceedingly heterogeneous—not only in the biblical sense that some seeds fall on stony ground, but in which to a mustard seed a worm cast is a mountain, a fallen leaf is a shade from light (or from the eye of a possible predator), a raindrop is a cataclysm."

The close correspondence between seed size and seedling vigor has been emphasized by agronomists working both with legumes (Black 1956, 1957, 1959) and with grasses (McKell 1972, Kneebone 1972). On the basis of his experience with species of *Agropyron, Andropogon, Bromus,* and *Elymus,* Kneebone (1972) concludes that "of all the selective criteria affecting seedling vigor . . . seed size is probably the most important." Consequently, successful adaptation to habitats in which seedling vigor is of prime importance inevitably involves natural selection for increased seed size.

Physiological factors affecting establishment

In addition to seed size, the most important factors that control the establishment of seedlings are physiological in nature, such as the action of growth substances in the promotion and breakage of dormancy, as well as in promoting the elongation of roots, hypocotyl, and other parts of the young seedling. Unfortunately, the comparative physiology and biochemistry of seed germination and seedling establishment is an almost completely unexplored field of research, and data that might be of value for evolutionary studies are completely unavailable. A recent volume on the germination of seeds (Mayer and Poljakoff-Mayber 1963) gives a thorough account of many of the factors and processes involved, but says nothing about their comparative efficiency in related species. Yet anyone who has had even the slightest experience with raising seedlings of various kinds is fully aware of the important differences existing between them. These differences are almost certainly related to the relative efficiency of such processes as digestion of stored food, absorption of water, translocation of materials through the seedling, and the speed and efficiency of those hormone-controlled processes that regulate cell division, protein synthesis, and cell enlargement. Some of these processes are definitely associated with seedling morphology. In maize and other plants, for instance, seedling establishment is greatly promoted by rapid cell elongation in the hypocotyl at the initial stages of germination. In other cereals, such as barley, the vigorous early growth of the seedling is based upon a rapid onset of both cell division and cell enlargement at the time of germination. In general, the action of growth substances in promoting such processes appears to be much more highly developed and efficient in most angiosperms than in other seed plants.

Indirect evidence of the importance of growth substances in the adaptation of seedlings to particular habitats comes from studies of the ecology of seed germination and seedling development in plants adapted to regions having seasonal drought. In *Yucca* (Arnott 1962), *Colchicum* (Fritsche 1955, Galil 1968), and *Marah* (Schlising 1969), extensive intercalary growth of the petiolar bases of the cotyledons (in *Marah*), the cotyledon (in *Yucca*), or both the cotyledon and the first leaf (in *Colchicum*) causes the young shoot to become deeply submerged in the ground during the moist season when germination takes place. Consequently, it is very well protected from drought during the subsequent dry season. This kind of growth would be possible only through the excessive activity of growth substances during the critical stages of development. Also essential is the presence of large quantities of stored food in the seed, providing energy for this extensive growth before photosynthesis begins in the seedling itself.

Drill mechanisms that aid in establishment

One of the most conspicuous kinds of adaptation for establishment is that in which accessory structures attached to the seed are drills having hygroscopic properties (M. Zohary 1937). In response to changes in atmospheric moisture they twist and untwist, thereby drilling the seed into the ground. The best known are the carpels and styles of the genus *Erodium* (Geraniaceae) and the lemmas of *Stipa* (Gramineae). In *Erodium,* the carpels are indehiscent, one-seeded, and provided with a hard, sharp-pointed base. From their upper end emerges a long, tough style, which becomes coiled when the carpel is ripe. Its top part is bearded with stiff hairs. The sharp-pointed carpel base, its stiffly hairy body, and the bearded style all aid in adhering to fur, feathers, or clothing, and so bring about external dispersal by animals. Once the fruit has been transported to a suitable place for germination, it can lodge in cracks in the soil, and, if its base points downward, hygroscopic movements of the twisted style can drill the fruit into the ground. The "fruit" of many species of *Stipa* is an exactly comparable mechanism. The pointed base and hairy or smooth body of the fertile scale or lemma are comparable to the carpel of *Erodium,* and the twisted awn is comparable to the style. The same mechanism has evolved in the genera *Chrysopogon, Heteropogon,* and *Trachypogon,* belonging to a completely different tribe of Gramineae (Andropogoneae). In all of these groups, the perfection of the drill mechanism is associated with an increase in size of the fruits and occupation of relatively dry habitats. In species of *Stipa* that occur in forests, meadows, and other more mesic habitats (*S. californica, S. Lemmonii, S. columbiana*), the lemma is broader and blunter based than in the xeric species that possess typical "drill-fruits" (*S. capillata, S. comata*), and the awn is shorter, more slender, and less strongly twisted.

Another form of drill mechanism has been described by D. Zohary (1959, 1960, 1963) in *Hordeum spontaneum.* In this species, as in all other wild species of *Hordeum,* the fertile spikelet, bearing a single large seed, is flanked by two sterile spikelets. At maturity, the fragile rachis of the spike breaks up into a number of dispersal units, each of which consists of a fertile spikelet flanked by two sterile ones, all of them being attached to an internode of the rachis that is sharp pointed at its base. This sharp base, plus the rough, linear awns of the sterile spikelets, aid in dispersal by animals, and later in the drilling mechanism that drives the unit into cracks in the soil. A similar mechanism exists in the large-seeded perennial species, *H. bulbosum.*

Seed Polymorphism and Its Significance

Harper, Lovell, and Moore (1970) have reviewed several examples of seed polymorphism, particularly those in which the same genotype produces two or more very different kinds of seeds or fruits. This condition greatly increases the flexibility of adaptation to highly variable environments. In some instances, two different kinds of fruits are adapted to different modes of transport. In *Hypochaeris glabra,* for instance (unpublished personal observations), the marginal florets of the capitulum bear achenes that are beakless, but have sharp-pointed bases, rough, adhesive surfaces, and a pappus of plumose bristles that forms a dense, cottony mass. These achenes adhere strongly to animal hairs or feathers, and so are adapted to external transport by animals. The achenes of the central florets, on the other hand, are beaked and have long, plumose pappus bristles that adapt them well for wind transport (Fig. 5-2).

In other examples, seed dimorphism in the Compositae, Leguminosae, and Cruciferae is a separation into one kind of fruit that possesses a mechanism for animal or wind transport and a second kind that has no obvious means of transport (Datta, Evenari, and Gutterman 1970, M. Zohary 1937). The presence of such dual mechanisms provides maximum assurance that some seeds will remain in the favorable spot where the mother plant grew, whereas others will be dispersed for varying distances and will colonize new habitats. A dimorphism having a similar adaptive value is the formation of accessory flowers at the base of the plant, which are usually cleistogamous so that they are automatically self-fertilizing. This adaptation is well known in species of *Viola* that inhabit the eastern United States (Fernald 1950), and is found in several genera of grasses such as *Danthonia* and *Stipa* in the

United States (Hitchcock and Chase 1950) and *Enneapogon, Cleistogenes,* and *Stipa* in Africa (Stopp 1958). In these examples, cross-fertilizing flowers borne on elevated stems or peduncles produce seeds that are more or less well adapted to transport by animals or wind, whereas the seeds formed by the accessory flowers at the base of the plant ensure replacement *in situ*.

Interactions of Factors

One approach to an understanding of the complexity of interaction between factors that affect seed and seedling ecology is to consider interactions between pairs of factors. The most important of these interactions will be reviewed in the present section.

Interactions between seed size and seed number

The interactions between seed size and seed number are of particular importance, since each of these characteristics has one overriding advantage and several disadvantages. Large seeds contain large embryos, large quantities of stored materials, or both. They can, therefore, produce seedlings that are superior in many kinds of environments in two ways. A large embryo produces a vigorous seedling with optimal competitive ability (J. W. Black 1956, 1957, 1959). An abundance of stored food enables a seedling to produce an extensive root system by relying on this food for its initial growth, so that it can be well established before producing large leaf surface, with the attendant hazard of excessive transpiration under dry conditions.

The disadvantages of large seed size are as follows. (1) If the amount of photosynthetic product that the plant can produce is limited by either drought, cold, low light intensity, or restriction of the growing season, increased seed size can be obtained only at the cost of reduction in seed number. (2) Other conditions being equal, large seeds take longer to develop than small ones; consequently, in many environments, restriction of the growing season may impose a limit upon seed size. (3) The growth pattern of large seeds inevitably includes a long early period of development during which the tissues are relatively soft and tender; they are, therefore, highly subject to attacks by predators unless they are provided with special means of protection from them. (4) Large, heavy seeds are less easily dispersed than small, light ones. Each of the last three disadvantages can be alleviated by various accessory adaptations. Much of the diversity of gynoecia and seeds in various groups of flowering plants is associated with these accessory, compensatory adaptations.

The great advantage of large seed number is the increased chance that, given random dispersal, some seeds will land in a spot that is favorable for seed germination and seedling establishment. In the words of Harper (1965), large seed number increases the chances that some seeds will find a "safe site."

The principal disadvantage of increased seed number is that, given a limitation on the length of the growing season or on the amount of organic material that the plant can produce, increased seed number can be obtained only at the cost of reduced seed size and its consequent disadvantages, which have already been enumerated.

Because of these divergent and often conflicting advantages and disadvantages of seed size vs. seed number, most plant species have evolved compromise evolutionary strategies that depend upon various conditions of their habitat, as well as upon their evolutionary ancestry.

A well-documented series of compromises that exists in the tropics is associated with insect predators. Janzen (1969) has examined 36 species of Leguminosae in Central America. In this region, legume seeds are prime objects of attack by insects of the family Bruchidae or "pea weevils." Two kinds of defense mechanisms against their attacks are described. In one group of species, belonging chiefly to the genera *Acacia, Mimosa, Leucaena,* and *Pithecellobium,* the seeds are relatively small (mean weight per seed 0.01–0.13 g) and are produced in very large numbers. These seeds are attacked by bruchids, which may destroy up to 90 or even 100 percent of the seed crop from a single tree. Nevertheless, the total crop of seeds produced by these species is so large that some seeds escape predation often enough to perpetuate the species. A second group of species, in the genera *Mucuna, Canavalia, Entada, Erythrina, Enterolobium,* and *Schizolobium,* are completely free from attack by Bruchidae, because of the high content of alkaloids in their seeds. The seeds produced by species of this group are, in general, larger than those of the first group, and they may be much larger (individual seed weight 0.14-24.07 g). Nevertheless, they are produced in such small numbers that, in spite of their larger size, the total seed crop per tree is much less in weight than is the seed crop of the first group.

These two groups of species apparently represent two kinds of compromise that have been evolved independently many times during the evolutionary history of the family. A noteworthy fact is that the genera in the first group belong to the subfamily Mimosoideae, in which the overall architecture of the inflorescence is associated with the production of large numbers of small flowers. On the other hand, the genera in the second group, except for *Enterolobium,* belong to the subfamily Papilionoideae, in which inflorescences containing few, large flowers are more common. This suggests that the particular compromise that may be adopted in response to selective pressure by bruchid beetles may depend to a large extent upon the inflorescence and flower structure of the populations in which the adaptive response begins.

Interactions between seed size and seed number that affect the dispersal of seeds and the establishment of seedlings are particularly easy to recognize in herbs. Because of their seasonal growth cycles, most herbaceous species are limited in the resources that can be devoted to seed production. Consequently, seed size can be increased only at the expense of seed number, and vice versa. As mentioned above, Harper, Lovell, and Moore (1970) and Salisbury (1942) have pointed out that seed size is strongly correlated with the ecological conditions of germination and seedling establishment. Depending upon these conditions, therefore, many different compromises have been evolved between the conflicting demands of seed size and seed number. The nature of these compromises has a great influence upon the structure of seeds and fruits because of the greater ease with which small seeds can become dispersed in comparison with large ones. If a plant produces large numbers of small seeds, these may be scattered widely by the wind, or may easily become embedded in small bits of soil that adhere to the feet or other parts of animals. Furthermore, if large numbers of seeds are scattered about in this fashion, the chances that at least some of them will land in a favorable site are high. Consequently, plants that produce large numbers of small seeds have, as a rule, evolved few specialized structures of seeds and fruits that aid in dispersal. On the other hand, a plant that produces small numbers of large seeds will be perpetuated only if the chance that these seeds will be transported to favorable habitats is relatively high. In such plants, therefore, mechanisms favoring seed dispersal are relatively highly developed. As has already been pointed out, the development of these mechanisms may involve, successively, modifications of the gynoecium itself, of the calyx, and finally of various structures of the inflorescence.

A number of structural and physiological modifications of the fruiting structures have evolved in response to partly conflicting selective pressures that favor protection of immature, developing fruits and seeds from animals, but promotion of access to animal vectors when seeds are ripe and ready for dispersal. Selection pressures of this sort may be responsible for the establishment of complex interactions of hormones that take place during the ripening of fruits (J. P. Nitsch 1965, van Overbeek 1962).

Another kind of interaction between plants and seed-dispersal agents is responsible for seasonality of fruiting in the tropics. Smythe (1970) has shown that the fruiting periods of rain-forest trees in Panama, particularly those having relatively large seeds, coincide with the period of greatest activity on the part of mammalian vectors. D. W. Snow (1966) has suggested that a similar factor is responsible for the diversity of sympatric species of *Miconia* and other genera in the American tropics.

Adaptive Shifts with Respect to Dispersal Mechanisms

Given the large number of possible compromises between the various selective pressures that act upon seed development, seed dispersal, and seedling establishment, we might expect that adaptive shifts from one compromise to another would play a large role in differentiation at every level, from the species to the family and even the order. Within the larger families, differentiation of genera may well be based to a considerable extent upon adaptive radiation, involving many different adaptive shifts. Unfortunately, the comparative biology of dispersal mechanisms is so poorly known that no examples of adaptive radiation involving entire genera or families are available. Nevertheless, some isolated examples of adaptive shifts either have been recorded or can be deduced from monographic data, so that a few of the principal kinds of shifts can be recognized. The most common ones in temperate regions are from wind dispersal, or from generalized seeds or fruits having no obvious adaptations, to various adaptations for animal dispersal.

Adaptive shifts between wind and animal dispersal

An example is in the genus *Geum* (Gajewski 1959, 1964). This genus of 56 species is widespread in temperate regions. It belongs to a small tribe, Geeae, of the Rosaceae, which is closely related to the tribes Dryadeae and Cercocarpeae. All species belonging to the two latter tribes are shrubby, and have fruits that are wind dispersed by means of greatly elongated, plumose styles (Fig. 5-3, *A*). The species of the tribe Geeae are herbaceous, and in several characteristics, including the basic chromosome number 7 rather than 9 as in the other two tribes, are more specialized than the Dryadeae. Of the three genera comprising the tribe Geeae, the small genera *Waldsteinia* and *Coluria* are forest-loving herbs that bear fruits with deciduous styles, and are dispersed by animals (Fig. 5-3, *B*). In *Geum* itself, the most primitive species have wind-borne fruits with persistent, plumose styles as in the Dryadeae, whereas the more specialized species have fruits with deciduous stigmas and specialized styles that develop a hook or "rostrum" in their middle portion, below a specialized abscission layer that is responsible for the removal of the stigma while the fruit is ripening. The hook enables the fruit to cling to the fur or feathers of animals, and so is essential for fruit and seed dispersal.

This example is complicated by the fact that among the species of *Geum,* the more specialized ones having deciduous stigmas are all hexaploid or 12-ploid, and their diploid ancestors appear to be extinct. Nevertheless, the inference appears clear and inescapable that in the tribe Geeae the condition of wind-borne fruits with persistent, plumose styles is primitive, and that the adaptive shift to deciduous styles or stigmas and animal dispersal has taken place, perhaps more than once, in association with adaptation to forest conditions, where winds are less strong and animals are more abundant.

Shifts from wind dispersal to external dispersal by animals have taken place in many other groups, but they are not so well documented. Several examples exist in the large family Compositae. In most species of this family, the fruit bears a varying number of hairs or thin scales known as the pappus, which aid in wind

Fig. 5-3. (*A*) Mature ovary and styles of *Geum* (*Sieversia*) *montanum,* a wind-dispersed fruit; (*B–E*) stages in the development of the fruit of *Geum rivale* L., representative of species having fruits that are dispersed by adhesion to animal vectors. (Original.)

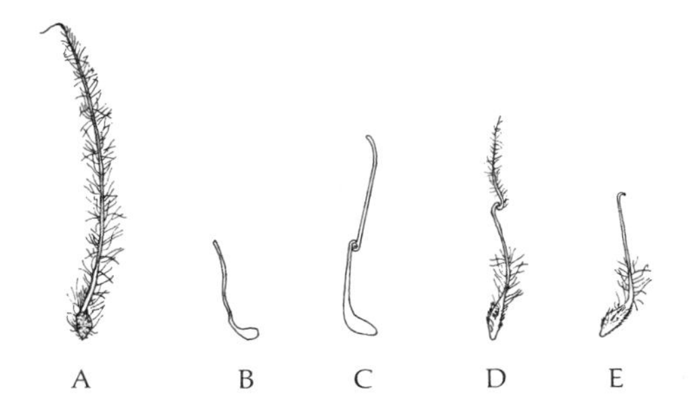

dispersal. In some genera, however, such as *Bidens,* the pappus appendages have become thickened and beset with rough barbs that cling to fur and feathers. In other genera, the pappus has disappeared, and the fruits are enclosed in bracts or phyllaries that bear either hooks or barbs (*Franseria, Xanthium, Arctium*), adhesive mats of woolly hairs (Inuleae, subtribe Filaginae), or sticky glandular hairs (Heliantheae, subtribe Madiinae).

In the Umbelliferae, many genera, such as *Hydrocotyle, Berula, Cicuta,* and *Ligusticum,* have fruits without any obvious means of dispersal. Most of these genera live in wet places, and the transport of their seeds through the mud that adheres to the feet of waterfowl is a reasonable probability. Other genera have evolved more specialized means of seed dispersal, such as greatly flattened fruits bearing broad thin wings that aid in wind dispersal (*Angelica, Heracleum, Pastinaca, Lomatium, Cymopterus*), or fruits that bear variously shaped hooks or spines that aid in external dispersal by animals (*Eryngium, Sanicula, Caucalis, Daucus, Osmorhiza*). The genera bearing winged fruits grow, for the most part, in open habitats where winds are frequent, whereas those with animal-dispersed fruits are either forest loving or relatively low growing. More careful studies of fruit and seed ecology of this family are clearly needed, but superficial impression suggests that much of its differentiation into genera is based upon adaptive radiation with respect to fruit characteristics that are associated with dispersal, and probably also with seedling establishment.

Adaptive shifts to internal dispersal by animals

Adaptive shifts to internal dispersal by animals can be recognized in several families of temperate regions, the most conspicuous examples being in the Rosaceae, Ericaceae, and Liliaceae. In the Rosaceae, the genus *Potentilla* bears dry, one-seeded achenes that are probably transported by adhering to the mud on the feet of animals. In the related and more specialized genus *Fragaria,* on the other hand, the fruit is eaten by animals because of its enlarged fleshy and red-colored receptacle. The large and diverse genus *Rubus* is probably another offshoot of a primitive potentilloid stock that has acquired animal dispersal by means of developing a fleshy, variously colored carpellary wall. *Potentilla* and its relatives occur chiefly in open country, whereas both *Fragaria* and *Rubus* occur predominantly in woodlands or forests. In both genera, therefore, the shift from generalized dispersal to internal dispersal by animals has been associated with the change in adaptation from open to forested habitats.

In the Ericaceae, the tribes Rhododendreae, Ericeae, and most genera of the Andromedeae possess dry, dehiscent capsules and small seeds that are dispersed by wind, whereas the tribe Arbutoideae, the genus *Gaultheria,* and the subfamily Vaccinoideae have fleshy fruits that are eaten by animals. Since the divisions between these tribes and subfamilies are rather sharp, and genera transitional between them no longer exist, the actual adaptive shifts that have taken place are difficult or impossible to reconstruct.

Nevertheless, the fact that in other families of Ericales that have relatively unspecialized floral characteristics, such as the Clethraceae and Pyrolaceae, the capsules are dehiscent and wind dispersal prevails, whereas both the Gaultherieae and Vaccinoideae have the specialized condition of pseudoinferior or actually inferior ovaries, indicates that internal dispersal by animals is a derived condition in the Ericaceae also. The genera having wind-borne seeds occupy a great variety of habitats, as do the Arbutoideae among the berry-forming genera. The Gaultherieae and Vaccinoideae, on the other hand, occur chiefly as undershrubs or trailing plants of forests. In the Ericaceae, therefore, the shift to internal dispersal by animals in association with adaptation to forest conditions has probably taken place, but it has not always followed this kind of adaptation. Furthermore, some of the genera, particularly those of the Arbutoideae, probably acquired edible fruits under other circumstances.

The family Liliaceae contains a number of genera, chiefly in the subfamilies Convallarioideae, Asparagoideae, Ophiopogonoideae, and Smilacoideae, having fleshy, brightly colored berries that are eaten by animals. These genera all occur in forests, as do many other genera of Liliaceae having less specialized gynoecia. In the Liliaceae, therefore, the shift to internal dispersal by animals apparently occurred several times in evolutionary lines that had already become adapted to life in forests. The Asparagoideae, which are often adapted to open, dry habitats, and have reduced, sometimes xeromorphic phyllodes, probably were derived from baccate, forest-inhabiting ancestors that secondarily became adapted to more open, drier conditions.

Adaptive shifts to active dispersal by animals

Some examples can be recognized of the shift from wind dispersal to active dispersal by animals. As has been pointed out by Sernander (1906) and Berg (1958), ant-dispersed plants occur mostly in forests. In the case of some examples, such as *Primula acaulis,* the related species (*P. elatior, P. veris*) occur in open country and have seeds that are dispersed at least in part by wind, being shaken out of their capsules by movements of the slender peduncle of the inflorescence. The ant-dispersed species of *Melica* are probably also derived from wind-dispersed ancestors, since in both the Old and the New Worlds the majority of the species of this genus occur in open country and have light, papery lemmas that can be easily blown by the wind. On the other hand, the ant-dispersed species of *Trillium* (Berg 1958) are probably derived from other species of the same genus that have red, fleshy berries that are eaten by animals.

The shift from wind dispersal to active dispersal by mammals has probably taken place in the Betulaceae. Most of the genera of this family (*Betula, Alnus, Carpinus, Ostrya*) have wind-dispersed fruits. *Corylus,* on the other hand, has large nuts that are gathered by squirrels and may be carried for long distances. The occasional seeds that escape being eaten probably serve as the principal means by which species of this genus are spread. That *Corylus* is one of the most specialized genera of Betulaceae is evident from the great differences that have evolved between its staminate and pistillate flowers, and the development in the latter of a prominent involucre, formed by the "fusion" of two leafy bracts.

Shifts from animal to wind dispersal

In some groups of grasses, adaptive shifts from animal dispersal to wind dispersal have taken place. An example is the genus *Sitanion,* of the tribe Hordeae. Its two recognized species are so closely related to *Elymus glaucus* that in the F_1 hybrid between *Elymus* and *Sitanion* the parental chromosomes pair almost perfectly at meiosis, and occasional back-cross progeny can be obtained (Stebbins 1957*b*, Stebbins, Valencia, and Valencia 1946). In *E. glaucus,* the rachis of the spike is stiff and continuous, and the individual lemmas bearing seeds, which break off separately, can adhere to animals by means of their stiff, bent awns. In *Sitanion,* on the other hand, the rachis of the spike is fragile, and easily breaks off of the plant at maturity. Furthermore, the sterile glumes are long awned, and have in some instances developed extra awns, so that either the entire spike or the individual complex of spikelets at a node is easily blown for long distances, "tumbleweed" fashion, by the wind. This adaptation has been acquired in association with the shift from the woodland habitat occupied by *E. glaucus* to the open grassland, steppe, or alpine habitats in which the species of *Sitanion* may be dominants. Similar shifts have occurred in other groups of grasses such as the genera *Aegilops* (D. Zohary 1965), *Pennisetum* (*P. villosum,* unpublished personal observations), and *Spinifex* (van der Pijl 1969).

At least one clear case exists of the trend from internal dispersal by animals to wind dispersal. This is the genus *Fraxinus.* In most of the other genera of the Oleaceae, the fruit is an indehiscent, edible berry. The fruit of *Fraxinus,* on the other hand, is a dry, indehiscent samara, which is dispersed by the wind in a manner similar to the fruits of maples (*Acer*). Although the majority of the genera of Oleaceae are evergreen shrubs or vines that inhabit forests or dry scrub formations, most of the species of *Fraxinus* are deciduous trees that form part of the canopy of deciduous forests in cool temperate regions. The shift from animal dispersal to wind dispersal, therefore, has been associated in *Fraxinus* with the evolution of the growth habit of a deciduous tree. Wind dispersal via indehiscent gynoecia or carpels is a very common method of dispersal in trees that form the canopy of deciduous forests, such as *Acer, Liriodendron, Betula, Platanus,* and *Ulmus.* Moreover, as van der Pijl (1969) has pointed out, indehiscent, wind-borne fruits have evolved in several genera of Leguminosae (*Dalbergia, Derris, Pterocarpus*) that form part of the canopy of tropical forests. Though the type of dispersal that existed in the ancestors of these genera is not known, the prevalence of animal dispersal via edible seeds in most tropical woody Leguminosae suggests that they are also examples of the shift from animal to wind dispersal.

Application of General Evolutionary Principles to Adaptations of Seeds and Seedlings

In the review of pollination mechanisms in the last chapter, I pointed out that adaptation to different methods of pollination, as well as to different pollen vectors, involves changes in character syndromes, rather than in individual characters. The same generalization can be made with even greater emphasis regarding adaptive diversification with respect to seeds and seedlings. Moreover, the interrelations between characters are so complex that successful adaptive shifts usually lead to simultaneous and harmonious changes in many parts of the plant, some of which may involve complex developmental patterns. This means that any adaptive compromise that has become highly successful is likely to be retained in large part even after the disappearance of the environmental conditions that originally exerted the selective pressure favoring that particular compromise. Modifications of such patterns by quantitative changes in sizes and numbers of parts are more likely to take place than are complete reorganizations of the pattern.

For example, the placentation of the ovules in the gynoecium is often used by taxonomists to distinguish families and orders of angiosperms, because it is regarded as relatively conservative. This conservatism is probably a real phenomenon, and may well depend upon the fact that each type of placentation results from a complex developmental pattern that is different in many respects from the pattern which leads to another type of placentation. Shifts from one to another type require many more genetic readjustments than do quantitative changes in the size of the gynoecium, the size of the ovules and seeds, or the number of seeds per gynoecium.

A common and widespread form of placentation is the parietal arrangement, in which the ovary has a single locule, even though the number of styles and stigmas present suggests that its derivation is from an ovary that contained several locules or carpels. In families such as the Cistaceae and Hypericaceae, which have regular, generalized forms of flowers, parietal placentation is associated with the formation of many small seeds. Consequently, selection pressure from increased seed number may well have played an important role in the evolution of parietal placentation. Once this condition was acquired, however, reversion to an ovary containing many locules may have required so many adjustments of the developmental pattern that it rarely or never occurred. On this basis, gynoecia containing few ovules with parietal placentation would be explained as a consequence of two successive evolutionary steps: an initial development of parietal placentation in association with selection for increased seed production per flower, followed by reduction in ovule number in response to selection for either increased seed size or general reduction of flower size associated with a more rapid reproductive cycle. A further discussion of this and similar points is presented in Chapter 11.

This type of argumentation is, frankly, speculative, but it is advanced to show

how intricate the selection pressures may have been that gave rise to major evolutionary trends in higher plants, and to suggest observations and experiments that might lead to a better understanding of the processes involved.

In my opinion, the most significant advances in our understanding of evolutionary trends in angiosperms will result from comparative studies on the ecology and physiology of seed development, seed dispersal, and seedling growth and establishment. Not only is this a field that has long been neglected and is ready for careful research on an experimental basis; in addition, the results obtained will have a more direct bearing upon the problem of the origin of angiosperm genera and families than will those derived from any other kind of research. This is because the differences with respect to gynoecia, seeds, and seedlings tend to be constant within genera and families, and to differ in a regular fashion between these categories more than any other kinds of characters. Many large families can be named in which great diversity with respect to perianth and androecium is accompanied by relative constancy with respect to gynoecium and seeds: Leguminosae, Scrophulariaceae, Apocynaceae-Asclepiadaceae, Compositae, Iridaceae, and Orchidaceae, to name only a few. In other families, such as Cruciferae, Umbelliferae, Cyperaceae, and Gramineae, differences between genera are based to a large extent upon gynoecial and seed characters. Here lies one of the most significant frontiers of knowledge for botanists who are interested in angiosperm taxonomy and evolution.

Summary

The complexity and diversity of adaptations for seed reproduction are partly caused by the three separate and to some extent conflicting demands that must be met—those for seed development, for seed dispersal, and for establishment of seedlings. The very great differences that exist between taxa of angiosperms with respect to both seed size and seed number reflect in part the diversity of these adaptations. Selection for increased seed number or for larger seeds may bring about profound alterations of the gynoecium and other parts of the flower.

Adaptations for protection of developing seeds may involve changes in the structure of flowers and their surrounding bracts. Other modifications are associated with various methods of seed dispersal by mechanical means—water, wind, and animals. Animal dispersal is regarded as more effective for most plants than wind dispersal; since longer distances can often be covered and since the animal usually travels from one region to another having the same habitat it is more favorable for seedling establishment.

Factors affecting the establishment of seedlings involve the amount of stored food present and the succession of hormone-controlled processes of cell division and cell enlargement, and in some examples the development of versatility is aided by phenotypic polymorphism, such that the same plant produces two kinds of dispersal units. These may be adapted to two different methods of dispersal, or one of them may be so poorly adapted for dispersal that it ensures the retention of the populations in the spot where the parental individuals grew.

Particularly in plants that are adapted to climates in which seasonal or other limitations are imposed upon growth, there may be competing demands for seed size and seed number, so that survival depends upon the establishment of a suitable compromise. Among the factors that affect this compromise are animal predators and the evolution of effective defense mechanisms against them.

The phylogenetic significance of variation in seed and fruit characters is evident from the frequency with which they are used in diagnostic keys to major plant taxa. When individual families are examined, examples can be found of adaptive shifts from wind to animal dispersal, from generalized fruits to those differentiated for wind dispersal, and in other genera of the same families for animal dispersal. Reverse shifts from animal to wind dispersal also exist. The conservatism of complex adaptive patterns is such that many of their features will persist after the selective pressures that brought them into being no longer exist.

6 / Gene Action, Development, and Evolution

The evolutionist cannot trace directly the alterations of genes and gene-controlled processes that have been responsible for evolutionary trends. He can observe only the outcome of these changes in terms of alterations in the morphology and reactions of the adult organism. Hence, unless he has some conception of the relation between gene action and ultimate form, he cannot interpret the course of evolution at the genetic level on the basis of comparisons between adult organisms, living or fossil.

Because of this fact, our ignorance of the pathways from genes to characters is one of the chief obstacles to our understanding of evolutionary trends. Fortunately, this ignorance is no longer complete. Recent research in developmental biology enables us to formulate some generalizations about developmental pathways. The most significant of these will be reviewed briefly in the present chapter. The most important message which this review is intended to convey is that the pathways from genes to characters are much longer and more complex than most biologists, including geneticists, formerly believed that they were.

The nature of this complexity is indicated by Fig. 6-1. The main object of this diagram is to emphasize the general occurrence and basic significance of three phenomena: pleiotropy, multiple-factor inheritance, and feedback mechanisms. Pleiotropy is expressed by the radiation of lines from the same primary gene product to several adult characters. In higher animals or plants, the complexity of gene action is such that pleiotropy may be expected to be universal.

Two examples of extreme pleiotropy are the hooded gene of barley (Stebbins and Yagil 1966) and the radiation-induced mutants known as *eceriferum*-g (Zeiger and Stebbins 1972). In this series of mutations, which are all at the same locus, the production of wax on the leaf surfaces is largely suppressed simultaneously with a marked alteration of the cellular pattern of the stomatal complexes. Since all mutations induced at this locus have this pleiotropic effect, whereas waxless or *eceriferum* mutations which were induced at other loci do not alter the stomatal pattern, there is little doubt that *eceriferum*-g mutations all involve widely different pleiotropic effects of the same gene.

Gene Action at the Level of Molecules and Cells

The complexity of gene action exists at several levels: the individual protein molecule, which reflects the structural information of the gene itself; supramolecular complexes such as membranes and organelles, constructed of protein molecules that bear precise relations to one another on the basis of their gene-coded structures; individual cells, the growth and division of which must be precisely integrated during morphogenesis; and finally tissues and organs. Since this complexity has been well described in textbooks and semipopular works, such as those of Watson (1970), Monod (1971), and others, no further discussion of it is necessary here. One conclusion is, however, worth mentioning. This is that the genes that determine morphological characters most probably code either for regulators or

Fig. 6-1. Diagram showing the complex interrelations between genes and characters that result from the existence of genes having multiple or pleiotropic effects, from the multiple-factor basis of most characteristics, and from the frequent presence of feedback interactions. (From Stebbins 1966. Reprinted by permission of Prentice-Hall, Englewood Cliffs, New Jersey.)

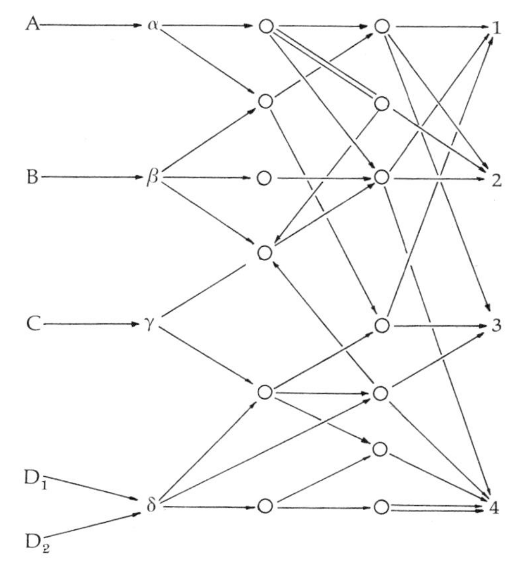

control systems that affect the action of other genes, or for structural proteins that can function properly only in conjunction with other proteins that are coded by different genes (Stebbins 1969). This is the probable basis of the fact that was emphasized in Chapter 1: differences between populations, species, and higher categories with respect to visible characteristics are inevitably controlled by many different genes. The evolution of new characteristics occurs not by the establishment of single mutations having large effects, but by extensive reorganizations of gene pools.

The Morphogenetic Basis of Differences in Form

The bridge between gene action as understood by developmental geneticists and an understanding of the genetic basis of morphological trends that is needed by the evolutionist must be provided by studies of morphogenesis. The basic postulate for such studies, as mentioned above, is that morphological characteristics are determined by complex sequences of gene action that are controlled by many genes, acting upon the cells of developing tissues. On the basis of this postulate, evidence must be sought that will permit evolutionists to decide what kinds of changes are accomplished with relative ease, what changes are more difficult from the point of view of gene action, what changes can be reversed with relative ease, and which ones are intrinsically irreversible or nearly so.

Unfortunately, morphogenesis has not provided firm and complete answers to these questions. Nevertheless, some tentative answers can be obtained on the basis of present knowledge, and will now be reviewed.

Meristematic capital and numbers of parts

A few years ago (Stebbins 1967), I suggested that the number of similar organs or parts that are produced in a particular whorl or series can be represented by the quotient $A^n = a^m/a^i$, where A^n is the final number of parts, a^m is the total number of meristematic cells that are capable of producing an A-type part, and a^i is the number of meristematic cell initials that are needed to produce one A-type part. This formula will apply equally well to the number of lobes in a compound leaf, of flowers in a strongly determinate inflorescence, such as a capitulum or umbel, and of sepals, petals, stamens, or carpels per flower. If the time factor is added, so that the length of time is considered during which primordia having a particular kind of cellular determination are differentiated, the principle can be extended to cover the development of a structure with more indeterminate growth, such as the number of flowers in a raceme or spike.

Four examples will be presented to show that changing interactions between cell proliferation and differentiation can be responsible for either increases or decreases in the numbers of organs or parts. The first is the experimental alteration of the number of protoxylem points in a pea root produced by Torrey (1955, 1957). He found that when 0.5 mm of the distal portion of pea roots, containing only visibly undifferentiated cells, was isolated and cultured *in vitro*, the great majority of the cultures produced roots having the triarch condition, which is normal for these roots (Fig. 6-2). About 2 percent of the cultures, however, which were tips of relatively small size, produced at first diarch roots, which later reverted to the triarch condition.

If to the culture he added indole acetic acid at a concentration of 10^{-5} molar, he obtained a greater proliferation of the meristematic cells from which vascular tissues are differentiated. As a result he converted the triarch condition, which existed in the roots from which the tips were excised, to the hexarch condition in the cultured and treated roots. The number of protoxylem points could, therefore, be increased or decreased, depending upon the amount of meristem present when procambial differentiation took place.

A second example is provided by developmental studies of the difference between wild, small-fruited varieties and species of the tomato (*Lycopersicon*) and the cultivated varieties of *L. esculentum* which have been selected for large fruit size (Fig. 6-3). Species such as *L. pimpinellifolium,* as well as the smaller, cherry-fruited varieties of *L. esculentum,* have the floral formula typical of the Solanaceae: pentamerous calyx and corolla, 5 stamens, and 2 carpels. On the other hand, the large-

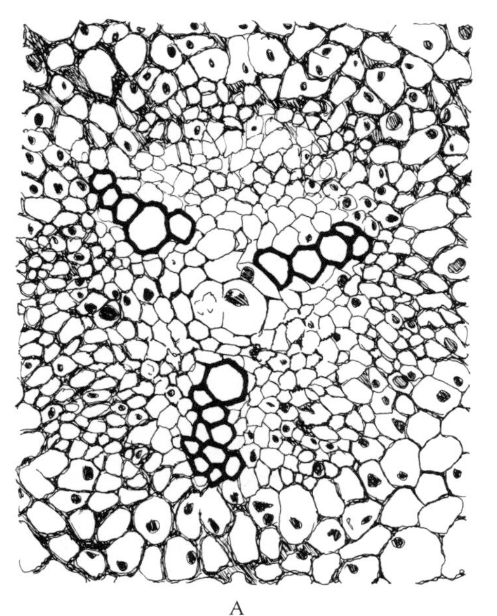

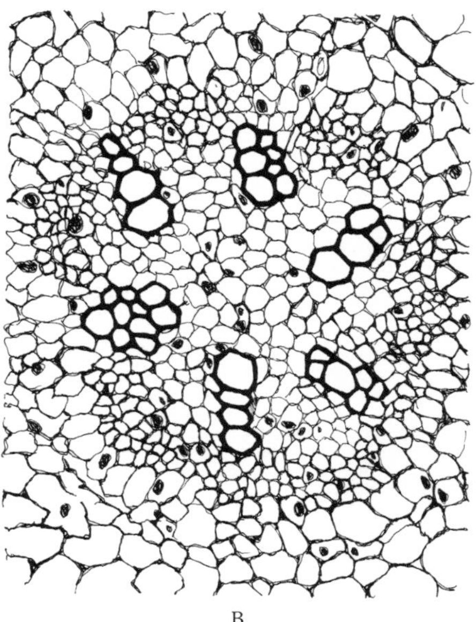

Fig. 6-2. Tracings from photographs of two sections of the same pea root, decapitated and treated with indole acetic acid at 10^{-5} M: (*A*) regenerated root tip immediately distal to the region of decapitation, showing the original triarch arrangement of the mature xylem elements (*heavy lines*), and extensive cell proliferation in the pericycle region; (*B*) tracing of a section cut distally to (*A*), showing hexarch arrangement of xylem elements that developed following extensive proliferation of undifferentiated cells. (From Torrey 1957.)

Fig. 6-3. Effects of natural and artificial selection for differences in fruit size in *Lycopersicum* (tomato): (*A*) undifferentiated floral meristem, (*B*) ovary at anthesis, and (*C*) mature calyces of *L. pimpinellifolium* var. "red currant"; (*D*) undifferentiated floral meristem and (*E*) ovary at anthesis of *L. esculentum* var. "red peach"; (*F*) mature calyces of a similar variety, "red cherry," showing the usual pentamerous condition and a hexamerous calyx which is occasionally found; (*G*) undifferentiated floral meristem and (*H*) ovary at anthesis of *L. esculentum* var. "Bonny Best," a large-fruited cultivar, produced relatively recently by artificial selection for increased fruit size; (*I*) calyces of *L. esculentum* var. "Marglobe," a variety similar to "Bonny Best." Note that selection for increased fruit size has resulted in an increase in the number of calyx lobes, as well as in fusions of lobes in some calyces. (*A, B, D, E, G, H* from Houghtaling 1935; *C, F, I* from plants in collection of Department of Vegetable Crops, University of California, Davis.)

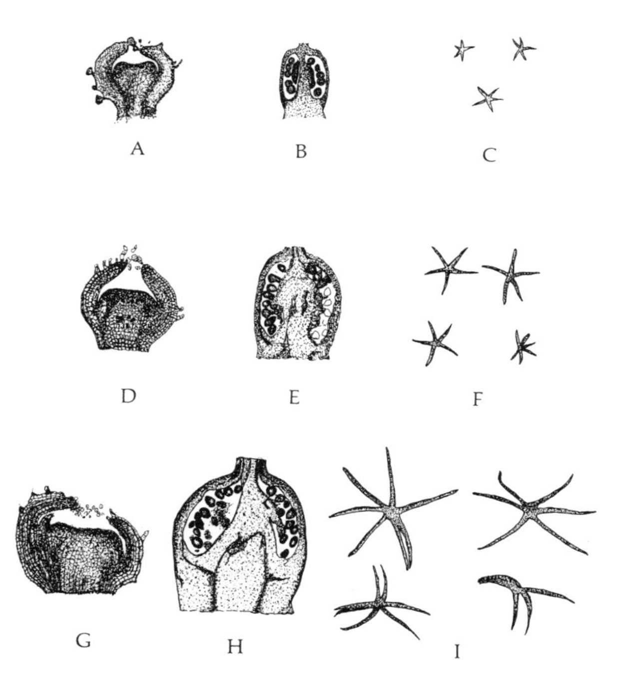

fruited varieties of *L. esculentum* usually have 6 or 7 calyx and corolla lobes, and 1 or 2 extra, imperfectly developed ovary locules. The development of the flower in different varieties of tomato was studied many years ago by Houghtaling (1935). She found that even before the calyx and corolla are differentiated, the floral meristem of the large-fruited varieties contains more cells than does the corresponding meristem of small-fruited varieties. Apparently, some of the selected genes or gene combinations favoring increased fruit size accomplish this result by increasing the size of the very young floral meristem. Comparable increases in the numbers of floral parts are found in large-fruited cultivars of other species, such as the eggplant (*Solanum melongena*), pepper (*Capsicum annuum*), and pomegranate (*Punica granatum*).

A third example, that of *Linanthus androsaceus* (Huether 1968, 1969) is described in Chapter 2.

A fourth example is provided by developmental studies of double flowers, some of which, as in *Matthiola incana*, are determined by a single gene (Saunders 1928). In this instance, also, the gene apparently exerts its effects by increasing the number of cells in the floral meristem, but it has an additional effect. The fact that all of the numerous primordia that are differentiated on the meristem after the sepals give rise only to petals is best explained by assuming that the gene for doubleness in some way blocks the normal succession of gene activation. The genes responsible for stamen and carpel differentiation never become activated, and those for petal differentiation persist in their activity. The gene for doubleness can, therefore, be characterized as producing an *allochronous deviation* from the normal course of development. The timing of activation of stamen and carpel differentiation genes is correspondingly inhibited.

Examples of allochronous deviation without change in numbers of parts are pistillody in certain hybrids of *Paeonia* (Saunders and Stebbins 1938) and the teratological form of *Trillium grandiflorum* described by Hall (1961), in which the sepals, petals, and stamens are leaflike. An example in *"Pulsatilla"* (*Anemone*), in which apparently "new" kinds of appendages appeared in the progeny of a hybrid between a polypetalous mutant and one having laciniate petals, has been described by Zimmermann (1964). He postulates that it results from an upset in the timing of action of various factors that control development—an interpretation that is essentially similar to the present one. Such allochronous shifts in the relative timing of action of different gene systems are probably important features of many evolutionary trends, particularly in the inflorescence. They are discussed in this context in Chapter 11.

The most reasonable hypothesis to explain the successive production of different kinds of appendages (sepals, petals, stamens, carpels) during floral development is that the cells of the undifferentiated apical meristem undergo regular changes in their capacity for determination, under the influence of control systems that successively activate and inactivate specific groups of genes. Evidence in favor of this hypothesis has been obtained from experiments in which floral primordia have been split surgically into two halves at various stages of development (Cusick 1956, Soetiarto and Ball 1969, Hicks and Sussex 1971). These experiments have shown that regeneration after the split is confined to organs of which the primordia have not yet been differentiated when the surgical operation is performed. Further evidence has been obtained from extractions of proteins (Barber and Steward 1968), which have shown that the differentiation of each kind of primordium is associated with the appearance of different proteins in the embryonic floral apex.

On the basis of this hypothesis, allochronous shifts can be explained by assuming that the factors which determine the number and positions of appendages act independently of the factors which determine the kind of appendage that will be produced.

The problem of phyllotaxy

One of the most vexing problems of plant morphogenesis is the mechanical and physiological basis of the regular phyllotactic spirals of leaf arrangement that plants so persistently display. The extensive literature in this field will be reviewed only to the extent necessary to establish three points, as follows: (1) The nature of the phyllotactic spiral is determined by forces acting within the meristem itself, rather than by influences emanating from the already determined leaves. Hence the genes that control it act on undifferentiated, actively dividing meristems.

(2) As in the examples of the number of parts in a flower, the magnitude of

the spiral, whether 1/2, 1/3, 2/5, 3/8, or higher, probably depends upon a particular relation between the number of cells existing in the apical meristem and the number required to form a primordium. (3) The forces that control the position where a new primordium arises are such that, even in a condensed shoot like the floral axis, rapid shifts in the magnitude of the phyllotactic spiral cannot be expected to occur. Consequently, the relation of a structure to the phyllotactic spiral is a significant fact bearing upon its homology to other structures.

Experiments with microsurgery have shown that the shoot apex is independent of influences from older parts of the plant (Snow and Snow 1948, Snow 1955, Ball 1950, Wardlaw 1957*a,b,* Loiseau 1959, 1960). When deep cuts are made in the apex, which completely sever the procambial connections between it and the older appendages, the phyllotactic spiral continues to develop in a normal fashion. In *Lupinus albus,* completely isolated central portions of the shoot apex develop a typical succession of leaf primordia. On the other hand, cuts made in critical parts of the "free area" of the meristem from which new primordia would normally arise affect greatly the position of these primordia, and may suppress them altogether.

Evidence that the magnitude of order of the spiral depends upon size relations between apex and primordia comes from three sources: comparisons between apices of different plants, studies of developmental sequences in a single plant, and surgical experiments. Wardlaw (1957*a*) has pointed out that in ferns, which have a relatively high order of phyllotaxy, the apex is broad and shallow, and several growth centers for new primordia can be accommodated on it at one time (Fig. 6-4, *A*). At the other extreme is the situation in grasses and in some species of *Magnolia* and *Peperomia* (Hagemann 1960; Fig. 6-4, *B*), in which the undifferentiated apex is not only small but also narrow and high, and many cells participate in the differentiation of primordia. Because of this situation, more than half of the "meristematic capital" available as undifferentiated cells of the apex is used up in the formation of a single primordium. As a result, only one primordium is differentiating at any one time, and the interval between successive differentiations is relatively great. Following the principle originally established by Hofmeister, and supported by recent surgical experiments (Snow 1955), the position of the new primordium is in the middle of the maximum available space, which in the grass apex is just opposite the position from which the previous primordium emerged. This combination of timing and position results in the distichous condition, or a 1/2 phyllotaxy, the lowest order possible.

If, on the other hand, the cells that participate in the formation of a primordium include about half of those available in the apex, the remaining available cells can participate simultaneously in the formation of a primordium on its opposite side, and two opposite appendages are produced. The decussate arrangement, with each successive pair appearing at right angles to the previous one, is the expected result of the operation of the principle of maximum available space, since the new primordia appear at a position equidistant between those occupied by the two previously formed opposite primordia.

Spiral phyllotaxies represent various situations in which the formation of a new primordium requires less than half of the meristematic capital present in the apex at one time. In most seedlings of dicotyledons, the first true leaves above the cotyledons are opposite each other or nearly so, and form a decussate arrangement with respect to the cotyledons themselves. The spiral arrangement may appear almost immediately, or it may be somewhat delayed.

The conservative nature of phyllotactic spirals during development is an inevitable result of the operation of the principle of maximum available space. On the basis of this principle, we would expect that in a rapidly expanding apical meristem a transition from a 2/5 to a 3/8 and a 5/13 order would occur during the emergence of a relatively small number of primordia. On the other hand, a sudden shift from 2/5 to 5/13 and back to 1/3 would be entirely unexpected, and inexplicable on the basis of any known morphogenetic principle. This point will be reemphasized in the discussion of the homology of stamens in Chapter 10. It is particularly significant with respect to flowers that have five sepals, five petals, numerous stamens, and five or fewer carpels. If in these flowers each stamen is regarded as a separate appendage, two radical shifts in phyllotaxy are required. If, on the other hand, a bundle of stamens is regarded as the basic

Fig. 6-4. (*A*) Shoot apex of *Liriodendron tulipifera,* a plant having relatively large leaves and a low phyllotactic spiral; the asymmetrical appearance of the apex is due to the fact that the single developing primordium has been formed from a high proportion of the undifferentiated tissue of the apical dome. (From Hagemann 1970.) (*B*) Shoot apex of a plant having relatively small leaves and a high phyllotactic spiral, *Elodea canadensis.* (From preparation provided by E. M. Gifford, Jr.) (*C-H*) Outlines of the shoot apices of six different species of angiosperms, showing the extent of variation between them. (From Wardlaw 1957*b*.)

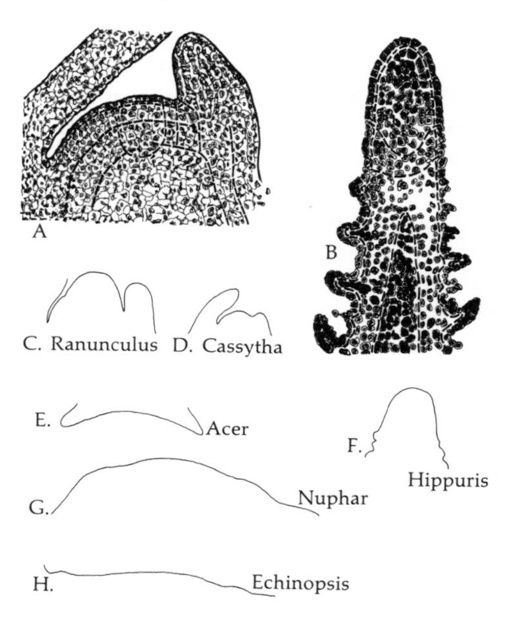

appendicular unit, a normal phyllotactic progression can be postulated. The latter viewpoint is supported by evidence from several other directions, as is explained in Chapter 10.

Intercalary Meristems and Their Significance

The development, intensification, and redistribution within the plant of intercalary meristems form some of the most important kinds of developmental changes, with respect to both the differentiation of angiosperms from other groups of seed plants and the differentiation of the various groups of angiosperms from one another. Such changes have affected stems, leaves, and floral organs. Botanists living in the modern world are accustomed to thinking of plants having elongate stems, bearing leaves separated by well-developed internodes, as the normal form of architecture for a vascular plant. In those plants having short internodes, with the leaves close to each other on the stem, the internodes are said to be "compressed," and the condition is often regarded as secondary in all of the groups in which it occurs. This way of thinking probably came about because the commonest contemporary plants are angiosperms, most of which have elongate internodes. It is probably the result of uncritical transfer of Goethe's archetype concept to evolutionary morphology.

A little reflection, accompanied by a review of the growth forms that exist in most groups of vascular plants other than angiosperms, will serve to show that in these other vascular plants elongate stems and internodes are by no means the rule. On the contrary, they are quite exceptional in all groups of vascular plants except for angiosperms and Gnetales. Ferns are typically rosette plants, and elongate internodes are found in them only in the rhizomes of some species (*Pteridium*), in the stems of climbers such as *Lygodium,* and in the sporangial stalks of such genera as *Ophioglossum* and *Botrychium* (Peterson and Cutter 1969*a,b*). The Lepidophyta (*Lycopodium, Selaginella,* and *Isoetes*) all have the leaves crowded on the stem, with essentially no internode elongation, as do also the cycads, the extinct cycadophytes, and most conifers. Where internode elongation exists in them, as in creeping species of junipers, and in some Araucariaceae and Podocarpaceae, it is associated with specialized growth forms or leaf shapes, and so must be regarded as derived. We can hardly escape the conclusion, therefore, that elongate stems and internodes, produced by the activity of intercalary meristems, are derived originally from stems having crowded leaves and short internodes, although reverse trends toward the rosette condition must have occurred frequently. The form of leaf development that occurs in angiosperms, consisting as it does of "plate meristems" that are intercalary in nature, is also largely confined to angiosperms, and so must be regarded as a derived condition.

Even more important than internode elongation through the development of intercalary meristems is the development of generalized intercalary meristems in the basal portions of various organs, by means of which the continued differentiation of separate parts is suppressed. This phenomenon, illustrated in Fig. 6-5, is responsible for the phylogenetic trends designated "union" or "fusion" of parts. These terms are misnomers, and their use has led to much confusion in anatomical literature as to the nature of the trend involved. Some authors have maintained that, since there is no union of previously separate parts, the free distal regions of parts such as calyx lobes cannot be regarded as homologous with the otherwise comparable organs, namely, sepals, which are completely separate from one another. This argument has been used most often with reference to the origin of the single cotyledon of the monocotyledons (Chapter 13).

In order to eliminate any confusion in discussing this matter, I substitute in this book for the terms "union" or "fusion" the term *intercalary concrescence,* which means growing together through the activity of an intercalary meristem. Whether this cumbersome term should be substituted for the older and simpler term "union" is a matter for botanists in general to decide. Nevertheless, whatever term is used, discussions of the evolution of concrescent parts will be meaningful only if the discussants have clearly in mind the alterations of developmental processes that are involved.

Intercalary concrescence of similar parts has given rise to synsepalous calyces, sympetalous corollas, the stamen tube of Malvaceae, the concrescence of stamens in the more advanced members of the Family Leguminosae subf. Papilionoideae, and the syncarpous gynoecium. Intercalary concrescence of different portions of the same organ has given rise to the closed, tubular leaf sheaths of many Gramineae (grasses), Cyperaceae (sedges), and other monocotyledons; to the leaf sheaths of Polygonaceae (buckwheats); to the "pitcher"-like leaves of Sarraceniaceae and Nepenthaceae; to the peltate leaves of *Tropaeolum,* Nymphaeaceae (water lilies), and other groups; and probably to the closed carpels of many apocarpous groups.

Important trends in the condensed reproductive axis of angiosperms are intercalary concrescences of superimposed organs of a different nature. These trends are termed *adnation,* a term that will be retained in the present book, since its developmental meaning is hard to misinterpret. Its simplest form is the adnation of stamens to a sympetalous corolla. The much more complex forms of adnation involved in the origin of perigynous flowers and particularly epigyny, or the inferior ovary, are discussed fully in Chapter 12, which is devoted to trends of evolution in the flower.

Fig. 6-5. Vegetative shoot apex of *Kalanchoe Blossfeldiana* after treatment with a 10-ppm solution of triiodobenzoic acid (TIBA) in six cycles of 11 hours each. The leaf primordia have become "fused" by means of intercalary concrescence, so that they form a cup. (From Harder and Oppermann 1952.)

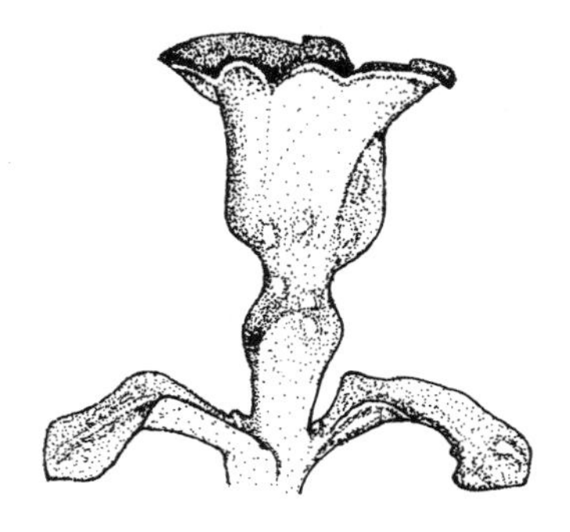

Intercalary meristems and the action of growth substances

The evidence now available indicates strongly that the activity of intercalary meristems is governed by the concentration, the distribution, and the activity of growth substances, particularly auxin (IAA) and the gibberellins. When growing plants are treated with 2,3,5-triiodobenzoic acid (TIBA), the action of apical meristems tends to be suppressed, and that of intercalary meristems is stimulated (Denffer 1952, Harder and Oppermann 1952, Wardlaw 1953, Heslop-Harrison 1967, Kiermayer 1960). Various lines of evidence indicate that this substance acts chiefly to bring about a redistribution of auxin in the plant (Leopold 1955, Heslop-Harrison 1967).

The intercalary meristems which are stimulated by the action of TIBA often produce the "fusion" of parts (Fig. 6-5). In a plant having opposite leaves, such as *Kalanchoe,* the two leaves of a pair may be "fused" to form a cup. In the tomato, the leaves may be replaced by entire, ascidia-like structures. In addition, side branches may become "fused" with the main axis, and ring-fasciations containing extra vascular cylinders may be induced in the stem (Gorter 1951). These effects are obviously teratological and highly inadaptive. Nevertheless, they can be regarded as extreme examples of phenomena which, on a much smaller scale and more precisely regulated, could be responsible for important evolutionary trends in angiosperm morphogenesis. On this basis, we can postulate that those trends of angiosperm evolution which involve the change in location and the *de novo* activity of intercalary meristems are based upon mutations that alter either the locations where growth substances are synthesized, the way in which they move through the plant, or both of these factors.

One feature of intercalary meristems which must be emphasized in connection with evolutionary trends is that their regulation and integration with other growth processes must require a much more complex system of regulatory factors than those needed for the apical meristem. As already mentioned, the latter persists throughout the active life of a shoot or shoot system, and its size and apical position change either gradually or not at all during shoot growth. On the other hand, intercalary meristems repeatedly arise *de novo,* and the timing of their activity must often be adjusted very precisely so as to produce organs of a particular size. The evolution of the regulatory mechanisms required for these precise adjustments must itself require very complex and equally precise interactions between genes. On this basis, we can postulate that intercalary meristems in general represent a derived and more highly evolved form of meristem than apical meristems. Hence structures that develop largely through the activity of intercalary meristems are more highly evolved than those formed directly by the apical meristem. The bearing of this complexity of regulation on the concept of evolutionary irreversibility is discussed in the next chapter.

Recapitulation, Embryonic Similarity, and Neoteny

Discussions of development and embryology in relation to evolution usually place their greatest emphasis upon the concepts of recapitulation, embryonic similarity, and neoteny. In contrast to the concepts and principles that have been discussed previously in this chapter, these three concepts are based upon comparisons between entire sequences of development from the morphological and descriptive point of view, with little attempt to analyze their genetic basis. Both their formulation and most of the discussions about them were made before any information was available about the role of genes in development. Hence they need to be reevaluated in the light of our present-day knowledge.

The concept of *recapitulation,* or the "biogenetic law," dates from the speculations made by Ernst Haeckel almost a century ago. In a review of this concept as applied to animals, De Beer (1950) concluded that it is more misleading than helpful. I reached the same conclusion with respect to plants (Stebbins 1950), and this conclusion has been either explicitly or tacitly accepted in most discussions of development and evolution since then. More advanced, specialized forms rarely if ever recapitulate the *adult* condition of their ancestors at any stage of their development.

The question of embryonic similarity is more complex. In many instances, embryonic or seedling stages of specialized plants resemble corresponding stages of unspecialized, primitive forms more closely than the adults of the forms in question resemble one another. This phenomenon is based upon the general principle that genes which affect later stages of development are less likely to upset the entire sequence than are those which act at early stages. Hence late-acting genes, if they confer an equal adaptive advantage with respect to some particular characteristic, have a greater chance of becoming established in the population by the action of natural selection than have genes that act early in development.

Nevertheless, the generalization of embryonic similarity is by no means universally true, and it cannot be applied uncritically to all situations. In higher plants, it must be considered in two contexts: (1) the succession of leaves produced by a seedling or sporeling and (2) the development and differentiation of individual primordia or groups of primordia into organs of the adult plant.

With respect to leaf succession, embryonic similarity is best exemplified by *seedling heterophylly.* The successive leaves of a seedling differ from each other as well as from those of the mature plant (Fig. 6-6). In an analysis of seedling heterophylly among species native to California (Stebbins 1959*b*), I distinguished two types: *elaboration heterophylly,* in which the earliest leaves are simple in outline and those pro-

Fig. 6-6. Succession of leaves on a seedling of *Adenostoma fasciculatum* (Rosaceae). The simple, linear, hard-textured, needlelike leaves shown at the right are the only form found on mature shrubs. (Original.)

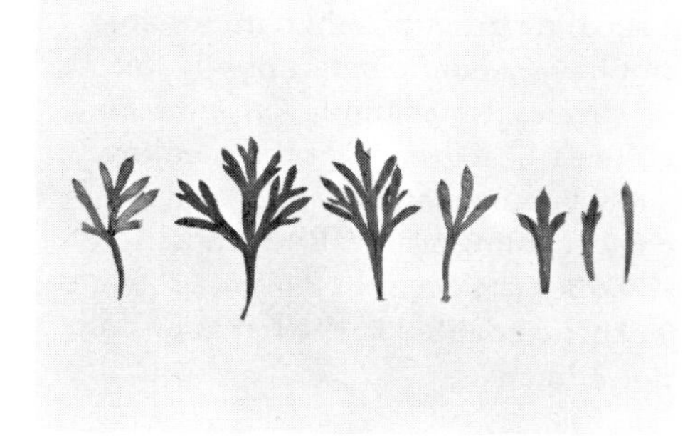

duced later have successively more complex outlines, and *reduction heterophylly*, in which the earliest leaves are relatively complex and the adult ones are simpler. These two types are associated with completely different ecological adaptations. Elaboration heterophylly is associated with a progressive increase in leaf surface, and is usually found in mesophytes. When it does occur in plants adapted to arid or semiarid situations, their seedlings are mostly quick-growing herbs that take advantage of brief periods of high available moisture. Reduction heterophylly in the California flora is almost entirely restricted to shrubs that grow in dry sites, in regions having mild, moist winters and hot, dry summers. The thinner, lobed or dentate leaves produced by the seedling adapt it to rapid early growth when light intensity is low and moisture availability is relatively high. The thick, simple leaves on the adult plant are adaptations to the hot, dry summer conditions.

Shrubs such as *Adenostoma fasciculatum*, which are well adapted to the Mediterranean-type, dry-summer climatic regime, have pronounced reduction heterophylly. On the other hand, their relatives that are adapted to the cold steppe climate found in eastern California and regions to the eastward, in which winters are so cold that seedling growth is then impossible, do not have seedling heterophylly. In these species, for example *Purshia tridentata*, the earliest growth of the seedling occurs when insolation is high and moisture is becoming scarce. Reduction heterophylly does not occur in *Purshia* or in any other shrub adapted to the Great Basin steppe climate.

For these reasons, reduction heterophylly must be regarded as at least in part an adaptation to a particular climatic regime. Both kinds of heterophylly, when they occur, may reflect an ancestral trend of evolution, though this is not necessarily so. On the other hand, the absence of heterophylly can never be taken as evidence that the form concerned did not evolve from one having leaves of a different kind.

Embryonic similarity and the development of primordia

The development of leaves from an axis having indeterminate growth is very different from any of the kinds of developmental sequences found in animals, from which von Baer's concept of embryonic similarity was derived. Much closer analogies to animal development can be found in the development of appendage primordia from the shoot apex, particularly in the modified, determinate shoots that give rise to flowers. In them embryonic similarity between primitive and more advanced forms may occasionally exist, but it is by no means universal. When present, it is usually based either upon the development of intercalary meristems or upon allometric growth.

Intercalary concrescence of sepals, petals, or carpels usually follows the appearance of separate primordia, which later form the calyx lobes, corolla lobes, or separate stigmas of a compound ovary. When the concrescence or "union" is weakly developed, the primordia may reach a considerable size before meristematic connections between their bases become evident (Fig. 6-7, *A*). In strongly concrescent parts, on the other hand, the intercalary meristem may appear almost or quite as soon as the primordia are differentiated, so that the compound structure, even in its embryonic condition, never resembles the primordia of the presumed ancestral form (Fig. 6-7, *B*). A frequent feature of the evolutionary trend toward concrescence is, therefore, the initiation of the intercalary meristem at earlier and earlier stages of development. This relatively gradual modification of gene action toward greater precocity would be less likely to bring about disharmonious development than would an abrupt shift in gene action that produced in a single step a highly active intercalary meristem at an early stage of primordial development. During the course of the evolutionary trend toward greater precocity, selection could establish in the population modifying genes which could counteract any harmful side effects that the more precocious inception of the intercalary meristem might bring about. The morphological progression in phylogeny of increased "fusion" or intercalary concrescence of parts is regarded as based upon *increasing precocity of gene action.*

We can expect, therefore, that, if specialization is accompanied by the appearance of a new kind of intercalary meristem, embryonic similarity might be recognizable during the early stages of the evolutionary trend, but almost or completely absent from the development of the most highly evolved structure. This difference is clearly evident from a comparison between early stages of development in a species such as *Agapanthus umbellatus,* in which the corolla lobes are two or three times as long as the tube, and *Kniphofia,* in which the corolla lobes are very short.

The second basis on which embryonic or primordial similarity may exist between forms having widely divergent adult structures is through the initiation of divergent directions of growth at an early stage of development. This phenomenon can be detected by means of the allometric growth constant. Sinnott (1960, Sinnott and Kaiser 1934) showed that fruits having different shapes at maturity

Fig. 6-7. Development of sympetalous corollas. (*A–C*) *Apocynum cannabinum,* in which the "fusion" of petals is only moderately developed; (*A*) early floral development, showing initiation of the corolla as separate petal primordia; (*B*) intermediate stage, showing beginning of intercalary concrescence; (*C*) mature corolla. (*D–F*) *Centranthus ruber* (Valerianaceae), a highly specialized flower: (*D*) earliest stage, showing beginning of intercalary concrescence simultaneously with the appearance of petal primordia; (*E*) intermediate and (*F*) nearly mature corolla, showing strong development of intercalary concrescence. (From Payer 1857.)

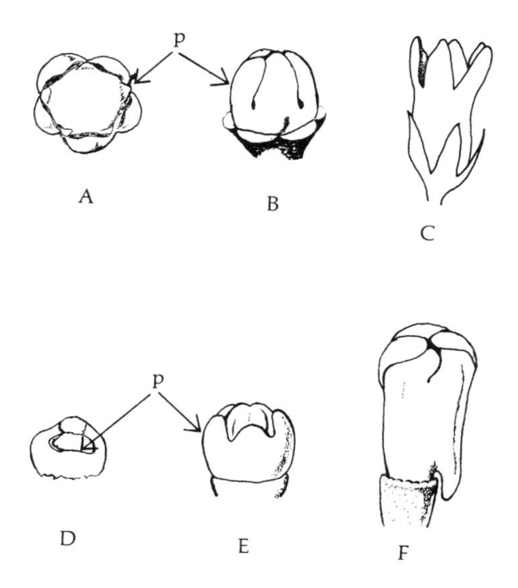

may in some instances also have primordia of different shapes, so that the growth constant *k*, which expresses the relation between growth in length and growth in width, is constant throughout development. This is the situation in *Cucurbita*, which, consequently, gives little evidence of embryonic similarity in its fruit development. On the other hand, the development of the elongate fruits of *Capsicum annuum* involves a change in the value of *k* at an intermediate stage of development. As a result, long- and short-fruited varieties of this species resemble each other closely in the development of their ovary primordia and of young fruits until shortly after flowering. The fruits then diverge, and the difference between them continues to increase as development progresses. As in other situations, embryonic similarity is not inevitable. It may be easily recognized in some groups and is nearly or quite absent in others.

The hypothesis of neoteny

A number of zoologists, particularly Koltzov (1936), De Beer (1951), and Hardy (1954), have suggested that major evolutionary transformations of development can come about via *neoteny* or paedomorphosis. Takhtajan, who has suggested that it plays an important role in plant evolution, defines neoteny as follows (1959:25): "Under extreme conditions for existence the process of evolution leads to the situation, that certain organs and tissues are arrested in their development, so that a definitely precocious conclusion of development can be demonstrated. Ontogeny breaks off, so to speak, and the earlier stages of development become definitive. The best known animal example is a salamander, the axolotl (*Ambystoma*). In various species of this genus a shortening of the period favorable for development brings on precocious sexual maturity in a form which otherwise has the appearance of a larva." Transferring the concept to plants, Takhtajan cites plants adapted to alpine meadows (p. 26), as well as the carpel as a neotenous leaf (p. 89) and the embryo sac as a neotenous female gametophyte (p. 143).

According to the concept of neoteny, the transition from one complex pattern of development to another occurs via the elimination of the later stages of the existing pattern, after which new and more complex later stages can be added to the essential earlier stages, which have been retained. The validity of this concept as applied to animals will not be discussed, since it is not relevant to the theme of this book. There are, however, good reasons for questioning Takhtajan's application of the neoteny concept to comparisons between developmental patterns in plants. His three examples will be discussed in turn.

In order to apply the concept of neoteny to plants of alpine meadows, one would have to show that the vegetative development of alpine races, up to the initiation of flowering, is essentially similar to the development of their lowland relatives. As has been shown by a number of comparisons between alpine and lowland races of the same species grown under the same controlled conditions, this is not the case. Differences between the timberline, middle-altitude, and coastal races of *Potentilla glandulosa* are evident in young seedlings, and persist throughout the life of the plant (Clausen, Keck, and Hiesey 1940). The same is true of *Achillea lanulosa* (Clausen, Keck, and Hiesey 1948). Adaptation of these races to high altitudes has not consisted merely in shortening their vegetative cycle, leaving the early stages unaltered; on the contrary, these adaptations have affected almost equally all stages of the cycle.

The principle of neoteny might be applied to the evolution of the carpel from a supposed leaflike ancestral megasporophyll, if this hypothetical course of evolution was actually followed. In one of the most primitive angiosperms, *Drimys Winteri* var. *chilensis*, the ontogeny of both the foliage leaf (Gifford 1951) and the carpel (Tucker 1959) have been investigated in detail. As one can see by comparing Figs. 6-8, *A* and *B*, the early stages of development of these two organs are very similar. One could imagine, therefore, that the carpel evolved from a leafy meristem, followed by the origin of new plate or intercalary meristems, which in the carpel give rise to the bent or incurved carpel wall.

This interpretation of the derivation of a closed carpel from a flat megasporophyll is not, however, the only possible one. An alternative interpretation could be based upon the hypothesis of increasing precocity of gene action, outlined in the next chapter. One could assume that the first stage in the transformation was the appearance of asymmetrical growth at a late stage of meristematic activity, which would have produced a sporophyll with a flat central portion and an incurved margin. A shift in the appearance of this asymmetrical growth to successively earlier stages of ontogeny could then produce the completely closed carpel.

Carlquist (1962) has suggested that in some herbaceous plants, as well as in oceanic-island species that are secondarily woody, "paedomorphosis," which must be regarded as a form of neoteny, is responsible for the presence in them of a more primitive kind of secondary xylem than might be expected on the basis of their otherwise rather specialized condition. In plants having relatively specialized secondary xylem, the length of the cambial cells becomes sharply reduced between the formation of primary and of secondary xylem, becoming somewhat longer again at a later stage of development. On the other hand, in the insular species studied by Carlquist, the length of cambial cells and of the xylem that they produce differs little or not at all from

Fig. 6-8. Stages in the development of the leaf and the carpel in *Drimys Winteri* var. *chilensis*, showing histological similarity at early primordial stages, and great morphological divergence at maturity: (*A–C*) leaf (from Gifford 1951); (*D–F*) carpel (From Tucker 1959).

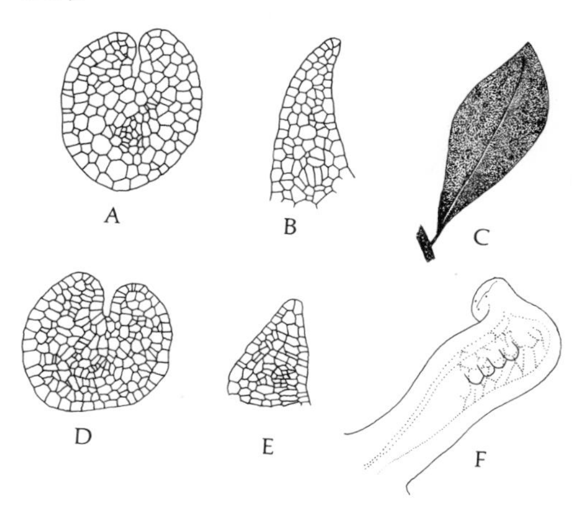

Fig. 6-9. Alternative explanations of the origin of a new character, united petals, on the basis of the hypothesis of neoteny and of increasing precocity of gene action: (*A*) developmental sequence, separate petals; (*B*) sequence of transitional stage, according to the neoteny hypothesis; (*C*) sequence of intermediate stage, according to the precocity-of-gene-action hypothesis; (*D*) developmental sequence, strongly united petals. (Original.)

primary to secondary xylem. The resulting secondary xylem is, therefore, regarded as an example of paedomorphosis, since it retains the cell length that is characteristic of the more juvenile tissue.

This example can also be interpreted on the principle of preferential establishment by selection of mutations that affect later stages of development. The mutations concerned are those that suppress the usual shortening of cambial cells between the formation of primary xylem and of secondary xylem.

In discussing the example of the embryo sac, Takhtajan (1959:143) places great emphasis upon the shortening of its development, as compared with the gametophyte of gymnosperms. He has not, however, produced any evidence which would suggest that this shortening has occurred through the elimination of later developmental stages, leaving the earlier stages unaltered. On the other hand, certain features of the early development of the angiosperm embryo sac, particularly the migration of its first two nuclei to opposite ends of the nucellar cavity in which the embryo sac develops, must be regarded as innovations rather than as retentions of an original developmental pattern.

All five of the examples that have just been discussed, as well as many others, can be explained on the basis of my hypothesis of increasing precocity of gene action as well as or better than on the concept of neoteny. Fig. 6-9 shows diagrammatically the difference between the two concepts.

In my opinion, the use of such generalized concepts as recapitulation, embryonic similarity, and neoteny to explain the complex interrelations between developmental patterns and their alteration through processes of evolution is no longer a fruitful approach to the problem. Such an approach is basically typological, since it attempts to apply one apparently simple term to a complex succession of events. The modern approach to an understanding of evolutionary change in terms of altered developmental sequences must be basically synthetic. It must begin with careful, quantitative analyses of developmental patterns in related species and genera. These must be carefully chosen so as to represent the kinds of trend that have been responsible for the differentiation of families and orders. Until a much larger number of such analyses have been made than are yet available, hypotheses concerning the nature and reversibility of evolutionary trends must be regarded as tentative and unproven.

Summary

The complexity of gene action in development leads to the phenomena of pleiotropy and multiple-gene inheritance. Probably the majority of genes that affect form exert their primary action on control systems or on structural proteins. The number of parts that are produced by a meristem, such as floral organs of a particular whorl or flowers of an inflorescence, is a function of the ratio of the total amount of meristem produced of a particular kind to the amount of meristematic tissue needed to produce one part or organ. Genes are known in several plants that increase the number of parts by increasing the amount of available meristem. Similar considerations, applied to the continuous development of a shoot meristem, can explain different orders of phyllotaxy. Intercalary meristems are specialized tissues that are particularly frequent in angiosperms as compared to other vascular plants. Their strong development is a derived condition. "Union" or "fusion" of parts, here termed intercalary concrescence, results from meristems that replace the growth and differentiation of separate organ primordia. The development of intercalary meristems probably requires elaborate control mechanisms for the production, distribution, and regulation of the activity of growth substance, and so is most likely a derived condition.

Though the Haeckelian concept of recapitulation is rejected, von Baer's concept of embryonic similarity is accepted with reservations; it is explained by assuming that mutations affecting developmental patterns are most likely to be adaptively advantageous if they affect later stages of development. Once they are established, however, their effects can be gradually transferred to earlier stages through the action of newly acquired modifying genes. This concept of increasing precocity of gene action is regarded as a more satisfactory explanation of many phenomena that other biologists have explained on the principle of neoteny.

7 / Methods for Recognizing Evolutionary Trends

In the preceding chapters of this book, I have tried to show that the most fruitful working hypothesis to explain evolutionary trends in flowering plants is that they have been guided by natural selection because of successions of complex interrelations between the plants and their environment. In the following chapters, the most probable trends will be outlined, together with reasons why my postulates differ from those commonly accepted. The present chapter is an outline of the principles that have guided the deductions made in subsequent chapters.

The first and most important of these principles is that, especially when the fossil record is scanty or nonexistent, hypotheses concerning past trends can never be proved in the way that experimental biologists prove working hypotheses about processes in living organisms. In dealing with phylogeny, we can never subject hypotheses to precise predictions and then test these predictions by repeatable experiments. We must always be satisfied with decisions as to which of several possible sequences of events has the greatest probability of being the correct one. No judgments or hypotheses about past events can be regarded as final; one must always be ready to modify them as new evidence comes in. I agree with Davis and Heywood (1963) that the construction of a definite and final phylogenetic tree for angiosperms is a hopeless task. Nevertheless, I believe that significant trends of evolution can be determined, if an effort is made to assemble all available evidence about them and to evaluate this evidence as carefully as possible.

In spite of these difficulties, botanists

can never be persuaded to give up their attempts to construct phylogenies, and there is no reason why they should be. Constructing phylogenies is a stimulating mental exercise. If it is done carefully and deliberately, and based upon all pertinent data, this operation can often provide insights into the nature of living plants that one cannot acquire in any other way. Constance (1964) has expressed the situation very aptly: "Why cannot phylogenies, or even phylogenetic trees, be considered under the heading of poetry, or visual aids, or metaphors, or analogies or hypotheses, all devices designed to lead us to a better understanding? To condemn a given phylogeny as speculative is as inappropriate as to damn a poet or composer for being imaginative."

There are two reasons why definite unalterable phylogenies cannot be constructed. First, the trends of evolution that gave rise to modern forms took place only once. The sequences of events involved will never be repeated, and they cannot be reproduced under experimental conditions. This situation contrasts strongly with that offered to the experimental biologists, including those evolutionists who are dealing with processes of microevolution, such as the origin of racial diversity and the distinctness of species. Even if the exact origin of any particular race or species cannot be precisely reproduced, models can be constructed in population cages and gardens in which the natural course of events can be simulated to a greater or lesser degree. Reconstructing sequences of events that took place in the remote past resembles the detective work that is required to solve a crime more than it does the precise scientific verification of hypotheses about contemporary biological processes.

Second, in groups of organisms like flowering plants, of which the early fossil record is so scanty that it can provide only uncertain clues about the appearance of ancestral forms, we must base our conclusions entirely upon indirect or circumstantial evidence. As contrasted to direct evidence that is obtained from observations of actual ancestors or the repeatable experiments on modern organisms, circumstantial evidence relies for its validity upon the correctness of certain assumptions, which must be made in order to establish connections between different lines of evidence. Too often, biologists working in this field mention these assumptions only casually or not at all.

The Evidence from Fossils

In order to provide direct evidence from the fossil record, the evolutionist must first be certain that the fossils being studied represent the direct ancestors of the living forms whose ancestry he is attempting to unravel. The fossils must also exhibit clearly the morphological characteristics that are used as guides to the relationships between modern forms. With respect to both of these criteria, the fossil record of early angiosperms, as well as of seed plants that might be regarded as their ancestors, is so scanty as to be almost useless. In Chapter 10, comparisons are made between the reproductive structures of Mesozoic seed plants belonging to the orders Caytoniales and Glossopteridales and those of angiosperms that are regarded as primitive. From these comparisons, the conclusion is drawn that plants belonging to these extinct orders are more similar to the probable ancestors of the angiosperms than most paleobotanists now believe them to be. Nevertheless, all fossil evidence concerning the origin of angiosperms is circumstantial.

Evidence from Comparative Morphology and Anatomy

The following assumptions about morphological and anatomical evidence are believed to be justified:

1. Single mutations that produce drastic alterations of the developmental pattern will always have a negative adaptive value by themselves, because they disrupt harmonious developmental sequences. Unless they are accompanied by modifying factors that restore the harmony, they will be eliminated by natural selection. The reasons for making this assumption are set forth in earlier chapters. Its consequences are that the character complexes of new genera, families, or orders never arise in a saltational manner, by single mutational steps.

2. There is no evidence to indicate that more advanced morphological conditions can arise by repeated mutations that alter morphological characteristics consistently in a particular direction, that is, by mutation pressure alone, in the absence of selection. On the contrary, much evidence from both the nature of gene action in development and the comparative morphology of related groups is against the assumption of such orthogenetically directed trends. As was pointed out in Chapter 6, internal factors can orient the direction of evolutionary trends, but these factors are concerned with internal selection of mutations on the basis of their harmonious interrelations with other genes, rather than with mutation pressure alone.

3. The consequence of the last assumption is that different specialized characteristics will be associated with one another only if their association has a functional basis. The assumption made by Sporne (1948, 1956, 1959, 1969) that correlation with other characteristics shown to be primitive is valid evidence by itself of the primitiveness of a character state is, therefore, unwarranted. Mosaic evolution (Takhtajan 1959), in which different characters become specialized to different degrees in the same group, is common rather than exceptional. Examples of mosaic evolution are easy to discover. In the Chloranthaceae, a very primitive woody anatomy (Metcalf and Chalk 1950) is associated with highly reduced and specialized reproductive structures. In the Alismataceae, on the other hand, the reproductive structures are among the least specialized of any herbaceous angiosperms, but the vascular anatomy and pollen are relatively specialized compared with other monocotyledons (Cheadle 1942, 1953). Mosaic evolution extends even to different parts of the same flower. In the genera *Delphinium* and *Aconitum* of the Ranunculaceae, the perianth and stamens are highly modified and specialized, whereas in most species the carpels retain the most unspecialized condition found in the family Ranunculaceae. In contrast, *Anemone* and *Clematis* have evolved highly

specialized carpels while retaining very generalized conditions in their perianth and stamens.

In addition, advancement with respect to one characteristic may actually retard advancement in others. Such retardation or inhibition is expected on the assumption that evolutionary advancement is based upon shifts in adaptive character complexes or functional syndromes. In many instances, the primitive state of one character can be associated in an adaptive syndrome with advanced states of other characters.

Adaptive shifts that include mosaic evolution could explain Sporne's conclusions with respect to flowers hermaphroditic vs. unisexual (character 10). On the basis of correlation, Sporne regards unisexual flowers as primitive, in contrast to most interpretations, which maintain that the primitive angiosperms had hermaphroditic flowers.

As stated in Chapter 4, I regard unisexual flowers as part of a syndrome that is adaptive for wind pollination, since in such flowers the chances for accidental deposition of pollen on flowers of the same plant are much reduced. Three of the six other "primitive" states with which, according to Sporne, unisexual flowers are correlated—woody habit, scalariform side walls, and unstoried wood—can be explained functionally on the assumption that, particularly in temperate mesophytic floras, wind pollination has a higher adaptive value in woody plants than in herbs, since trees and tall shrubs are more likely to be exposed to wind than are low herbs. A fourth character, actinomorphic flowers, has an expected functional correlation, since zygomorphic flowers are adaptive in correlation with pollination by specialized insects and other pollen vectors. A fifth character, ovules crassinucellate, is to some extent associated with large seeds, which have a higher adaptive value in woody plants than in most herbs (Salisbury 1942). We would expect, therefore, that a shift from bisexual to unisexual flowers would be more likely to become established in woody plants than in herbs, because it would more often have an increased adaptive value. Nevertheless, very many examples exist of trends from bisexual to unisexual flowers within individual angiosperm families as well as genera, including both woody plants and herbs. Among the best known of such trends are *Thalictrum* (Ranunculaceae), *Rumex* (Polygonaceae), *Atriplex* (Chenopodiaceae), *Lychnis* (Caryophyllaceae), *Sedum roseum* (Crassulaceae), *Zanthoxylum* (Rutaceae), *Pistacia* (Anacardiaceae), *Celastrus* (Celastraceae), *Rhamnus* (Rhamnaceae), *Vitis* (Vitaceae), *Fraxinus* (Oleaceae), *Coprosma* (Rubiaceae), and numerous genera of Compositae. To maintain, therefore, that unisexual flowers must always be primitive, and are, consequently, a reliable guide to the primitive status of a family, is completely at variance with the facts. The trend from unisexual to bisexual flowers must have occurred in the early ancestry of the angiosperms, and one cannot completely rule out the possibility that some primitive angiosperm families have flowers that are primitively unisexual. Nevertheless, the reverse trend, from bisexual to unisexual flowers, has obviously occurred many times during the evolution of flowering plants, and is easily explained by the loss of either the androecium or the gynoecium. Hence the safest course is to regard bisexuality vs. unisexuality as a reversible character, at least with respect to the most primitive angiosperms. In more advanced, modern orders, unisexuality is usually and perhaps always a specialization.

This example has been discussed in detail because, in my opinion, it indicates clearly the mistakes that can be made if one attempts to use statistics uncritically as a means of determining phylogenetic primitiveness or advancement. The computer can do nothing more than perpetuate and amplify the correct judgments, as well as the mistakes, that are fed into it.

4. An additional corollary of these assumptions is that no parts of the plant can be regarded as innately conservative relative to other parts. The doctrine of innate conservatism is accepted tacitly by systematists when they rely almost entirely and uncritically upon reproductive characteristics as diagnostic criteria for higher categories, and is expressed specifically in some of the older literature (Jeffrey 1916). It has been justly criticized by Davis and Heywood (1963). Nevertheless, the recognition of mosaic evolution implies that conservatism exists with respect to certain characteristics in particular lines of evolution. We should not be surprised, therefore, to find that certain characteristics behave conservatively, that is, are relatively little subject to change, in angiosperms as a whole. This conservatism, however, probably has a functional basis, since certain structures or conditions are more resistant to selective pressures than others. On the basis of the principle of modification along the lines of least resistance, set forth in Chapter 2, such functional conservatism would be expected. One of the most conservative of all characteristics in seed plants is the organization of the aerial parts into shoots that consist of an axis that bears lateral appendages, with branches emerging from the axils of the appendages. This conservatism is probably based upon the highly adaptive nature of this kind of organization in land plants and in plants that live in quiet water, with their roots in the ground. Departures from it are found in families such as the Podostemaceae, which inhabit fast-running streams, and the Lemnaceae, which both are small in size and float on the surface of water.

Another general basis for conservatism is the adaptive value of harmonious, highly integrated developmental sequences. Such sequences characterize the development of structures having highly determinate growth, such as the more complex and specialized flowers. As has been stated previously, genes affecting early stages of development of such structures are less likely to be modified successfully than are genes that affect later stages. Consequently, we can expect that the early stages of development in highly complex, determinate structures are likely to be conservative. Since the procambial strands of vascular bundles are usually laid down during these early stages, the supposed conservatism of the vascular system of the plant, to the extent that it exists, is probably based upon these circumstances. As Hagemann (1963) has suggested, and as was already noted by

Goebel (1931), the course of the vascular bundles in an organ may reflect its course of development.

5. The inevitable consequence of the last assumption is that vascular anatomy cannot be accepted uncritically as an intrinsically better indication of phylogenetic trends than other characteristics. Carlquist (1969*c*) has recently reviewed this entire subject, and has concluded that vascular strands appear only in response to the functional needs of the particular species in which they are found. He believes, therefore, that evidence from vascular anatomy is less reliable as a criterion of phylogenetic trends than that from external appearance.

As indicated above, my opinion on this matter stands between that of Carlquist and those of such authors as Eames (1961) and Melville (1960, 1962, 1963), both of whom attach great importance to the priority of vascular anatomy, only to arrive at conclusions that are diametrically opposed to each other. Evidence from vascular anatomy cannot be disregarded altogether, but it must be weighed critically in comparison with other evidence, particularly that from development. This topic is discussed further in Chapter 10.

6. The generalized or specialized condition of a character cannot be determined on the basis of its frequency within a group. The assumption that the characteristics possessed by the largest number of members of a group are, therefore, the most generalized and primitive characters of the groups was advanced many years ago by Willis (1922) and his follower Small (Small and Johnston 1937) for genera within families. It has been revived recently by Wagner (1961) for morphological characteristics as well as by Taylor and Campbell (1969) for biochemical characteristics. The fallacy of this deduction lies in the hidden assumption that the present members of a group are representative of that group as it was originally formed. This assumption is unwarranted. During the early stages of evolution of any successful group, rapid alterations occur with respect to many characteristics, and numerous species become extinct because they have not achieved the optimum adaptations possible for the group. Consequently, the most common characteristics of a group will be neither the most primitive ones, which were predominant in its earliest representatives, nor the most advanced ones, which are confined to certain specialized derivative lines. They will, rather, be those characteristics found in the members of the group that first achieved widespread success.

Numerous examples can be cited to support this point of view. These examples are uncommon states which are primitive for the group in which they occur. In the family Gramineae, for instance, the commonest number of style branches is two per ovary and of lodicules also two per floret. Nevertheless, these characters are not primitive. In all of the families of monocotyledons related to the Gramineae, the flowers are trimerous. Hence, the relatively uncommon condition of three stigmas per ovary and three lodicules per flower, found in the bamboos and a few other tribes, is rightly considered by agrostologists as the most primitive condition in the grass family. In the Compositae, the commonest character combination is a receptacle that lacks paleae and achenes that bear a capillary pappus. Nevertheless, the morphological link between the capitulum of Compositae and the inflorescence found in related families is provided by those genera of the tribe Heliantheae that possess bracts or paleae on their receptacle and a pappus of a few scales that resemble calyx lobes. In the genus *Crepis,* the largest number of species possess capitula having two distinct and well-differentiated series of involucral phyllaries, and a basic chromosome number $x = 4$. Yet, as Babcock (1947) has clearly shown, those species that provide connecting links between *Crepis* and other genera of the tribe Cichorieae possess involucres with three or more imbricated series of relatively undifferentiated involucral phyllaries, and a basic chromosome number $x = 6$. These species are very few in number; their characteristics are uncommon in the genus.

Examples of this kind could be multiplied indefinitely. The following generalization has relatively few exceptions. Characteristics that can be regarded as primitive because they provide connecting links between a particular genus, family, or other taxon and its nearest relatives are usually found in only a few species of the taxon being studied. These connecting species are likely to occur in relatively stable habitats or in habitats that for various reasons can be supposed to have been subjected to relatively low selective pressures. The characters most widespread in a group form a part of certain highly and broadly adaptive character complexes. They are usually neither the most primitive nor the most specialized characters of the group.

7. The best way of determining whether a particular specialized character, found in a number of related groups, has had a single origin (monophyletic) or has arisen several times in separate evolutionary lines (polyphyletic) is to determine the degree of association between that character and other characters found in the various groups in which it appears. If a particular characteristic is associated in various separate groups with the same or similar combinations of other characters, both primitive and advanced, that have little functional relation with one another, the probability that such character combinations have arisen independently more than once is very low. In such situations, similarity with respect to a particular specialized character can be regarded as indicating descent from a common ancestor. Thus, in the family Compositae, the association of milky juice or latex with a capitulum consisting entirely of ligulate corollas, anthers having basal appendages of moderate length, and styles and stigmas of a relatively homogeneous and unspecialized kind is characteristic of a large number of genera that form the tribe Cichorieae. In all probability, this combination of characters evolved only once, so that latex formation appeared only once in the line that led to the tribe Cichorieae. On the other hand, achenes that are attenuated into a long, slender beak appear in many different species, such as those belonging to the genera *Hypochaeris, Tragopogon, Picris, Lactuca, Taraxacum, Crepis,* and *Agoseris,* which differ from one another in most other characteristics except for those that are diagnostic of the tribe Cichorieae. Furthermore, each of these genera or species having beaked achenes resembles species with unbeaked achenes with respect to many other char-

acteristics. Thus *Picris echioides,* which has beaked achenes, resembles in most other characteristics *P. hieracioides,* which does not; species of *Lactuca* and *Crepis* having beaked achenes can be matched with respect to other characteristics by species of these same genera having unbeaked achenes; *Tragopogon* resembles species of *Scorzonera* with unbeaked achenes; *Taraxacum* resembles *Dianthoseris; Agoseris* resembles *Nothocalais,* and so on. We can, therefore, conclude that beaked achenes arose many times independently during the evolution of the tribe Cichorieae.

Evidence from Development

The nature of the developmental evidence that is regarded as most valuable for determining phylogenetic trends was reviewed in Chapter 6. There I expressed the belief that valid evidence can best be obtained by comparing developmental patterns in an empirical fashion, without reliance on more abstract and idealistic conceptions, such as recapitulation and neoteny. Two assumptions, which were discussed in terms of gene action, appear to be well enough supported that they can be used as guidelines for determining phylogenetic trends (see Fig. 6-9):

1. Mutations that affect late stages of development are less likely to disturb harmonious interrelations between gene-controlled processes than are mutations that produce comparable effects on early stages. Hence we can expect that radical alterations of the morphogenetic pattern are likely to appear initially at relatively late stages of development. They can be shifted to successively earlier stages by the accumulation of modifying genes that increase the precocity of gene expression.

2. Mutations that affect early stages are most likely to be integrated into harmonious developmental patterns if their effects on these stages are relatively slight. They can, however, exert relatively strong effects upon the final pattern by altering the rate of a continuing process, so that their initially small effects at early stages are amplified into much larger effects at later stages.

3. The kinds of gene action mentioned in assumptions 1 and 2 would both result in embryonic similarity. Hence, the generalization that young stages in the development of a particular organ will resemble one another more than they do adult stages of the same organ will hold except for examples in which great divergence in development at an early stage has an adaptive or functional value.

The problem of "vestigial characters"

A phenomenon that is related to both morphological and developmental evidence is that of so-called "vestigial characters." In animals, these characters usually consist of structures that themselves have a fairly complex and distinctive developmental pattern, such as the gill slits of the vertebrate embryo and the vermiform appendix. In plants, however, the concept has most often been applied to much simpler situations. A number of them can be found in books on comparative anatomy such as that of Eames (1961). One of the commonest kinds of example is of vascular bundles that are believed to have no function and to represent the vestiges of structures that have disappeared.

From the developmental point of view, the problem of "vestigial" vascular bundles is of an entirely different order of magnitude from that of vestigial animal organs such as gill slits and the appendix. The procambium cells that form the xylem and phloem of these bundles are probably differentiated from meristematic cells during a single mitotic cycle (Olson, Tibbits, and Struckmeyer 1969). More important, the epigenetic sequence responsible for the formation of these bundles is an exact repetition of a course of events that occurs in many other parts of the plant; only the position where it occurs is distinctive. Wetmore and Rier (1963) have shown that vascular tissue can arise in cultured callus tissue as a result of relatively slight alterations in the nutritive medium. Consequently, the appearance of a bundle in an unexpected position requires only a slight shift in the distribution of nutritional factors or in the balance of hormonal interaction within the developing organ system.

Two developmental situations can be mentioned that might be responsible for the appearance of "vestigial bundles." One of them might be associated with reduction; the other definitely is not. In his correlated study of the vascular anatomy and development of the androecium in various species of Malvales, van Heel (1966) showed that in several instances vascular bundles which in the mature flower were not associated with any recognizable structure nevertheless appeared in a position where small stamen primordia could be recognized in early stages of development. These primordia later became enveloped by the growth of the surrounding tissues, presumably produced by persistent intercalary meristems. These examples could be regarded as terminal stages of reduction series, but such an interpretation is by no means the only one possible.

The second kind of example is associated with changes in numbers of parts, as discussed in Chapter 6. There the hypothesis is suggested that the number of parts in a particular compound structure—that is, the number of flowers per inflorescence, of sepals, petals, or stamens per flower, or of leaflets per compound leaf—is a function of the total amount of meristem of a particular kind divided by the amount of this meristem needed to make one part. In many instances, the quotient expected from this division includes a fraction, that is, there may be $2\frac{1}{2}$ or $3\frac{1}{3}$ parts. In that case, the "fraction" could be expressed in development as a part that is small and rudimentary, and does not complete its development. Since the presence of a fraction in the quotient would be equally likely in a sequence evolving toward an increase, as in a reductional series, the existence of a rudimentary structure would in this case be no indication whatever of the direction of evolution. The rudiment might well contain a vascular bundle, developed from a procambial strand that had become differentiated early in its development. As mentioned above, such rudiments are otherwise often difficult to detect, owing to later growth of the tissue that surrounds them.

Essentially this interpretation was presented many years ago by Murbeck

(1914) to explain variation in number and vasculation of the calyx lobes in two species of the family Rosaceae. In *Comarum palustre* (*Potentilla palustris*), the normal number of calyx lobes is 5, but occasional flowers may have either 4 or 6 calyx lobes. In addition, some flowers having either 4 or 5 calyx lobes may have one lobe that contains one or more extra vascular bundles. In *Alchemilla vulgaris* a similar situation was found, except that in this genus the normal number of calyx lobes is 4, so that the deviants associated with reduction and with increase have, respectively, 3 and 5 lobes, one of which contains extra vascular bundles (Fig. 7-1).

The hypothesis just discussed can be used equally well to explain another "vestigial structure" that has been postulated by a number of anatomists (Eames 1961). This is the rudimentary dome or shoot apex in flowers such as those of the Leguminosae that have a single carpel. In many flowers of this family, the formation of the carpel apparently uses up all of the meristematic tissue present in the embryonic floral apex, so that the carpel arises terminally (Thompson 1929). In other flowers, as in species of *Acacia* (Newman 1936), a rudimentary dome appears on one side of the carpel. The presence of this dome is regarded as evidence that the carpel is basically a lateral organ. In my opinion, the determination of a lateral or

Fig. 7-1. Drawings of the perianth of several flowers of *Alchemilla vulgaris* (*sens. latiss.*) showing similar anomalies of gross structure and vascular pattern in association with both increase and decrease in numbers of parts: (*center right*) a typical tetramerous perianth; (*top row*) abnormal flowers, selected from several thousand that were examined, showing abnormalities in association with increase in number of parts up to the pentamerous condition; (*bottom row*) similar abnormalities associated with the decrease to the trimerous condition. (From Murbeck 1914.)

terminal position for the carpel depends upon the position and size of the meristem which is capable of differentiating into a carpel as compared with that which actually undergoes the process of differentiation. If the processes of differentiation do not affect all of the meristematic cells, a rudimentary dome will appear, and the carpel will apparently develop as a lateral organ. If, on the other hand, all of the meristematic cells that form the final apex of the flower become capable of differentiating into carpel tissue, the carpel will develop as a truly terminal structure. There is no reason for excluding the assumption that during the course of evolution shifts in the position of differentiating cells relative to ground meristem could cause a more primitive form bearing carpels in a lateral position to give rise to a descendant bearing carpels in a truly terminal position. Evolution in a reverse direction would, of course, also be conceivable. On this basis, the following hypothesis may be of general application:

When meristic series of similar structures are being differentiated, rudimentary structures may be formed as a result of certain kinds of relations between a particular capacity for differentiation of meristematic cells and the actual differentiation of these cells into functioning structures or into rudiments. The presence of such rudiments, therefore, cannot be taken as a guide to the direction of evolution.

The need for continuity of examples

When attempting to interpret phylogenetic trends on the basis of adult structures, morphologists and taxonomists usually obtain as continuous a series of forms as possible before making their interpretations. On the other hand, interpretations of phylogeny on the basis of developmental studies often rely on a relatively small number of forms that are distantly related to one another. This procedure might be justified on the grounds that the time required to investigate complete developmental sequences in a large number of forms is excessive. Nevertheless, this approach has serious flaws. In many instances, developmental patterns may become so drastically altered during evolution that the relations between them cannot be seen unless the development of intermediate forms has been studied. This often results in the mistaken conclusion that two structures bear no relation to each other because their developmental patterns are completely different, when in fact they are phylogenetically related via intermediate forms that have not been studied. In my opinion, the need for examining a continuous series of forms is even greater with respect to developmental evidence than with respect to morphological and anatomical evidence.

The Evidence from Chromosomes

Three kinds of evidence about evolutionary trends can be obtained from comparative studies of chromosome number and form. The first of these is the association of certain alterations of the karyotype, or the morphology of chromosomes at somatic metaphase, with phylogenetic trends. This evidence is discussed more fully elsewhere (Stebbins 1971) and will only be summarized here. The great majority of vascular plants possess karyotypes that are relatively symmetrical, that is, that have chromosomes of similar sizes within the same karyotype, and chiefly median or submedian centromeres. In many evolutionary lines, however, trends occur toward greater asymmetry, including both increasing difference in size between the chromosomes of the same karyotype and shifts of the centromere from median to subterminal or even terminal position. These trends are most often observed among the different species of a genus, as in *Crepis* and other genera of the tribe Cichorieae of the Compositae (Babcock 1947, Stebbins, Jenkins, and Walters 1953, Stebbins 1958). They can, however, involve different genera of the same tribe, or, more rarely, all of the genera in a particular family. No such trends have been observed at the level of the differentiation of plant families from one another. A reverse trend, toward greater symmetry, also exists. The best example recorded in plants is the family Podocarpaceae (Hair and Beuzenberg 1958). This trend is characterized by the "fusion" of entire chromosomes having terminal or nearly terminal centromeres to form new chromosomes having median centromeres. This process of "fusion," contrary to the processes that give rise to greater asymmetry, does not alter either the size or the number of chromosome arms. It merely converts a larger number of rodlike chromosomes into a smaller number of V's. Such "centric fusion" is rather common in animals, particularly in certain groups of insects, but is encountered much less frequently in plants.

The second kind of chromosomal evidence is obtained from examples of chromosome doubling, or polyploidy. At the level of species or subgenera, analysis of the cytogenetic characteristics of the numerous polyploids that are of hybrid origin can provide the surest evidence of a particular phylogenetic ancestry that a botanist can obtain (Stebbins 1950, 1971). Moreover, even at the level of subfamilies and families, polyploidy can provide indirect evidence regarding the nature of past events. It can do this in three ways. First, many subfamilies or even families have apparently originated through adaptive radiation and speciation in the descendants of a hybrid polyploid derived from crossing between the primitive ancestors of two other subfamilies. The subfamily Pomoideae of the Rosaceae, which apparently originated from hybrid polyploids involving primitive representatives or ancestors of the subfamilies Spiraeoideae and Prunoideae (Stebbins 1950), is the best-known example. Another probable example is the subfamily Oleoideae of the Oleaceae (Stebbins 1950). Where examples of this kind exist, the phylogeny of the family concerned must be regarded as at least partly reticulate.

A second kind of evidence about past evolutionary events comes from the existence of families that have such high chromosome numbers in all of their genera that the most probable inference is a polyploid origin for the entire family as we know it. Classical examples of such families are the Salicaceae ($x = 19$), Platanaceae ($x = 21$), and Magnoliaceae ($x = 19$). More recently, several more examples have been added, such as the Bombacaceae ($x = 36$, Baker and Baker 1968), Austrobaileyaceae ($x = 22$, Takhtajan 1966), Eucommiaceae ($x = 17$, Takhtajan 1966),

Didieraceae ($x = 75$, Takhtajan 1966), Melianthaceae ($x = 19$, Takhtajan 1966), and Bromeliaceae ($x = 17, 25$, Marchant 1967). In the family Winteraceae, which is of critical significance with respect to the origin of the angiosperms, two widely different basic numbers exist, $x = 13$ and $x = 43$ (Takhtajan 1966, Ehrendorfer *et al.* 1968). The significance of these examples is that for most of them an origin by chromosome doubling from hybrids between ancient, now extinct diploids is the most probable explanation of their chromosome numbers. This means that the modern forms must be the last survivors of groups that formerly were much richer in genera and species. Consequently, we cannot postulate that plants closely similar to the modern representatives of these families were the ancestors of other, more advanced families. In some of their basic vegetative and floral characteristics, the ancestors of the angiosperms may have resembled the modern Magnoliaceae, Winteraceae, and Austrobaileyaceae. Nevertheless, the hypothesis that these ancestors were much like any modern genus of these families with respect to either external morphology, anatomy, ecological preference, or geographic distribution is not in accord with the evidence from their chromosomes, or with other kinds of evidence.

The situation with respect to the much larger number of families in which the lowest basic number is $x = 12, 13$, or 14 is more problematical. The hypothesis that these families, also, are originally of polyploid origin was advanced many years ago (Stebbins 1950). On the basis of more recent evidence, this still appears to be the most probable explanation of their basic numbers (Stebbins 1966, 1971). If it is correct, then we must extend the reasoning developed in the last paragraph to many more families, including all of those placed in the Magnoliales by Hutchinson, Takhtajan, and Cronquist, except for the Annonaceae. The significance of this conclusion is discussed further in Chapter 10.

The third way in which evidence from polyploidy can be useful is in relation to theories about the ecological conditions under which angiosperms originated and began their evolutionary diversification. Recent comparisons between the geographic distribution of diploids and their polyploid derivatives (Stebbins 1971) indicate that the spread of polyploids is favored chiefly by the presence of extensive new ecological habitats which they can occupy, as was the case during the retreat of the Pleistocene ice sheets and is likely to exist for plants that have arrived on oceanic islands that have been formed recently and have not yet acquired mature soils and stable plant communities. To the extent that this is true, we can expect polyploids to have different geographic and ecological distributions from those of their diploid ancestors. This is true for most modern polyploid complexes, particularly those in which the polyploids have become much more abundant than the diploids. If we extrapolate from this situation to that which probably existed when the relictual polyploid families mentioned in the preceding paragraphs were at the height of their success, we can suggest that the extinct diploid ancestors of these polyploids probably occupied different geographic regions and ecological habitats from those of their modern descendants. This extrapolation leads to the following generalization: if a modern genus or family having supposedly primitive characteristics is represented only by species having such high chromosome numbers that they are probably of polyploid origin, then the present habitat of the group is usually different from the conditions under which the group originated.

The third kind of evidence from chromosomes comes from the existence of aneuploid series of chromosome numbers. As I have explained elsewhere (Stebbins 1950, 1971), such series most often are produced by a series of unequal reciprocal translocations. Genera that have such series, such as *Astragalus, Astroloma, Boronia, Crepis, Haplopappus, Leucopogon, Limonium, Primula, Salvia, Saxifraga,* and *Verticordia,* are almost exclusively adapted to pioneer habitats in open or semiarid situations. The chromosomal translocations that bring about their aneuploidy are probably favored in these situations because they tie together by genetic linkage adaptive combinations of genes (Stebbins 1971). Colonization of new pioneer habitats depends upon the origin of new linked combinations, and hence of new translocations. As I have suggested elsewhere (Stebbins 1965) and will discuss more fully in Chapter 10, the earliest angiosperms are believed to have been shrubs or subshrubs, and so may have occupied ecological sites similar to those now occupied by shrubby species of *Haplopappus* (*sens. lat.*) in North America or of *Astroloma, Boronia, Leucopogon,* and *Verticordia* in Australia. The original aneuploid series of basic numbers which they possessed may have originated under these conditions.

Biochemical Evidence

During the last few years, refined methods have become available for estimating similarities and differences between groups on the basis of various organic molecules associated with metabolism. Three groups of molecules have received major attention: nucleic acids, proteins, and secondary metabolic products of various kinds. Each of these groups has its advantages and disadvantages, as summarized below.

Secondary metabolic products

The first secondary metabolic products to receive major attention were the phenolic compounds, mainly flavonoids (Alston and Turner 1963). When extracted and separated by two-way paper chromatography, or by thin-layer chromatography, these compounds form patterns that in many groups of plants are highly diagnostic of individual species or races and can be used to recognize interspecific hybrids or their polyploid derivatives. More recently, other compounds, such as alkanes (Douglas and Eglinton 1966) and terpenes (Forde and Blight 1964, Weissmann 1966, Mirov 1967), have been separated by gas chromatography and used in a similar fashion. In much of the early work, the compounds were recognized by such properties as R_f values, which could easily be determined by paper chromatography, by their fluorescence in ultraviolet light, and by some diagnostic color reactions. Their chemical structure was unknown. Now, however, the more progressive research in this field deals with several classes of molecules whose structure is determined. The anthology of research papers assembled by Swain (1966) included reports on the majority of these, and a critical review of the field has been presented by Merxmüller (1967).

All these groups of compounds have the same advantages and disadvantages as morphological characters. In many instances, they possess the advantage of relatively easy, definite, and objective determination on a particular specimen, and a clear-cut distinction between taxonomic groups. One of the most striking examples of their use at a higher level of the taxonomic hierarchy is the recognition of the ten betacyanin-producing families that belong to the Caryophyllales or Centrospermae, and their separation from other families (Caryophyllaceae, Molluginaceae) that have been placed with them on morphological grounds but possess anthocyanins or their derivatives rather than betacyanins (Reznik 1955, 1957, Mabry 1964, 1966, Wohlpart and Mabry 1968). If additional evidence, both chemical and morphological, can be produced to show that the Caryophyllaceae and Molluginaceae are more closely related to other families than to the betacyanin group, a major taxonomic revision will have been carried out primarily on the basis of a biochemical characteristic. At the level of the family, Ruijgrok (1966) has obtained biochemical evidence in support of a realignment of genera of Ranunculaceae that had previously been suggested on the basis of morphological and chromosomal evidence.

The chief disadvantage of these secondary compounds is the sporadic occurrence of many of them, which apparently results from repeated parallelisms and convergences with respect to biosynthetic pathways (Hegnauer 1966, Harborne 1966). Parallelism or analogy exists through the possession of similar compounds by unrelated plant groups, such as anabasine in Chenopodiaceae and Solanaceae, and glycoflavones as well as the glycoxanthone, mangiferin, in Leguminosae and Iridaceae. In these examples the biosynthetic pathways by which the compounds are produced are different, or are not recorded. Convergence is exemplified by the occurrence of sparteine in *Lupinus* (*Leguminosae*) and *Chelidonium* (Papaveraceae), the biosynthetic pathways for this compound being identical in these two entirely unrelated genera. Such situations should provide a warning to biochemists who would suggest relationships between genera and families on the basis of biochemical similarities, when no other evidence for such relationships is at hand. An example is the postulated relationship between *Cornaceae, Escallonia, Daphniphyllum, Eucommia,* and *Scrophulariaceae,* based on the presence of iridoid compounds in these otherwise very disparate groups (Bate-Smith and Swain 1966). At present, there is no reason for believing that similarity with respect to molecules such as phenolics, flavonoids, alkaloids, terpenes, and other secondary metabolic products is any more reliable a criterion of genetic and evolutionary relationships than is similarity with respect to visible morphological characters. As the genetic bases of the biosynthetic pathways leading to these compounds become better known, their reliability or unreliability may become clearer.

Proteins

The use of proteins in plant systematics began many years ago, with the serological research of Mez and Siegenspeck (1926). This work was conceived on a grand scale, but it attempted too much. Its results could not be consistently repeated in other laboratories and were not generally accepted. Consequently, the method lost favor for some time. More recently, it has been revived on a more modest scale, using more precise techniques, and is showing considerable promise for determining the relationships within genera of problematical families, such as the Gramineae, Ranunculaceae, and Solanaceae (Fairbrothers 1966, Hawkes and Tucker 1968, Jensen 1968*a*, Smith 1969). In addition, it has provided insight into relationships between specific families. Johnson (1954) and later Johnson and Fairbrothers (1965) have obtained serological evidence that supports certain lines of morphological evidence indicating that the genus *Illicium* clearly belongs to a different family, and probably a different order, from the Magnoliales. Fairbrothers and Johnson (1964), on the basis of similar evidence, have supported those taxonomists who recognize the Nyssaceae as a separate family from the Cornaceae. Jensen (1968*b*) has obtained evidence in support of the recent opinion of taxonomists that the Papaveraceae are related to the Ranunculaceae. The difficulty with serological or immunological work, however, is that the individual proteins that are responsible for the recorded reactions cannot usually be identified. Its value as a phylogenetic tool depends upon the validity of the assumption that all of the proteins, or at least those that are extracted and present in the samples used, are evolving at approximately the same rate. If certain proteins were evolving much more rapidly or more slowly than others, the method could produce misleading results. Moreover, since individual proteins cannot be identified, the significance of serological divergence cannot be associated with either gene change or the adaptive differences that are affected by natural selection.

An example of this kind of difficulty is presented by the work of Kloz, Turkova, and Klozova (1960) on storage globulin proteins in the seeds of legumes. They found that these proteins differ much more from one taxon to another than do the proteins of the leaves, stems, and hypocotyls. They interpreted this difference to mean that storage proteins are phylogenetically older than metabolic proteins. This is not in accord with current genetic theories. A geneticist would predict that storage proteins, if they consist merely of accumulations of amino acids that are useful to the plant after the protein has been digested, would accumulate many random mutations, and so would come to differ from one another both between species and among the individuals of the same species. The constancy that Kloz found to exist within species would not be expected. On the other hand, if the structure of globulin proteins is related to the conditions under which they are digested, including the specificity of the digestive enzymes and the temperature, moisture conditions, and other factors that contribute to the efficiency of seed germination, then one would expect constancy within species, as well as differences between species that differ with respect to the physiology and ecology of the highly sensitive stage of seed germination and seedling establishment. Distinctive and constant characteristics of this kind have been recognized by Boulter's group (Boulter, Thurman, and Derbyshire 1967) in globulins of legumes by means of disk electrophoresis.

These difficulties are alleviated by methods involving extraction of proteins and their separation by means of gel electrophoresis. This method has been used in two ways. Some workers have extracted all the soluble proteins from seeds or other plant parts, have subjected the extracts to electrophoresis, and have obtained patterns of bands known as electrophorograms by using some generalized stain that detects all kinds of proteins (Hunziker 1969). More commonly, gels have been stained selectively for enzyme activities so that comparisons are made with respect to the allozymes ("isozymes") of a single enzyme (Scandalios 1969, Sing and Brewer 1969). Each of these alternatives has its advantages and disadvantages. Comparisons of total proteins by electrophorograms resemble serological comparisons in that several stainable proteins may be present in each band, and so are compared without regard to their genetic basis or metabolic function. Its advantage over serological methods lies in the fact that each group of proteins can be analyzed separately, so that dissimilarities which might be due to unusually rapid or slow evolution of one or two particular proteins could, presumably, be recognized. A possible disadvantage lies in the fact that only about a quarter of the differences between proteins are reflected in their electrophoretic mobility (Shaw 1965). Comparisons based only upon this method might well produce a bias in favor of differences with respect to certain amino acid residues that are particularly influential in altering the electrostatic charge of the molecule or in modifying its tertiary or quaternary structure in such a way that the electrostatic properties of individual residues are either revealed or hidden.

Serological methods have been combined with electrophoresis in a method known as immunoelectrophoresis. This method has been used successfully to supplement morphological characteristics in determining the relationships between the species of a genus. An example is the analysis by Smith (1972) of relationships between annual species of the grass genus *Bromus.*

Analysis of allozymes ("isozymes") by means of electrophoresis is proving to be particularly valuable for determining the amount of genetic variation in populations of both animals and plants (Prakash, Lewontin, and Hubby 1969, Selander, Yang, and Hunt 1969*a*, Allard and Kahler 1972). Moreover, since the metabolic function and kinetic properties of individual enzymes can be determined by other biochemical methods, allozyme comparisons provide the best method of correlating biochemical differences with adaptive properties. Furthermore, electrophoretic comparisons between enzymes that play key roles in metabolism, such as amylases associated with seed germination, various enzymes that participate in the photosynthetic and glycolytic pathways, and peroxidases that are associated with plant growth (Galston and Davies 1969) and meristematic activity (van Fleet 1959), will probably provide us with highly significant information about phylogenetic trends.

The most complete analysis of protein molecules is the determination of their amino acid sequences. When this has been done, as with hemoglobin (Jukes 1966) and cytochrome C (Margoliash and Fitch 1968, Boulter *et al.* 1970*a,b*), the number and kinds of mutations required to convert one molecule into another can be accurately determined. The drawbacks of this method are that present techniques permit sequencing only of proteins that have a relatively low molecular weight, which are easily isolated in sufficient quantity in a pure form. Nevertheless, as more data accumulate and better analytical techniques become available, amino acid sequencing may provide us with reliable data on the genetic and evolutionary basis of protein differentiation in contemporary taxa.

Nucleic acids

From the genetic point of view, accurate comparisons between deoxyribonucleic acids (DNA) would be the ultimate in precision, since they represent the genes themselves. Direct comparison of nucleotide sequences is still impossible, however, since no practical method of determining the nucleotide sequences of a particular nucleic acid molecule is available. Nevertheless, some degree of overall similarity with respect to the nucleotide sequence can be estimated by "hybridization" studies (Hoyer, McCarthy, and Bolton 1964). This method takes advantage of the fact the DNA is a double helix, the individual strands of which match each other precisely in a complementary fashion. By means of carefully controlled heat treatments, the two strands can be separated from each other, and single-stranded molecules or relatively large segments of them can be trapped in agar. If radioactively labeled smaller fragments of molecules, either single-stranded DNA or RNA, are added to this suspension, new double helices will be formed within the agar, and their amount can be measured accurately. The method is quantitative, since double helices between single strands that match each other completely are formed quickly and irreversibly, whereas the thermostability of the molecule becomes progressively lower, the more the nucleotide sequences of the separate strands differ from each other.

The advantage of this method is obvious. To the extent that it is accurate, it provides the best indication that can be obtained of genetic similarity or dissimilarity. It has, however, certain defects. In the first place, recent studies (Britten and Kohne 1968, Britten and Davidson 1969) have shown that all higher organisms examined possess regions of DNA that are replicated hundreds or even thousands of times, and other regions that are present only once in a particular genotype. Since both the speed of hybridization and the thermostability of its product can be greatly affected by the presence of such duplication, the comparison of nucleic acids by means of hybridization can be accurate only if corrections can be made to allow for this effect. Methods by which this could be done are now being worked out, so that this difficulty may not be insurmountable.

Another difficulty with the method is that it appears to be not very discriminatory when nucleic acids of distantly related species are compared with each other. For instance, in the study of Hoyer *et al.* (1965), it discriminated very well between human DNA and that of other primates, but could not provide information on whether human DNA is more similar to that of the hedgehog or to that of the mouse.

General comments on biochemical characteristics

From this brief review, the fact should be evident that many different biochemical comparisons can be made that will help us to understand phylogenetic relationships and evolutionary trends. Each of these has its advantages and disadvantages; no one of them can be regarded as sufficient in itself. Our eventual understanding of relationships will be best explained by developing all of these methods, and, hopefully, by using all of them on the same material. This application will be particularly useful if biochemical, morphological, and developmental characteristics can be correlated with one another.

Furthermore, biochemical comparisons involve one difficulty in common with most morphological comparisons of contemporary forms: we can never know what were the biochemical characteristics of the actual ancestors of modern forms, which are now extinct. For this reason, many of the estimates that have been made of evolutionary rates on the basis of biochemical comparisons are, to say the least, open to question. The method used is to compare two protein or nucleic acid molecules by means of the methods just mentioned, and then to see whether the differences between them are proportional to the times since they diverged from a common ancestor, according to such fossil evidence as is available. If, for instance, a particular protein differs by twice as many amino acid substitutions between species A and C as between species B and C, then the ancestor of A and C is presumed to have lived twice as long ago as that of B and C. If paleontological evidence, however scanty, appears to support this expectation, the conclusion is reached that the protein molecules concerned evolved at a uniform rate over millions of years. Their supposed rate of evolution is then expressed mathematically by the ratio of the total time elapsed since the supposed common ancestor to the number of amino acid differences between contemporary forms. Aside from the inaccuracies involved in recognizing and dating the supposed common ancestor, this method is an oversimplification of what were undoubtedly very complex evolutionary pathways. A glance at such charts as that for rates of evolution in lungfishes, obtained by Westoll and reproduced by Simpson (1953:Fig. 4), should serve to illustrate this point. As more correlations between biochemical and paleontological evidence become possible, particularly those involving species that have diverged from each other relatively recently and by relatively simple evolutionary pathways, the validity of biochemical evidence for determining evolutionary rates may be more accurately assessed.

The Synthesis and Interpretation of Different Lines of Evidence

As Dobzhansky (1968*a,b*) has so clearly stated, evolutionary problems must be solved by a succession of two different kinds of approach. First, the problem must be reduced to its separate elements, or to separate lines of evidence that can contribute toward its solution, and each of these aspects must be attacked individually. Subsequently, the different lines of evidence must be compared and synthesized to form the complete picture. These two kinds of studies are equally important and are different in nature. One cannot expect that a synthesis will emerge automatically once the separate elements have been analyzed.

The nature of any such synthesis will depend upon the overall philosophy of the biologist who is making it, since this philosophy will determine, either consciously or intuitively, the relative values that he will assign to different kinds of evidence. In my opinion, any evolutionist who is attempting to determine phylogenetic trends by accumulating and evaluating evidence of many different kinds must be clear in his own mind as to the nature of this philosophy, and must express it clearly to his readers. I shall attempt to do this in the remainder of the present section.

My own philosophy can be summarized by referring to the hypotheses and principles that have already been stated in earlier parts of this book. I believe that evolutionary trends have been the result of a succession of complex interactions between organisms and their environment, acting through the mechanism of natural selection, which is based on the superior reproductive success of organisms having certain combinations of genes. Furthermore, within any gene pool there exist several possible and alternative gene combinations that might adapt the organism to any new environment. The particular adaptive combination that will become established will depend largely upon the nature of the gene pool already present, acting through the principles of evolutionary canalization and selection along the lines of least resistance.

The consequence of this philosophy is that two different kinds of guidelines are used for comparing and evaluating evidence. One of these is based upon consideration of environmental factors, and the other upon morphological and particularly developmental factors, insofar as they affect evolutionary canalization.

Environmental criteria for evaluating evidence

The chief result of an emphasis upon the organism–environment interaction is that it leads inevitably to the conclusion that rates of evolution will be very unequal, both for related groups of organisms in different environments and for different parts of the same organism. The latter situation, mosaic evolution, has already been discussed. The problem of widely divergent evolutionary rates has been discussed by many authors, particularly Simpson (1953), on the basis of fossil evidence from animals. The three classes of evolutionary rates which he recognizes—horotelic, or the normal range; tachytelic, or exceptionally fast; and bradytelic, or exceptionally slow—can also be recognized in plants. The evolution of conifers during the Mesozoic era, as determined by Florin (1950, 1953) on the basis of fossil evidence, is an example of a horotelic rate. Tachytelic rates are characteristic of many groups of annual species in modern times, as in the genera *Gilia* (Grant 1963) and *Clarkia* (Lewis 1966), as well as of epiphytic Orchidaceae (van der Pijl and Dodson 1966). Among the well-known examples of bradytelic genera can be mentioned *Lycopodium*, *Selaginella*, and *Marattia* among spore-bearing plants, *Taxus*, *Torreya*, *Sequoia*, and *Metasequoia* among conifers, as well as *Magnolia*, *Cercidiphyllum*, *Platanus*, *Engelhardtia*, and many other genera among angiosperms.

Two further facts about differential rates of evolution must be emphasized. First, the rate of evolution of any particular line can vary greatly during different parts of its evolutionary history. This is very well illustrated by the lungfishes (Dipnoi), worked out by Westoll and reviewed by Simpson (1953:23). These fishes evolved very rapidly during the first 50 million years of their existence, then at slower rates for about 100 million years, and have remained static, or bradytelic, for the remainder (175 million) of the 325 million years that they have been in existence. The modern lungfishes are excellent examples of a principle stated by Simpson. Modern bradytelic species, far from being basically primitive, were advanced and highly specialized at the time when they first became bradytelic. Most of them had previously undergone a stage of rapid evolution. Among modern forms, they may appear to be primitive, but this is only because they have remained static for millions of years while other evolutionary lines have undergone extensive change.

We must, therefore, distinguish between truly primitive organisms and those that are merely archaic, that is, the products of bradytely following the attainment of a high degree of specialization in past geologic eras. In my opinion, most of the families that systematists have in recent years regarded as primitive angiosperms—Magnoliaceae, Winteraceae, Degeneriaceae, Himantandraceae, and Austrobaileyaceae—are archaic rather than primitive. They all possess various combinations of primitive and highly specialized characteristics. These are discussed in Chapter 10.

Second, since rates of evolution depend upon organism–environment interactions, we can expect that the environment that favors the evolutionary stability and persistence of archaic, bradytelic species is different from the one that in previous geological eras has promoted their earlier rapid evolution and specialization. This topic is discussed fully in Chapter 8. The conclusion reached there is that rapid evolution can be expected in regions that are marginal or limiting with respect to some essential factor of the environment, such as moisture or cold. Regions having an equable climate, or those that are harsh and extreme, such as deserts, mountain summits, and the arctic regions, are unlikely to be the "laboratories" in which new adaptive complexes originate. Diversification in mesic tropical groups is most likely to consist of elaborate modifications that are adaptations to the complex biotic factors that prevail in such regions.

Morphological and developmental criteria

Two sets of alternative concepts have dominated the thinking of biologists with respect to phylogeny: homology vs. analogy and monophyletic vs. polyphyletic origins. These will be discussed in turn.

The use of homologies. The determination of homologies is the method by which relationships between entire organisms can be analyzed in terms of their individual parts. Structures are regarded as homologous if they are descended from the same structure in the common ancestor of a group of modern forms regardless of whether the modern structures perform the same or different functions. The most significant comparisons are often those between homologous structures that have different functions.

In the absence of fossils representing the common ancestor of modern forms, the determination of homologies is beset with the same difficulties as the determination of relationships based upon entire organisms. Moreover, the botanist encounters greater difficulties in dealing with plant structures than those that face the zoologist who is dealing with animal structures. The three principal criteria upon which homologies are based are similarity in position of origin of a structure, similarity in anatomical and histological characteristics, and similarity in developmental pattern. In animals, these are usually reliable criteria. Their patterns of development are highly determinate and carefully programmed. In them, determinism of pattern is essential if a functioning organism is to be produced. A functioning vertebrate or arthropod must have two eyes placed on the front side of its head. An angiosperm flower, on the other hand, can exist functionally at various positions on the plant, and even in different individuals of the same population can have varying numbers of stamens or carpels. Because of this relative indeterminism with respect to both the number and the position of plant parts, these criteria are much less diagnostic of homology in plants than in animals.

The unreliability of position as a criterion for homology is shown most clearly by the examples of flowers that emerge from the midribs of leaves, as is discussed in Chapter 2.

The unreliability of anatomical details, particularly the course of individual vascular bundles, for interpreting homologies, has been clearly pointed out by Carlquist (1969*c*). Botanists who rely on this criterion as one of major importance fail to realize the great evolutionary plasticity of developmental patterns. This is the principal weakness of Melville's "gonophyll theory." As an example, compare his (1963) discussion of the vascular anatomy of the flower of *Caltha* with the observations and discussions of Tucker (1966) regarding the same object.

Developmental pattern is also a much less reliable criterion of homology in plants than in animals. This is because patterns of development of individual organs are, in general, much simpler in plants than in animals. Consequently, parallel and similar but independent evolutionary modifications of structures as adaptations to similarly changed functions, the resemblance that Lankester (1870) called homoplasy, are much more common in plants than in animals. If a complex organ like a vertebrate appendage becomes modified independently in two separate evolutionary lines in association with similar changes of function, the resulting structures are so different that their independent origin can usually be recognized. The modification of forelimbs for flying has been so different in birds from that in bats that, even apart from other considerations, no anatomist would ever regard them as homologous modifications that indicate true relationship. The same is true of the modifications of forelimbs for digging in unrelated fossorial animals such as moles, gophers, and badgers.

On the other hand, parallel but independent modifications of plant structures are much more difficult to recognize. Botanists such as Bessey (1915), who constructed the first modern phylogenetic trees, tended to regard such modifications as sympetaly and epigyny as so "fundamental" that they could not be expected to occur many times independently. Consequently, they grouped most of the families having these advanced characters in the same circle of affinity. As more and more additional characters have been studied, evidence has increasingly favored the hypothesis that these changes in floral structure have occurred many times in different, independent evolutionary lines, so that sympetalous or epigynous families cannot be regarded as closely related unless they resemble one another with respect to many other characteristics. For interpreting phylogenetic relationships in higher plants, individual homologies are being subordinated to overall resemblances.

Serial homology and concepts of "fundamental morphological units." A particularly controversial criterion in plant phylogeny is the resemblance that Lankester (1870) termed serial homology. This is the resemblance between organs having similar positions on different parts of the plant. The best examples in higher plants are the foliar appendages: leaves, bracts, sepals, petals, stamens, and carpels. The obvious structural and developmental resemblances between these structures have led some botanists to the hypothesis that all are descended from similar leaflike structures, that is, that the original vascular plant had only one kind of appendage, and that differentiation of successive appendages from each other has been characteristic of progressive evolution. The fact that many ferns produce only one kind of frond or leaf, which may either bear sori and sporangia or lack them, appeared to be evidence in favor of this hypothesis. Fossil evidence, however, has shown that both leaves and reproductive structures ("sporophylls" or "sporangiophores") in contemporary ferns and seed plants were derived originally from branch systems. Furthermore, vegetative and reproductive shoot systems have been modified independently of each other and at different rates. When we compare the most advanced leptosporangiate ferns (Polypodiaceae) with forms that are regarded as more primitive (Schizaeaceae, Osmundaceae), we find that the difference between vegetative leaves and sporophylls is greater in the more primitive than in the more advanced forms. We can conclude from this fact that during the course of evolution serially produced organs may either become progressively more strongly differentiated from each other, as is the case in most angiosperms, or more similar to each other, as has apparently been the trend in leptosporangiate ferns. Consequently, little or no reliable information about phylogeny can be obtained by comparing the structure and development of serially produced structures on the same plant. We must always compare structures that have similar positions in different but related taxa.

The attempts to interpret phylogeny by comparing serially produced organs are a legacy from the idealistic morphology that has dominated the thinking of many botanists ever since the time of Goethe. Even before they believed in evolution, botanists looked upon plant parts as modifications of certain "fundamental units," such as leaf, shoot, and root. The discoveries of paleobotanists, which have shown clearly that flattened, dorsiventral leaves, as well as reproductive branches or appendages of various sorts, have been derived in phylogeny through modification of dichotomous branch systems, have led some botanists to seek for other "fundamental units" of morphology. This search has reached its climax in the "telome theory" (Zimmermann 1959, 1965). According to this theory, leaves, bracts, sporangiophores, sporophylls, and all of the other aerial parts of the plant must be looked upon as modifications of dichotomously branching shoot systems. A corollary to the theory is that dichotomous branching, wherever it occurs, is an indication of primitiveness.

As a vehicle for understanding the evolution of the shoot system in vascular plants, this theory has two weaknesses. In the first place, it directs the attention of investigators to alterations of adult structures, and does not emphasize sufficiently the need for understanding the succession of developmental patterns that bring these structures into being. Second, it assumes, without supporting evidence, that a certain trend in the mode of branching of vascular strands—from dichotomy to other patterns—is irreversible. There is no doubt that the evolution of both flat and cylindrical structures from dichotomous branch systems was the way in which the shoot system of modern vascular plants evolved from the dichotomous branch system that existed in their primitive ancestors. The concepts of overtopping ("Uebergipfelung"), webbing, and anastomosis of vascular strands are, moreover, useful for describing the ways in which the adult structures became modified. They do not, however, even begin to analyze the causal factors involved, or to associate these changes with modified gene action that could be the result of new mutations and gene recombinations. A first step in the direction of such an analysis would be to describe and compare carefully the developmental patterns that give rise to each of the different kinds of adult structures, in terms of the distribution and activity of meristematic regions, cell division as compared to cell enlargement, polarized growth of both cells and tissues, and the times and methods of procambial cell differentiation. Such descriptive studies should be followed by altering development experimentally through changing the balance of growth substances, as discussed in Chapter 6. Only upon the completion of such analyses would the botanist be in a position to state whether or not trends away from dichotomy can be reversed. At least until such studies have been carried out, comparisons between such distantly related structures as the branching system of early spore-bearing plants and the leaf of angiosperms belonging to the family Ranunculaceae, such as were made by Zimmermann (1965:143), are fruitless and can be very misleading. Zimmermann rightly points out that the trend from dichotomous to netted venation must be regarded as potentially reversible. Nevertheless, his belief that in response to the aquatic habit the leaf of species of *Ranunculus* subg. *Batrachium* has reverted to the condition that prevailed in the most primitive, now extinct, Ranunculaceae, and can therefore be used as a guide to tracing the phylogeny of the complex array of leaf types found in the family, is unfounded speculation. Equally useless are comparisons between the branching stamen or "microsporophyll" of an advanced form such as *Ricinus* and the dichotomous sporangiophores of primitive seed ferns.

Other ways in which the terminology of idealistic morphology has been used to describe evolutionary trends can be equally misleading unless both the author and his readers understand clearly that the terms are employed in a purely figurative rather than a literal sense. For instance, many of the early discussions of comparative morphology were cluttered up with such terms as the "splitting" ("dédoublement") of stamens as a means of increasing the number of stamens per flower. Violent arguments raged as to whether or not such "splitting" could take place. In our day, such arguments seem futile and outmoded. Of course, stamens do not "split" in evolution; neither do they "fuse." This is not even true of the meristematic regions that give rise to them. Such terms certainly have no relation at all to modifications of gene systems that are ultimately responsible for the morphological changes. We do not yet know how modified gene action can give rise to increased or decreased numbers of parts, or to the activity of intercalary meristems that are responsible for so-called "fusion." Until we do, botanists will be best understood by both students and other research workers if they use terms that have the least possible interpretive connotation, that are as direct as possible.

One kind of semantics that is particularly likely to produce confusion is a failure to distinguish between the words "is" or "consists of" and the words "is homologous to" or "is derived from." When we say that the ovary locule with its associated style and stigma "is" the carpel of a syncarpous ovary, we are not implying that it is in every respect the same as the carpel of an apocarpous gynoecium. During the process of "fusion," involving as it does the initiation of an intercalary meristem in a new position, the entire pattern of development may be altered. Nevertheless, in favorable groups, such as the genus *Saxifraga* and the family Rutaceae (Gut 1966), so many intermediate stages exist that the phylogenetic homology between an individual carpel of apocarpous gynoecia and the "carpellary unit" of the syncarpous ovary is clear.

A good review and analysis of the "universal categories" of idealistic morphology, and of the difficulties in semantics and interpretation that such concepts have generated, has been presented by Sattler (1966). He has brought forth a number of clear-cut examples to show that no empirical evidence exists in support of such categories. He has suggested that in place of the valueless notion of "universal categories" and "essentialistic homology" a concept of "semiquantitative homology" should be recognized. Botanists should not ask the question, "Is organ X homologous to A (a leaf) or B (a stem)?" They should ask, rather, "Is organ X more homologous to A than to B, or vice versa?"

On the basis of the fact, which has been noted by a number of botanists such as Cronquist (1968) and is mentioned repeatedly in this volume, that parallelisms and convergences have been very common in angiosperm phylogeny, I cannot accept Sattler's conclusion, and would like to suggest instead a further modification of the concept of homology. This is that botanists should think in terms of specific homologies. They should ask, for instance, "What is the degree of homology between the juvenile leaf of a palm as compared, on the one hand, to the leaf of a juvenile plant of the Winteraceae, and on the other hand, to the phyllode of *Acacia?*" If a number of such specific questions can be answered positively, new and more valid generalizations may emerge.

Mono- vs. polyphylesis. One of the much-discussed questions of phylogeny, particularly with reference to the angiosperms, is that of single vs. multiple ancestry. These discussions are often clouded by insufficient attention to their theoretical basis. In this connection, it is important to recognize the distinction between phylogenies that are based upon races or types ("Rassenphylogenie" Zimmermann 1959), and those that are based upon characters ("Merkmalsphylogenie"). In groups like angiosperms, which lack fossils, the race or type approach leads automatically to the selection of either one or several modern groups as "ancestral prototypes" of the phylogenetic line or lines in question. The question of mono- vs. polyphylesis is, therefore, answered by deciding whether one or several of these "prototypes" must be regarded as ancestral to the group in question. Botanists who regard the angiosperms as monophyletic often select the Magnoliales as the ancestral prototype, and attempt to derive all other orders from them. The difficulties of this interpretation are discussed in Chapter 10.

These difficulties consist in the probability that primitive as well as advanced characteristics are distributed almost equally among several different modern groups of angiosperms. This situation does not require us to believe that the angiosperms are polyphyletic, in the sense that they have been derived from several very different ancestral forms, and have acquired basic resemblances as a result of evolutionary convergence. On the other hand, it does point out a fundamental weakness in the type or racial approach to the problem. This approach excludes the possibility that a common ancestor existed which had a combination of characters not present in any modern group, and which, if living today, would be classified in a separate order from any living forms.

The high probability that all major existing categories had common ancestors of this kind is testified to by the fossil record of those groups of animals and plants in which the record is sufficiently complete that it can be used for interpretation. For botanists, the most instructive example is that of the conifers, as worked out by Florin (1950, 1953). Before his interpretation of the fossil record was available, many specialists on conifers attempted to explain their origin as monophyletic, on the basis of an ancestral group corresponding to a modern family. Some selected the Pinaceae as ancestral, others the Taxodiaceae, and still others the Araucariaceae. Florin showed clearly that none of these interpretations was correct, but that, nevertheless, the conifers are basically monophyletic. The fossil forms *Lebachia* and *Ernestiodendron,* which either represent or closely resemble the common ancestor of modern conifers, have combinations of characters that are completely different from those found in any modern family.

Consequently, the question of mono- vs. polyphyletic origin, at least in groups lacking a significant fossil record, must be answered, if at all, in terms of character phylogeny rather than racial or type phylogeny. We must ask not, "Is this or that order the ancestral prototype?" but, "Has this or that character (vessels, closed carpels, and so on) arisen once or several times in the evolutionary history of the group in question?" When we do this, we usually find that some of the characters that are diagnostic of the group have probably arisen only once, whereas others may have arisen several times. Furthermore, we have no reason for supposing that all of the diagnostic characters arose at the same stage of evolutionary history. In most instances, including the origin of the angiosperms, the different specialized character states originated at very different times. This point is discussed further, with references to the origin of the angiosperms, in Chapter 10.

Evolutionary reversibility and developmental complexity

The degree to which evolutionary trends can be reversed is one of the most important factors to be evaluated by anyone who desires to determine evolutionary trends. It has also most frequently been the victim of misinterpretation on the part of comparative plant morphologists and systematic botanists. Since higher plants have no fossil record that is significant for determining trends, all conclusions about this matter with respect to them must be based upon extrapolations from comparisons of contemporary forms. Consequently, the conclusions reached by vertebrate paleontologists, which are based upon studies of actual phyletic sequences, should be carefully considered before any interpretations of plant trends are made. For this reason, the statements made by Simpson in *The Major Features of Evolution* (Simpson 1953:310–311) are presented here:

> A group of organisms does not return entirely to the different condition of its ancestors. In broadest statement, that is the principle of irreversibility of evolution, sometimes called "Dollo's Law" . . . There has been much solemn dispute over whether an adaptive trend in some descriptive character can be reversed. Of course it can be and often is, witness the horses that became smaller although their immediate ancestors had a definite trend toward larger size . . . Some students have such a respect for law that they insist that apparent exceptions to their (or even to Dollo's) concept of Dollo's Law are inadmissible, for instance that a small animal cannot really be the descendant of a larger one, even if everything points conclusively that way, because that would be against the law. Such quibbles do not merit much serious consideration, nor does semantic argument whether, when lost structures do reappear (as in *tetraptera* mutants in *Drosophila* or the polydactylous guinea pigs of Wright 1934) they are "really" the *same* structure (e.g. Gregory 1936).
>
> There is nothing in the analyzable separate factors of evolution that prohibits reversion. Back mutations occur. Lost combinations can be reconstituted by recombination. Variation is practically always present on *both* sides of the mode. Selection can reverse its direction. As regards particular characters and trends, therefore, if genetic variation and selection really are essential factors of evolution, evolution should be reversible—and it is. Evolution is readily reversible for particular features, especially those like size that have a very broad genetic basis not necessarily or probably homologous in different groups . . . This does not alter the fact that evolution does not reverse itself exactly or for the whole organism . . .
>
> The statistical probability of a complete reversal or of essential reversal to a very remote condition is extremely small. Functional, adapted organisms noticeably different in structure have different genetic systems. They differ in tens, hundreds or thousands of genes, and such genes as are the same are fitted into differently integrated genetic backgrounds. A single structure is likely to be affected by many different genetic factors, practically certain to be if its development is at all complex.

The last statement deserves further elaboration. It leads to the generalization that the probability of reversal in the phenotypic character of a structure is inversely proportional to the degree of complexity and integration of the developmental reaction system which gives rise to that structure. In anticipation of discussions of specific trends that will appear in later chapters, some of the trends that morphologists and taxonomists have used most extensively as guides to angiosperm phylogeny will be compared with respect to the probability of reversal, on the basis of the principle of inverse correlation with developmental complexity, stated above.

1. *Characters that are easily reversible:* Size of a plant having indeterminate growth, that is, from shrub to tree and vice versa; number of branches; length of internodes; size of leaves, flowers, seeds, or other individual organs; number of flowers per inflorescence.

2. *Character trends that can be reversed, but with greater difficulty:* Numbers of serially or simultaneously produced parts of any

structure, that is, of leaflets in a compound leaf, of sepals, petals, stamens, or carpels per flower, and of ovules per carpel. With respect to sepals or calyx lobes, petals or corolla lobes, and carpels this reversibility has been demonstrated by selection experiments, as mentioned above (Huether 1968). It is based upon the relation between total amount of meristematic capital and the amount needed to differentiate a single primordium. Reversal of trends in number become limited chiefly when the number of parts becomes very small, particularly when it is reduced to one or two. Under these conditions, the reorganization of the developmental reaction system which a secondary increase in number would require may be so great that the adaptive value which might result from it can be more easily achieved by other means.

Another character trend that belongs in this category is the abruptness or gradualness of the transition from one kind of organ to another. Most morphologists have more or less implicitly assumed that the gradual transition from leaves to bracts, or from petals to stamens, is always a primitive condition, whereas abrupt transitions are always derived. This assumption has been questioned by Burtt (1961), in his review of *Morphology of the Angiosperms,* by A. J. Eames (1961). As long as one believes that only reductional trends exist in the shoot, the irreversibility of the trend toward abruptness is a logical corollary, since the fewer are the serially produced organs, the more abrupt is the transition. If, however, one believes, as is maintained in this volume, that both reduction and elaboration can occur with respect to the serial production of appendages upon a shoot axis, then the elaboration phase, involving the increase in number of appendages produced by the shoot, will often bring about a more gradual transition from one kind of organ to another. The production of such more gradual transitions by means of artificial selection in cultivated ornamentals is a familiar phenomenon. Most double flowers show this condition, particularly with respect to the transition from petals to stamens. Consequently, when gradual transition between these two kinds of organs occurs naturally in genera that otherwise have many specialized features, as in *Nymphaea, Victoria,* and other genera of Nymphaeaceae, the assumption that this condition is necessarily primitive is unwarranted.

3. *Character trends that can be reversed only with great difficulty. (a) The specialization of the vascular system.* One of the major trends of evolution in flowering plants has been the derivation of vessel elements from unspecialized tracheids and the differentiation of wood fibers, of wood parenchyma, and of the various specialized cells of the phloem (Bailey 1957, Metcalf and Chalk 1950). The irreversibility of this trend has been an unassailable dictum of comparative anatomy. As with all other trends, minor reversals might be expected from time to time. There is no reason for believing that the general trend from longer to shorter vessel elements and fibers can never be reversed. Occasional increases in length, without reversion to a more primitive wall structure, would be in accord with the general principle that size characteristics are the most easily reversed of

any. On the other hand, the trend from unspecialized vesselless wood such as that found in primitive Magnoliales to the highly specialized system of vessels, fibers, rays, parenchyma, and other elements found in specialized woods such as those of the Fagaceae and Leguminosae involves many harmoniously integrated adjustments of cellular structure and function. Hence a coordinated reversal of these changes is so improbable as to be essentially impossible.

(*b*) *The trend from woody plants to herbs.* Anatomically, the origin of herbs from woody plants is a continuation of the kinds of structural and functional adjustments at the cellular and tissue level that are responsible for the origin of specialized characteristics within woody plants themselves. The irreversibility of this trend is based upon the same considerations as those mentioned under (*a*). To be sure, shrubs and trees do sometimes evolve as specializations from herbaceous groups. The most conspicuous examples are monocotyledons such as palms, bamboos, and dragon trees (*Dracaena, Cordyline*). These, however, are comparable to the reversions of land vertebrates to aquatic life. Palms and bamboos are as different from primitive preangiospermous shrubs and trees as whales and seals are from fishes. More subtle reversals probably exist among dicotyledons. The arboreal Lobelioideae, Goodeniaceae, and Compositae of oceanic islands (Carlquist 1967, 1969*a,b*), and the arboreal members of such predominantly herbaceous families as Phytolaccaceae (*Phytolacca dioica*) and Chenopodiaceae (Wilson 1924) are probable examples. One anatomical feature that may be a distinctive clue to secondary woodiness is the presence of interxylary phloem. This condition, as it exists in the Myrtaceae, certainly reflects a different kind of cambial activity from that which gives rise to the usual and more primitive kind of woody stem.

(*c*) *The trend from perennials to biennials or annuals.* The monocarpic rhythm of flowering, in which the plant dies immediately after a single production of flowers and seeds, involves a complex readjustment of hormone balance and of other factors responsible for the translocation of materials through the plant. This specialized condition, therefore, is the end point of an irreversible trend. On the other hand, there is no reason for believing that the trend from monocarpic biennials to monocarpic annuals is irreversible. By altering their adjustment to seasonal conditions, spring annuals might be expected to evolve into winter annuals and monocarpic biennials almost or quite as easily into annuals.

(*d*) *From radial to bilateral symmetry.* The evolution of a bilaterally symmetrical organ from one having radial symmetry involves a complete readjustment of the pattern of growth. Complex changes in polarity and in the balance and distribution of growth substances must certainly be required. Consequently, it is not surprising that when a species having flat leaves is exposed to strong selection pressure for drought resistance and consequently reduction in leaf-surface area, the leaves never revert to the condition of radial symmetry that characterized the photosynthetic organs of the primitive vascular plants from which seed plants arose.

Instead, the leaves may become reduced in size and devoid of chlorophyll, so that the photosynthetic function is taken over by branchlets or leaf rachises, as in the genus *Casuarina.*

The essential irreversibility of the trend from radial to bilateral symmetry in the flowers of many advanced families (Scrophulariaceae, Labiatae, Orchidaceae, and so forth) is also based upon the principle that the reorganization of a complex developmental pattern is essentially irreversible. In these examples, however, an additional complication has arisen. Bilaterally symmetrical or zygomorphic flowers have the stamens and stigma arranged in such a way as to assure efficient cross-pollination. If a reversal in symmetry occurs with respect to one of these organs, as in the example of the peloric *Antirrhinum,* the relations between corolla nectaries, stamens, and stigma become so disturbed that successful pollination of the flower under natural conditions is difficult or impossible.

(*e*) *The architecture of the inflorescence.* In Chapter 10 a complete account is given of the evolutionary relationships between different kinds of inflorescences. The trends described there involve both the suppression of intercalary meristems and the origin of new ones. Also involved are changed relationships between the transformation of the shoot apex from the vegetative to the flowering condition and the differentiation of foliar appendages as well as branches. All of these changes are integrated into highly complex patterns, and so fall well within the range of trends which on the basis of the probability factor would be expected to be irreversible. As will be shown in Chapter 11, this essential irreversibility is well supported by the facts of comparative morphology.

(*f*) *"Fusions" and "adnations," especially in the flower.* These trends involve the development and regulation of intercalary meristems. As is pointed out in Chapter 6, intercalary meristems are much more strongly developed in angiosperms than in other vascular plants. Consequently, morphological structures that develop primarily through the activity of such meristems must usually be regarded as more specialized than structures that develop principally by apical growth. These conclusions support the widespread opinion of botanists that trends toward intercalary concrescence, exemplified by either "fusion" or "adnation," are rarely reversed. Nevertheless, reversals of such trends may occasionally occur, as has been demonstrated by Eyde (1946) for the Araliaceous genus *Tetraplasandra.* This and other examples should serve as lessons to tell us that we cannot expect to find absolute irreversibility of trends with respect to single characters.

Summary

In spite of the hazards and uncertainties involved, the construction of phylogenies is a worthwhile occupation because it gives insight into relationships between organisms. It must, however, be accompanied by a clear statement of the assumptions made by the biologist who is constructing them. Evidence must be obtained from fossils, comparative morphology and anatomy, morphogenesis, chromosomes, and biochemistry. The way in which each kind of evidence should be used has been discussed. The synthesis and interpretation of the different kinds of evidence must be made on the basis of clearly expressed criteria, which are based as much as possible upon known facts of biological processes, particularly the action of genes and growth substances, rather than idealistic morphology that confines its attention to adult structures. Reversibility in evolution is possible for all single character differences that are based upon simple alterations of the developmental pattern. The degree of irreversibility of a character condition depends upon the number and complexity of the separate factors that contribute to it. Using this criterion, various characteristics of angiosperms are grouped according to the probability of reversal vs. irreversibility.

Part II
Trends of Angiosperm Phylogeny

8 / The Ecological Basis of Diversity

In the first two chapters reasons were given for the hypothesis that adaptive radiation will be most strongly favored in habitats that present the maximum challenge and stimulus to the population for evolving new kinds of adaptations. In Chapter 5, as well as in the discussions of phylogeny presented in Chapters 10–13, the fact is emphasized that the character differences which most often separate genera, families, and orders of angiosperms are those associated with seed development, seed dispersal, and seedling establishment. Consequently, one would expect that the majority of adaptive radiations that gave rise to these major groups must have taken place in regions having the greatest possible diversity with respect to factors that would affect seed and seedling characteristics.

The regions concerned are those ecotones or marginal zones in which seasonal drought, cold, or both exist at levels of the borderline between sufficiency and insufficiency. In such regions, relatively slight differences between nearby sites with respect to the amount of precipitation or frost will affect greatly the composition of the plant community. Furthermore, in these transitional regions local edaphic factors, such as direction of exposure, steepness of slope, and depth and fertility of the soil, will tend to produce a mosaic of radically different communities within a single region. Moreover, as climates change over periods that are measured by the geological time scale, this mosaic will shift in kaleidoscopic fashion, so that all species that inhabit the region must migrate, evolve, or perish.

Effects of Limiting Factors on Diversity

The statements made in the last paragraph are supported by the changes in floristic composition that can be observed in Pacific North America. The coastal portions of this region having mean annual precipitations that range from 1500 mm (60 in.) to three times this amount are all covered by a mesic coniferous forest. The difference between these extremes is reflected in the luxuriance of the forest, but not very much in its species composition. On the other hand, regions having one-third the amount of precipitation, 500 mm (20 in.), are semiarid, and are covered by completely different plant communities. A reduction to a third of this, 167 mm (6.7 in.), brings on desert conditions and another complete alteration of the plant cover. However, the difference between mean annual precipitations of 150 mm (6 in.) and 50 mm (2 in.) is reflected relatively little in the species composition of the plant cover, which simply becomes sparser. In the regions of intermediate or limiting precipitation, for the most part those temperate climates having mean annual values between 200 and 1000 mm, the plant cover of hilly or mountainous regions is greatly affected by local edaphic factors.

The study made by Morison, Hoyle, and Hope-Simpson (1948) on vegetation and soil in the southwestern part of the Sudan showed that increased diversity of both vegetation and soils can be expected in tropical regions having intermediate amounts of precipitation, in the case of the Sudan from 800 to 1500 mm per year. The natural vegetation of this area is savanna woodland, but the diversity of communities is such that the authors could not produce a meaningful description of it by the use of the conventional concept of a uniform climax accompanied by various successional stages. Instead, they used the concept of the "catena," which they define as a grouping of soil-vegetation types that reflects topography and precipitation, and is repeated in different regions wherever the same conditions occur. My own experience with the vegetation of California in regions having annual precipitation between 250 and 1000 mm indicates that the application of the catena concept to these regions will produce a much more meaningful description of the situation than any attempt to use or modify the concept of uniform climatic climaxes.

Population structure in intermediate or ecotonal zones

The mosaic pattern of plant communities in intermediate or ecotonal zones has a profound effect on the individual species that occupy them. The distribution of such species cannot be continuous, but must be broken up into more or less widely separated colonies that occupy favorable habitats. Moreover, as climatic and topographic conditions change over geological epochs, the favored sites will alter their positions, and the relations between climate, topography, and soil will be subject to continuous change.

These conditions will enforce upon the species inhabiting ecotonal regions the optimal kind of population structure for rapid evolution. Wright (1940) has pointed out that such a structure consists of a large or medium-sized population that is divided into many small partly isolated subunits, which can exchange genes through occasional migration. This permits new gene combinations to become established in the individual subunits, both through natural selection and through random fixation of chance variants. At the same time, migration between colonies prevents their stagnation, and allows the population as a whole to draw upon a large supply of genes.

Fedorov (1966) has suggested that a similar population structure might be expected in tropical rain forests, since individuals of any one species in such forests are often widely separated from one another, and are likely therefore to be self-pollinated. Ashton (1969) has severely criticized Fedorov's conclusions, pointing out that many of the insect and bird pollinators of tropical trees are able to fly for long distances in a short time, so that the degree of isolation of these individuals is not as great as might appear at first sight. His conclusions are supported by the observations of Janzen (1971*a*), who has found that euglossine bees, which are among the principal pollinators of flowers in the American tropics, may fly as far as 23 km in a few hours.

Ashton has also analyzed patterns of the distribution of species and varieties in *Shorea* (Dipterocarpaceae), a dominant genus of the Malaysian forests, and has concluded from them that speciation in this genus is little if at all different from that found in temperate groups, except that hybridization is relatively rare, and slight differences between related, sympatric "subspecies" (probably sibling species according to the terminology of zoologists) may be maintained over large areas. In contrast to Fedorov's hypothesis, Ashton's observations indicate that, within an individual species or subspecies, variation between local populations is relatively slight. Hence the rather scanty data which are now available indicate that tropical rain forests do not appear to contain predominantly species having large gene pools which are broken up into genetically diversified subunits.

Whatever may be the outcome of this controversy, it is largely irrelevant to the main argument of this chapter. With respect to factors that affect characters of seed development and seedling establishment, the tropical rain forests are far more homogeneous than the semiarid areas surrounding them. Consequently divergence of populations with respect to these characters will always occur on a larger scale in the border areas than within the forests themselves, even if the latter contain populations having a genetic structure that is favorable for diversification.

Observations comparable to those of Ashton on tropical rain-forest species need to be made on species that inhabit the semiarid deciduous forests and thorn scrub of tropical regions, particularly in areas of rugged topography and soil diversity. From the observations of Ford (1964) on populations of Lepidoptera and other animals, we might expect to find in these regions a different pattern of variability from that which exists in the rain forest. Ford's observations on small, insular populations of Lepidoptera, as well as on populations which have periodically been reduced to a very small size, indicate that small size by itself is unlikely to promote the action of genetic drift so strongly that it can bring about genetic divergence in the absence of divergent selection. On the other hand, small populations can respond more quickly than large ones to radical changes in the environment, and can undergo more drastic alterations of genetic composition, since gene frequencies can be altered more rapidly by similar selection pressures. Ford points out in addition that the selective pressures which would be required to produce the changes observed in populations of Lepidoptera are far greater than those postulated by Wright.

Chromosomal diversification in ecotonal regions

Adaptive radiation in mountainous, ecotonal regions would also be favored by the effect of divergent selection on chromosomal differentiation and its role in promoting reproductive isolation. These regions have a maximum number of pioneer communities, inhabiting such habitats as steep slopes having a thin soil cover and areas previously denuded by landslides or volcanic eruptions. I have pointed out (Stebbins 1958) that in the Compositae, tribe Cichorieae, the species inhabiting pioneer associations are much more likely to differ with respect to karyotype morphology and chromosomal rearrangements than are species found in more stable communities. The same kind of chromosomal diversity exists in many other herbaceous plants that inhabit pioneer communities, such as the Compositae, tribe Madiinae (Clausen 1951), *Clarkia* (Lewis and Raven 1958), *Gilia* (Grant 1963), and genera of Dipsacaceae (Ehrendorfer 1965). The same holds for woody genera that inhabit the subtropical, semiarid regions of Australia (Smith-White 1955, 1960, Barlow 1959), and has been found to a remarkable degree in *Brachycome* (Compositae), an herbaceous genus that is widespread in Australia (Smith-White, Carter, and Stace 1970).

The basis of this chromosomal diversity in species inhabiting pioneer associations is probably selective pressure for linked gene combinations. The importance for adaptation of epistatic interactions between genes at different loci is well recognized by geneticists (Mayr 1963, Dobzhansky 1970). If a particular group of genes confers a high adaptive advantage upon a population because of epistatic interaction between them, any population in which such genes are so closely linked that they were inherited as a single unit will have an adaptive advantage over populations in which similar genes are scattered over the chromosomal complement and transmitted separately. Selection will therefore favor translocations that place the cooperating genes on the same chromosomal segment (Darlington and Mather 1949). The constant inheritance of a particular gene combination resulting from such linkage is of particular value to a population that is colonizing a new area, since, when it prevails, the maximum number of progeny are likely to retain the genetic advantage that enabled the original colonizers to become established.

On the other hand, in a region having a variety of diverse habitats, neighboring populations may be subjected to different selective pressures, so that different gene combinations have a maximal adaptive advantage. Consequently, different chromosomal rearrangements will be favored in various separate populations, resulting in chromosomal diversification.

The importance of such chromosomal diversification is that barriers of reproductive isolation often consist of differences in segmental arrangements of chromosomes. Hence in mountainous ecotonal regions natural selection will not only promote the diversification of populations, but favor the development of barriers of reproductive isolation between them. Some of these reproductively isolated populations will then be in a position to become the ancestors of new lines of evolution. The occurrence of such chromosomal diversity in the earliest angiosperms, the Mesozoic ancestors of the modern woody Ranales, is strongly suggested by the chromosome numbers that exist in the modern genera (Ehrendorfer *et al.* 1968).

In short, the climatically intermediate regions can be regarded as the laboratories in which most new adaptive complexes of plant groups are produced. The extreme regions, whether they are lakes and streams, the great forest belts, the arctic tundra and alpine fell fields, or the deserts, are more likely to be the museums in which extreme adaptations are preserved for shorter or longer periods of time.

The statements made in the preceding sections must not be interpreted as meaning that speciation ceases to occur when a group becomes fully adapted to highly favorable mesic habitats. Many large genera, such as *Ficus, Miconia, Inga, Viburnum, Acer, Fuchsia,* and *Begonia,* occur exclusively in such habitats. Ashton (1969) has recently shown that in the genus *Shorea* (Dipterocarpaceae) in the rain forests of Borneo, geographic speciation occurs according to a pattern similar to that found in those temperate genera of plants that have been intensively studied, as well as in most genera of animals. In some families, such as the Annonaceae, Lauraceae, Commelinaceae, Bromeliaceae, Zingiberaceae, and Orchidaceae, extensive differentiation of genera and tribes has taken place entirely in tropical rain and cloud forests. Much of this differentiation, however, has consisted of adaptive radiation with respect to methods of pollination, and has had little effect on the vegetative characteristics of the plant body, the periodicity of flowering and fruiting, or the establishment of seedlings. The hypothesis proposed in this chapter is that both the intensity of the environmental challenges and the nature of the responding gene pools greatly increase the probability that groups having primary adaptations to intermediate or marginal ecotones will evolve radiant derivatives that are adapted either to favorable mesic conditions or to the harsh environments of deserts, steppes, or arctic-alpine habitats. The chances are much less that groups which have become fully and exclusively adapted to one of these extreme habitats will give rise to descendants that are adapted either to marginal habitats or to an extreme habitat of a very different nature.

The observations and deductions made in this section are analogous to those made by E. O. Wilson (1961) as a result of his studies of the ant fauna of Melanesia. He concludes that expanding species occur preponderantly in marginal habitats, which he defines as those containing the fewest ant species, and consequently offering the smallest amount of direct competition to the expanding species. Ecological niches in the climax rain forest are occupied only by highly specialized and efficient adaptive radiants, which do not give rise later to radically different variants. Wilson's warning that his reasoning should not be extended to other groups of animals might have been even more emphatic if he had considered plants. Nevertheless, it appears highly significant to me that I had essentially completed the argument presented in this chapter before I became aware of Wilson's paper. The fact that very similar conclusions can be reached by two evolutionists working independently with entirely different organisms suggests that the principles involved may be of general application.

Relative stability of semixeric and mesic forest communities in the tropics

The great fragility and susceptibility to change that characterize plant communities of semiarid tropical regions were strongly impressed upon me during a recent brief visit to the island of Oahu, Hawaii. The southern and southeastern coastal regions of this island are semiarid. The gradient of precipitation increases sharply as one ascends the ridges from the outskirts of Honolulu to the crest of the mountains that form the "backbone" of the island. The lower slopes of these ridges, as well as outlying crests such as Diamond Head, are so dry that they can support only a low-growing shrubby vegetation. Originally, this vegetation consisted largely of endemic species that are now almost completely extinct. They have been replaced by introductions from North America, chiefly species of legumes or Mimosoideae (*Leucaena, Acacia, Prosopis*). In areas having intermediate amounts of precipitation, much of the native flora still persists, but it is intermixed with introduced woody species, such as guava (*Psidium*) and candlenut tree (*Aleurites*). Finally, the highest ridges still support a cloud forest consisting of native, chiefly endemic species, and has been altered relatively little during the past century.

The man-made disturbances that destroyed the vegetation of the lower slopes and altered that of the intermediate altitudes took place many years ago, before the present urbanization began. At that time, European cattle, sheep, and goats, introduced during the middle of the 19th century, roamed freely over the island. These animals, along with woodcutting and burning practices by the settlers, were responsible for the destruction of the communities adapted to semiarid conditions. They were, however, unable to alter significantly the dense, resistant cloud forests of the higher slopes.

This differential disturbance is representative of the kinds of change that man and his domestic animals have brought about in all tropical regions. Communities adapted to semiarid regions have, for the most part, been greatly altered, whereas rain and cloud forests have been relatively undisturbed. These man-made disturbances are, obviously, much more extreme than any that occurred before mankind evolved. Nevertheless, prehuman biological changes, such as erosion, landslides, and the evolution and migration of large herbivorous dinosaurs, probably had similar though much less pronounced effects on the vegetation. We might expect, therefore, that alterations in both climatic and biotic factors would bring about frequent replacements of one kind of plant by another in the communities adapted to semiarid climates, but would affect much less the composition of rain and cloud forests.

Diversity of Species Complexes in Marginal Regions

Since speciation may be the prelude to further adaptive radiation toward differentiation of genera and higher categories, as suggested in Chapter 2, some indication of the ecological conditions that would maximize differentiation of this kind may be provided by comparing species diversity of comparable groups adapted to various habitats within the same geographic region. Table 8-1 presents the results of such a comparison in the flora of the Pacific Coast of the United States (California, Oregon, and Washington). This flora was selected partly because it is familiar

Table 8-1. Distribution by habitat of genera of different sizes in the angiosperms of the Pacific Coast of the United States.

Genera are classified according to the habitat distribution of the majority of their species. (From Stebbins 1972.)

	Number of genera and species in genera with—														
	1–2 spp.		3–5 spp.		6–10 spp.		11–20 spp.		21–30 spp.		Over 30 spp.		Total		Number of species per genus
Habitat	Gen.	Spp.	Gen.	Spp.	Gen.	Spp.	Gen.	Spp.	Gen.	Spp.	Gen.	Spp.	Genera	Species	
Lakes, streams, swamps, bogs, marshes	73	109	29	116	4	32	5	75	1	25	2	96	114	447	3.8
Mesic woodlands	81	122	37	148	16	136	10	150	1	25	1	47	146	628	4.3
Mesic open country	45	68	20	80	17	136	12	180	5	125	9	456	108	1045	9.7
Mesic alpine areas	15	23	6	24	0	0	0	0	1	25	0	0	72	22	3.3
Semixeric open woods, grasslands, and shrub formations	114	171	40	160	40	320	37	555	9	225	23	1233	263	2664	10.1
Steppes and deserts	127	191	27	108	13	104	3	45	1	25	0	0	171	473	2.8

to me, but also because it offers a great variety of habitats, each of which (except for alpine habitats) covers an area not very different in extent from those of the others. The six generalized habitat categories represent the best classification permitted by the available data. In the case of many of the larger genera, such as *Astragalus, Eriogonum, Phacelia,* and *Senecio,* most habitats contain at least a few of the numerous species belonging to them. These genera were classified according to the habitat distribution of the majority of their species. Three of the larger genera, *Carex, Veronica,* and *Viola,* defied classification and were not tabulated.

As can be seen from Table 8-1, the habitats fall neatly into two groups with respect to the sizes of the genera that are principally adapted to them. In one group, the mean number of species per genus is low: from 2.8 to 4.3. This includes wet habitats, mesic woodlands, alpine habitats, and desertic regions. The remaining two habitats are mesic open country, chiefly fields and meadows, with 9.7 species per genus, and semixeric open woods, grasslands, and shrub formations, with 10.1 species per genus. The two habitats with high mean values are ecologically intermediate between the four having low values.

These data agree well with those compiled previously on the endemic species of the California flora (Stebbins and Major 1965). The species regarded as relictual, because they belong to small genera having no close relatives, are bimodally distributed. The largest concentration of them is in the moist forests of northwestern California, but a secondary concentration occurs along the margins of the deserts and in the desert mountains. On the other hand, the greatest number of examples of rapid speciation is in intermediate regions—the hot scrub lands or chaparral, the oak parkland, and the drier margins of the forest belt. Individual examples of rapid speciation have been found most often in these regions, as in *Delphinium* (Lewis and Epling 1959), *Layia* (Clausen 1951), *Clarkia* (Vasek 1964, 1968), *Gilia* (Grant 1963), *Penstemon* (Straw 1955), and *Stephanomeria* (Gottlieb 1971). In the Old World, the similar ecotonal regions of the Mediterranean also offer examples of rapid speciation (Ehrendorfer 1951, 1965), as do such regions in Australia (Briggs 1964).

These observations on the flora of the western United States agree well with those of Weimarck (1941) on the flora of the Cape Region of South Africa, which probably contains a higher concentration of endemic species than any other temperate flora in the world. He noted a number of centers and subcenters of concentration for these species, similar to those recognized by Stebbins and Major (1965) for California. The great concentration of these endemics is in ecologically intermediate habitats, such as forest margins and rocky outcrops. As in California, the flora of the climax forests as well as that of the open deserts contains many fewer localized endemics. Weimarck concluded, as I did for California, that some of the endemics of South Africa are ancient relics, whereas others are products of relatively recent speciation. He postulated that the extensive speciation which was necessary to produce these endemics took place in association with marked oscillations of the climate that occurred during the Pleistocene epoch.

Pertinent comparisons between semiarid and moist regions in the tropics are not known to me, and are probably not available. Nevertheless, comparisons between tropical and temperate regions as to numbers of families, genera, and species do not support the hypothesis that the well-known richness of tropical floras is due chiefly to extensive speciation in progress at the present time. Table 8-2 shows three such comparisons: two large areas of comparable size—the Philippine Islands (from Merrill 1923) and California (from Munz and Keck 1959); two relatively small areas—the Panama Canal Zone (from Standley 1928) and Marin County, California (from Howell 1949); and two comparable plant formations—the Brazilian cerrado (from Rizzini 1963) compared with the California chaparral (compiled by me from Munz and Keck 1959). The most striking fact that emerges from this table is that, though in each comparison the total number of species is greater in the tropical than in the temperate flora, the number of species per genus is in one comparison the same in the two regions, and in the two others is actually larger in the temperate, Californian example. The greater richness of the tropical floras is due chiefly to the larger number of genera found in them and to a lesser extent to a larger number of families. Since many of the large genera of the tropical floras, as in the Orchidaceae and Gesneriaceae, and such genera as *Ficus*, have undergone speciation almost entirely in response to the diversity of animal pollinators, with little or no dependence upon the diversity of the habitat, one can conclude that in the mesic or moist tropical forests speciation in response to habitat diversity and environmental changes is, if anything, less active than it is in semiarid temperate habitats. The large number of genera in the tropical floras could possibly reflect a greater

Table 8-2. Comparisons between tropical and temperate angiosperm floras as to numbers of families, genera, and species of vascular plants. (From Stebbins 1972.)

	Numbers of—				
Flora	Families	Genera per family	Genera	Species per genus	Species
Philippine Islands (Merrill)	186	7.9	1468	5.4	7858
California (Munz)	147	6.9	1021	5.4	5470
Panama Canal Zone (Standley)	97	6.5	627	2.2	1397
Marin County, Calif. (Howell)	92	4.1	380	2.5	961
Brazilian cerrado (Rizzini; woody plants only)	70	3.5	242	1.8	537
California chaparral (woody plants only)	40	2.1	83	3.0	250

amount of speciation in past geological epochs. More probably, however, it is due to the fact that these floras have acquired much of their richness by immigration from other floras combined with a lower rate of extinction.

This hypothesis is supported by the monographic study made by White (1962) on *Diospyros* in tropical Africa. He found that among the 90 tropical African species of this genus, 48 are taxonomically isolated. Of these 70 percent occur in rain forests, and several of the others in arid regions. Among the species that occur in semiarid borders of forests and savannas, 72 percent are components of superspecies or species groups that show indications of active speciation. Consequently, *Diospyros* in Africa resembles closely the larger genera of South Africa and the western United States in its distributional pattern.

The relictual nature of small genera found in extreme habitats

On the basis of the adaptive-radiation hypothesis, the isolated, small genera found in mesic or xeric habitats could be regarded as surviving radiants from ancient examples of adaptive radiation. The central species of the radiating complexes to which they belonged formerly occupied intermediate semiarid or open regions, but have become extinct there, having succumbed to competition from recently evolved more successful and dominant groups. Those radiants that managed to become adapted to the more extreme habitats became bradytelic species in the sense of Simpson (1953). In regions where the rate of speciation is relatively low, they have persisted long after their ancestors have become extinct.

The existence of two kinds of intermediate situations lends support to this hypothesis. One is the fact that each of the mesic or xeric habitats contains a few species belonging to genera that are predominantly adapted to intermediate habitats. Examples of such species in wet habitats are *Polygonum amphibium, Senecio hydrophilus,* and *Allium validum.* Among the numerous such examples in mesic woodlands are species of *Senecio, Aster, Solidago,* and certain radiants of *Antennaria* and *Ceanothus* that will be described in the next chapter. *Senecio, Crepis,* and *Antennaria* also provide examples of alpine radiants. In the fourth extreme habitat, the deserts and steppes, much of the flora is made up of radiants belonging to genera that are centered in semiarid regions. Many of these are drought-evading annuals, but truly drought-resistant perennials adapted to desert conditions occur in such genera as *Artemisia, Astragalus, Galium, Lotus, Oenothera, Penstemon,* and *Salvia.*

The second kind of intermediate situation is the presence of small genera that are entirely adapted to mesic or xeric habitats, but are specialized radiants belonging to families that are chiefly developed in the intermediate habitats. Examples of these, some of which can be found in Table 2-2, are as follows:

Wet habitats: *Berula, Chamaedaphne, Lilaeopsis, Phragmites, Subularia;*

Mesic woodlands: *Aruncus, Brachyelytrum;* also *Circaea* (Onagraceae) and *Adenocaulon* (Compositae);

Deserts: *Alhagi;* also many mono- and ditypic genera of Compositae;

Alpine habitats: *Smelowskia* (Cruciferae), *Podistera* (Umbelliferae), *Cremanthodium* (Compositae), *Raoulia* (Compositae).

Tropical Rain Forest: Cradle or Museum?

The argument presented in the preceding paragraphs leads inevitably to a question concerning the role of tropical rain forests in the origin of the angiosperms. The hypothesis that angiosperms originated and differentiated in moist tropical regions, either lowland rain forests or upland cloud forests, has been accepted by nearly every botanist of the 20th century who has considered the problem. Among these authors are Diels (1908), Hallier (1912), Bews (1927), Wulff (1943), Camp (1947), Bailey (1949), Corner (1949), Axelrod (1952), Cronquist (1968), Fedorov (1966), van Steenis (1969), and Takhtajan (1969, 1970). Is this the most probable hypothesis?

The reasoning of all of these authors is similar. It is based, at bottom, on the hypothesis of Willis (1922, 1940) that the region that contains the largest number of contemporary species belonging to a group is the region where that group originated. At first sight, this is a most plausible hypothesis. "Centers of diversity" are familiar to biogeographers. By mapping many distributional patterns, one often finds that the number of species belonging to a group diminishes gradually as one moves outward from its "center." This reduction of diversity may take place along radiant lines, rendering even more plausible the interpretation of these data as indicating a center of origin from which species have radiated.

This is, however, not the only interpretation that can be made of such patterns. It is based upon one of two assumptions. One might assume, as Willis did, that the proportion of angiosperm species that have become extinct is so small that extinction has had no substantial effect in altering distribution patterns. This assumption is patently false, and is justly rejected by modern plant geographers. Consequently, a second assumption is necessary if one is to conclude that modern distribution patterns are similar to those that existed when families of angiosperms were first differentiated from one another. This is that extinctions have been essentially equivalent in proportion to the total flora, at least in all parts of the world that have not been drastically affected by major upheavals, such as the mountain-building activity of the later Tertiary and the Pleistocene glaciations.

In my opinion, this second assumption is also unfounded. In the first place, most of the differentiation of angiosperm families took place during the Cretaceous Period, from 70 to 120 million years ago, when the earth's surface and climates were very different from their present aspect. Second, the fossil record tells us that differential extinction has occurred in regions which have not been exposed to orogenic or climatic upheavals. For instance, the coast redwood, *Sequoia sempervirens,* was common in Japan during the Miocene and Pliocene epochs (Miki 1941) and became extinct during the later Pliocene or Pleistocene, even though Japan was little affected by the glaciations and such relictual species as *Cryptomeria* and *Trochodendron* survived. Finally, and most important, the fragility of ecotonal and semixeric plant associations, which is discussed elsewhere in this chapter, as well as the small number of species that they can contain at any particular time, supports the hypothesis that over the millions of years that must be considered the proportions of extinctions were vastly greater in these ecotonal or transitional regions than in the more stable communities of mesic forests. If one combines the deductions made in this paragraph with those that have been made elsewhere in this volume, leading to the hypothesis that from the genetic and ecological point of view major categories of angiosperms are most likely to have become differentiated in the rich mosaic of diverse communities that characterizes ecotonal or border areas, then one inevitably reaches a conclusion diametrically opposite to that of the age-and-area hypothesis or any modifications of it. This is that the so-called "centers of origin" and "radiant lines of migration" that some authors have postulated with respect to the modern angiosperms that appear to be primitive are chiefly reflections of modern climatic and ecological conditions. They are the locations of those plant communities that have suffered the least disturbance during the past 50 to 100 million years, and so have preserved the highest proportion of archaic forms in an essentially unchanged condition. This is the *museum hypothesis,* which I wish to place in opposition to the currently held "center of diversity–center of origin" hypothesis.

Obviously, the only decisive evidence that could enable evolutionists to decide between these two hypotheses would be a representative fossil record, in which equal proportions of all kinds of plant communities were preserved. This, however, can never be obtained. Deposits of both macro- and microfossils are far more abundant in sites of deposition than in sites of erosion. Consequently, the floras of mesic forests that occupy lowlands are of necessity more strongly represented in the fossil record than are floras of the surrounding hills or mountains. Furthermore, with the exception of sporadic occurrences of fossil beds in the ashes of volcanic eruptions, fossilization requires the presence of lakes or large quiet streams, and so will inevitably be less frequent in semiarid areas than in those having abundant rainfall. Consequently, if the fossil record is taken at its face value, the bias that it contains reinforces the "center of diversity–center of origin" hypothesis with respect to mesic floras.

Indirect evidence might be provided by careful analyses of ecological distributions and relative levels of specialization found in neighboring semiarid and mesic areas, in both temperate and tropical regions. The hypothesis could be tested that, as predicted under the "center of diversity–center of origin" hypothesis, species that inhabit semiarid habitats are always, or at least usually, more specialized than their relatives that live in mesic habitats. This is, in fact, what most botanists believe. In the next chapter, examples are presented that have led me to question this widely accepted dogma.

The botanist who has presented the largest number of examples which he regarded as supporting the hypothesis of mesic tropical origins was J. W. Bews

(1927). Most botanists have accepted his evidence at face value. In the next section, reasons are given for not doing so.

A critique of the "evidence" presented by J. W. Bews

Although at first sight Bews's evidence appears impressive, upon closer scrutiny many flaws in it can be found. A case in point is the list of 70 groups in his "Appendix" on pp. 39–41. Twelve of these (nos. 2, 3, 10, 17, 18, 21, 35, 48, 50, 58, 69, 70) are relevant only to the postulate that temperate groups are derived from tropical or subtropical groups, since the groups listed as "advanced" are temperate and mesic. I agree completely with Bews on this point, and question only his belief that, among tropical and subtropical groups, plants adapted to semiarid conditions are all derived from the inhabitants of mesic rain and cloud forests.

Of the remaining 58 groups, 22 (nos. 1, 11, 13–16, 22–26, 32, 36–45) are poorly classified with respect to ecological adaptation. Although the "relatively primitive" families or tribes listed opposite these numbers are characterized as "hygrophilous or mesophytic," actually they contain many semixeric forms or even xerophytes, as in evergreen Fagaceae, Phytolaccaceae, Nyctaginaceae, Capparaceae, Mimosoideae, Caesalpinoideae, Sophoreae, Euphorbiaceae subf. Crotonoideae, and most tropical species of *Euphorbia.* In another five (nos. 46, 49, 53, 64, 68), the "advanced" groups are not regarded by modern botanists acquainted with phylogeny as derived from forms similar to those listed as "primitive" in the same example. For instance, Anacardiaceae, Celastraceae, and Staphyleaceae are not regarded as advanced derivatives of Icacinaceae and Sapindaceae, nor are Onagraceae believed to be specialized derivatives of Myrtaceae. In another six of Bews's examples (nos. 20, 27, 29, 32, 34, 59), the groups designated as "advanced" are, with respect to floral structure and other characteristics, less specialized than those designated as "primitive." For instance, *Clematis,* having its carpels reduced to one-seeded indehiscent achenes bearing elaborate plumose styles, is certainly not the most primitive genus in the Ranunculaceae. In the Rosales, the Saxifragaceae are not more advanced than the Pittosporaceae. The Hamamelidaceae, if they can be placed in the Rosales at all, would certainly qualify as one of its most advanced rather than primitive families. In the "Rutales" (or Sapindales), the Meliaceae are definitely not the most primitive family, since they have the stamens joined into a column and the seeds with little or no endosperm. Among the Rutaceae themselves, the most generalized floral structure, with carpels almost separate and seeds having copious endosperm, is found in many shrubby forms that inhabit semiarid habitats, whereas subfamilies such as the Flindersoideae, which consist entirely of mesic trees, have more specialized floral structure. Still another series of five examples (nos. 5–9) involve specialized parasites or hemiparasites, and so are peripheral to the problem of major evolutionary trends. The remaining 20 examples are also peripheral, since even the groups designated as "primitive" are already rather specialized, and in several instances (nos. 12, 28, 30, 31, 33, 54) represent relatively minor examples of adaptive radiation, which would be compatible with any theory of ecological evolution of angiosperms.

Since the work of Bews, botanists who support the hypothesis of angiosperm origins in the moist tropics use as their examples principally the Magnoliales or "woody Ranales," which in many respects are obviously primitive. Most of the members of this order are, however, relictual representatives of mono- or digeneric families that are only distantly related to one another (Bailey and Nast 1945*a*). The larger families are the Annonaceae, which with respect to anatomical characteristics are relatively specialized; the Monimiaceae and Lauraceae, which are relatively specialized with respect to both vegetative and reproductive characteristics; the Magnoliaceae; and the Winteraceae. The last two families consist chiefly or entirely of genera having a polyploid origin (Ehrendorfer *et al.* 1968), and their diploid ancestors are extinct. Since polyploids very often invade regions having different ecological conditions from those to which their diploid ancestors were adapted, we are by no means justified in concluding that the diploid ancestors of these families were of necessity adapted to moist tropical forests.

The last argument brings us back to the point raised at the end of Chapter 1. If modern phenotypes are used as the principal guide to hypotheses concerning the conditions under which angiosperms first diversified during the Mesozoic era, the hypothesis of mesic origins receives some support. If, on the other hand, the principle of genetic uniformitarianism is regarded as more significant than the distribution of modern relictual phenotypes, the hypothesis of semixeric origins appears more plausible.

Further evidence in favor of the museum hypothesis

Arguments in favor of the museum hypothesis that are based upon a review of the ecological conditions that would have favored the origin and early diversification of angiosperms are presented in Chapter 10. Three arguments of a general nature are given here.

In the first place, the diversity of tropical forest communities is maintained by the relations between the trees that dominate them and their insect predators (Janzen 1970, 1971*b*). Many of these trees are attacked by insects that live only on a single species or a few closely related ones. The insects will forage from the mature tree and destroy the seedlings of the same species that start to grow around it. This permits the establishment of seedlings that belong to different species or genera, and so are immune from attack. Consequently, when an old tree dies, space is formed that may accommodate any immigrant species that is adapted to the forest conditions, and the richness of the association in species may be increased.

Second, in contrast to the impression that is apparently held by many botanists, most of the woody species found in tropical rain and cloud forests do not belong to families that are generally regarded as primitive. On the contrary, these forests contain a complete spectrum from the most generalized to the most specialized forms with respect to their diagnostic reproductive characteristics. The bulk of the species that dominate the lowland tropical forests, according to descriptions by Richards (1952), Black, Dobzhansky, and Pavan (1950), Ashton (1969), and Poore (1968), belong to families whose phylogenetic position is intermediate between the most primitive and the most advanced groups. Such families are the Leguminosae, Lauraceae, Burseraceae, Meliaceae, Dipterocarpaceae, Euphorbiaceae, Melastomataceae, Sapotaceae, and Apocynaceae. The upland forests of Southeast Asia are dominated by Lauraceae and Fagaceae, families that are either intermediate or advanced in their phylogenetic position. Although species belonging to the "primitive" order Magnoliales occur in the forests that cover parts of the region characterized by Takhtajan (1970) as "between Himalaya and Fiji," these "primitive" angiosperms are always a minor element of the forest flora. They are equaled or exceeded in numbers of both species and individuals by representatives of the most advanced families of flowering plants, such as Verbenaceae, Bignoniaceae, and Rubiaceae.

In Chapter 10, reasons are given for rejecting the commonly held hypothesis that large trees, which constitute the dominant growth forms of tropical rain and cloud forests, are not the most primitive growth form that characterized the earliest angiosperms. My hypothesis is that in the evolution of angiosperms, shrubs preceded trees. Other growth forms that are common in tropical rain forests are obviously highly specialized. Examples are lianas, epiphytes, parasites, and saprophytes.

Finally, the families and even the genera of woody plants belonging to these tropical forest floras are for the most part very well marked and sharply distinct from one another. Transitional forms, which one might expect to be preserved occasionally if the major differentiation of these groups occurred in their present habitat, are almost nonexistent. This situation has led some authors, such as van Steenis (1969), to reaffirm the position maintained by Goldschmidt (1940), Willis (1940), and others, that families and genera originate through the occurrence of single mutations having drastic effects on many phenotypic characteristics. As is explained in previous chapters, current knowledge about the nature of genes and gene action reduces the probability of such assumptions almost to zero. The absence of transitional forms can be explained more plausibly by assuming that they formerly existed in the semiarid regions peripheral to the rain and cloud forests, and became extinct because their habitats were very sensitive to environmental changes.

According to this hypothesis, therefore, the modern distribution of angiosperm families and genera in no way reflects the ecological conditions under which they arose. It is, rather, a result of the extreme ecological plasticity of angiosperms. They can become genetically adapted to new habitats, which may be either drier or more mesic, with relative ease. In addition, open semiarid habitats undergo periodic revolutions in the species composition of their floras, whereas the great forest belts are much more stable. They are well fitted to be museums in which many archaic types can be preserved for millenia, but do not have the conditions favorable for the origin of radically new adaptive complexes.

Summary

The maximum environmental challenge for adaptive radiation is likely to occur in marginal or ecotonal regions in which precipitation is intermediate between complete sufficiency and continuous deficiency, and in which frosts are minimal but present. Such regions contain a mosaic of different changes over geological periods. Their population structures are particularly favorable for both genetic divergence and the development of reproductive isolating mechanisms. The plant communities of these transitional regions are more fragile than those of rain and cloud forests, and so are more likely to be altered by changing biotic factors, such as the entrance of new herbivores or insect predators.

Data from the Pacific Coast of the United States indicate that speciation in marginal or transitional habitats of this region has been more active than in either mesic forests or deserts. Comparison between numbers of species per genus and genera per family in temperate and in tropical regions indicates that in spite of the much greater richness of tropical flora in total numbers of species, genera, and families, there is no evidence that speciation is more active in them than it is in the varied communities of the marginal temperate regions, and it may even be less active.

This kind of evidence suggests that tropical rain and cloud forests are not the communities within which angiosperms originated and differentiated, but to a greater extent are museums in which representatives of most families have been preserved because of low rates of extinction. In addition to evidence for little extinction, the museum hypothesis is supported by the fact that tropical forests contain a complete spectrum from the most primitive to the most advanced families, and forms that provide transitions between families are either absent or at least not more common than they are in other kinds of communities.

9 / Adaptive Radiations and Ecological Differentiation

In earlier chapters, evidence is presented in favor of the hypothesis that, although trends of evolution in flowering plants have been based upon adaptation and natural selection, environment–organism interactions have been so complex and diverse that the adaptive responses leading to the differentiation of genera and families are often difficult to recognize.

The sedentary nature of plants gives rise to a further complication of their organism–environment interactions over long periods of time. This is the frequent reversal of trends with respect to ecological adaptations that can take place within a single evolutionary line. During successions of geological epochs, many parts of the earth have been exposed successively to climates having temperature and moisture conditions highly favorable for plant growth, alternating with much more severe conditions, imposed by drought, low temperatures, or both. With the deterioration of the climate, plant species either became extinct, migrated to regions where the climate was still favorable, or evolved adaptations to the more severe conditions. The effects of an ameliorating climate would have been similar except that the kinds of evolutionary response made by those groups that evolved *in situ* must have been entirely different. An ameliorating climate would make possible the immigration of many vigorous species from other parts of the earth having more favorable climates. Successful adaptation of drought- or cold-resistant species for survival in the lusher communities characteristic of more favorable environments requires increase in vigor, fecundity, and competitive ability.

The Kinds of Changes Fostered by a Deteriorating Environment

In a deteriorating environment, the surviving evolutionary lines evolve adaptations to drought or cold. In addition, the rhythm of flowering and seed production must become adjusted to coincide with the favorable growing season. Survival will often be achieved by: (1) reduction in the amount of growth and hence in the total size of the plant; (2) acquisition of physiological specializations, such as increased osmotic tension of the cells; (3) acquisition of hormonal mechanisms that bring about dormancy during the unfavorable season and quick recovery from this dormancy when favorable growing conditions are resumed; (4) differentiation of specialized organs, such as root crowns, rhizomes, corms, and bulbs, which facilitate this recovery; (5) development of protective coverings as well as of various internal substances (oils, resins, alkaloids, latex), which may serve as partial protection against desiccation, cold, and the attacks of herbivorous animals, particularly insects. Protection against animals has a much greater adaptive value in plants adapted to severe conditions than in those that live in more favorable habitats, because the unfavorable climate limits regeneration. This fact is familiar to everyone who has had experience with range or pasture management in regions having a long dry season. Palatable grasses or other plants that do not become dormant during the dry season can withstand very little grazing if they are not irrigated, but can tolerate much more if their environment is improved by summer irrigation. The only plants that can grow actively in a dry pasture that is being grazed during the rainless season are spiny, bitter, heavily glandular, or otherwise unpalatable, such as star thistle (*Centaurea solstitialis*), prickly lettuce (*Lactuca serriola*), tarweed (*Hemizonia, Holocarpha,* and *Madia* spp.), prickly pear cactus (*Opuntia* spp.), and turkey mullein (*Eremocarpus setigerus*).

Adjustment of the rhythm of flowering and fruiting to coincide with the favorable growing season is likely to involve particularly a shortening of various reproductive stages, and hence a reduction in the amount of growth that takes place in the reproductive organs. A likely hypothesis to explain on an adaptive basis the reductions and condensations of gynoecial structure that have been an important feature of the differentiation of angiosperm families and genera is that they have been adaptive responses to reduction in the length of the favorable season for fruit and seed development.

The timing of the favorable season, particularly its onset and cessation, may influence greatly the rhythm of flowering and seed maturation, and consequently the structure of the inflorescences, flowers, and fruits. For instance, in regions having a Mediterranean type of climate, with mild winters and hot, dry summers, the spring season may be very long, but the favorable portion of the summer is correspondingly short. Many of the plants that are adjusted to this rhythm, particularly those that do not have deep taproots, xeromorphic leaves, or other protective devices, possess elaborate inflorescences which mature over a long period of time, but small, few-seeded gynoecia, which mature their seeds relatively rapidly. Examples are the more mesic Leguminosae (*Trifolium*), the Umbelliferae, Hydrophyllaceae, Boraginaceae, Labiatae, Globulariaceae, and Dipsacaceae. On the other hand, tropical and subtropical regions which have a cool dry season and a warm wet season permit a variety of evolutionary strategies with respect to the rhythm of flowering and fruiting. One of the most highly successful of these, which is very unlikely to evolve in regions having a Mediterranean type of climate, is the presence of much-condensed inflorescences having numerous relatively small flowers, coupled with fruits that have many, sometimes relatively large, seeds. Such plants flower very quickly and abundantly when the favorable season begins, which may occur very suddenly. They then have a long favorable season over which to ripen their seeds. Examples of this rhythm are various genera of Leguminosae, subfamilies Mimosoideae and Caesalpinoideae, of Myrtaceae (*Eucalyptus, Melaleuca,* and so forth), and of Proteaceae.

The extent to which floral structures have become modified in response to the restraints imposed by such climatic rhythms can be determined only on the basis of comparative phenological data obtained on a large scale and analyzed statistically. Until such data are available, this kind of influence of the environment on the evolution of reproductive structures must be regarded as only a likely possibility. Nevertheless, it cannot be dismissed lightly when the problem of the adaptiveness or nonadaptiveness of differences between families and genera is being considered.

Changes Fostered by an Improving Environment

If the climate is becoming more favorable for plant growth owing to increasing moisture and reduction in the amount of frost, the selective pressures tend to be imposed less by the inanimate than by the biotic environment. In a mesic environment, particularly the rain and cloud forests of the tropics, keen competition between plants for light and space confers a great adaptive advantage upon vigor of growth. This is at least as true of the establishment and initial growth of the seedling as it is of the growth of the mature plant. Moreover, the floor of the forest is so dimly lighted that photosynthesis often cannot be carried on efficiently, so that seeds possessing a large amount of stored food in their endosperms or cotyledons are likely to have a great advantage over those that do not. This means that, other conditions being equal, those individuals of a population that produce the largest seeds will have a larger proportion of surviving progeny in the next generation than those that do not. Selection pressure of this nature thus favors genes that promote increases in the size of seeds. As is pointed out in other chapters, side effects of such genes can alter greatly the structure of the gynoecium, and even of the entire flower.

Another influence of an improving environment is to promote an increase in the number of seeds produced per plant. This effect need not be regarded teleologically, but from a simple statistical viewpoint. The principle involved is that if the subsequent generation arises from a random sample of the zygotes produced by individuals of the previous generation, those individuals that contribute the largest proportion of zygotes to the population from which the sample is extracted will transmit the largest proportion of genes to the subsequent generation. In other words, if two individuals are equal in their capacity for producing vigorous, well-adapted seedling progeny, the one that produces the most seeds will transmit more of its genes to the next generation than the one that produces fewer seeds. Since some of these genes are associated with increased seed production, the effect of this differential representation will be an increase in the mean fecundity of the subsequent generation. If sampling of surviving seeds is random, this kind of selective pressure will increase fecundity far beyond the reproductive needs of the species. This principle is exemplified by competition experiments with cultivated cereals. Suneson (1949) placed four barley varieties, Atlas, Vaughn, Club Mariout, and Hero, in competition with one another, initially in a mixture containing equal proportions of the four varieties. Each successive generation was sown from a random sample of the seeds produced by the previous generation. In 16 generations the Atlas variety increased in proportion to the others so that it constituted 88 percent of the mixture. To analyze the factors responsible for this differential, Lee (1960) placed Atlas and Vaughn in competition for a single generation, and followed the development of each individual plant in the mixture. He found that the competitive advantage of Atlas was not associated with a greater proportion of surviving seedlings, since in his experiment no seedling mortality occurred. On the other hand, the mature Atlas plants were more vigorous than those of Vaughn, and produced a larger number of seeds per plant. This differential seed production was great enough to account for Suneson's results, even on the assumption that seedling mortality did not occur, or that it was equal for the two varieties.

Selection for maximum fecundity will be exerted on all populations, whether they live in favorable or in unfavorable environments. In unfavorable environments, however, limits are often set upon the extent to which the plant can grow and the length of time over which it can flower and produce seeds. These limits require a compromise between the genetic advantage of fecundity and the limits on seed production imposed by the environment. Hence among a group of related species or ecotypes we would expect to find progressively higher fecundity and a longer flowering season in those populations that are adapted to increasingly favorable conditions. Unfortunately, observational or experimental data that might test this prediction are scanty. One example is the study by Johnson and Cook (1968) of *Ranunculus flammula.* In this species, populations occupying relatively high altitudes in the Cascade Mountains of Oregon have a shorter growing period and a lower production of seeds, both per flower and per plant, than populations inhabiting more favorable regions which have a longer growing season. These differences are maintained when the plants are grown side by side in a uniform environment, showing that they are based upon genetic differences between the populations concerned.

Another effect of an improving environment will be an increase in the potential animal pollinators. We might expect, therefore, that evolutionary lines which are exposed to improving environments, and which have the requisite gene pool, will undergo adaptive radiation for diverse methods of pollination. The diversity of genera and species in families like the Asclepiadaceae, Acanthaceae, and Orchidaceae, and of species in genera like *Aristolochia* and *Salvia*, is best explained on this basis. This subject was discussed in Chapter 4.

The Reversibility of Xeric Adaptations

Many botanists might reject the arguments presented above on the assumption that adaptations to xeric or markedly seasonal habitats require such highly specialized characteristics that the reverse evolution, from xeric to mesic adaptations, is difficult or impossible. Xerophytes are often regarded, *ipso facto*, as being irreversibly specialized. This belief has no foundation in fact. Originally, it is probably a holdover from an archetypal concept in which the ideal and original plant was conceived as similar to the majority of mesophytic plants growing in moist temperate regions where most botanists live. More recently this belief has apparently been confirmed by the fact that most macrofossils are relatively large leaves belonging to plants that were obviously mesophytes. In this respect, however, the fossil record is deceptive, since plants inhabiting mesic sites have a far greater chance of preservation than those living in drier sites.

Another source of the belief that xeromorphic specializations cannot be reversed is the great emphasis that is often given in descriptions of desert habitats to extreme specializations, such as stem-succulent cacti. The evolution of such highly specialized forms into less specialized mesic derivatives is, of course, hard to imagine. The majority of plants that inhabit semiarid and even desert regions, however, do not possess such extreme specializations. Their xeromorphic adaptations are those listed at the beginning of the present chapter. How easily can plants possessing these less extreme specializations give rise via natural selection to more mesic derivatives?

From the standpoint of developmental genetics, such trends must be regarded as not only possible, but actually more probable than the acquisition of xeromorphic specializations by derivatives of mesic groups. As explained in Chapter 7, irreversibility of evolutionary trends depends upon the inability of an evolutionary line to retrace a complex sequence of adaptational steps. The reversion from many kinds of xeromorphic to more mesic adaptations, however, can be accomplished in a relatively simple fashion. It requires only loss mutations, which inactivate the pathways leading to the elaboration of complex structures or the synthesis of special chemical compounds, plus increases in vigor and fecundity, which can be elicited with relative ease by selection for mutations that promote greater meristematic activity.

Origin of Mesic Cultivated Plants from More Xeric Wild Ancestors

One way of seeking more direct evidence on this point is to ask the question: Can relatively xeric wild species be converted into more mesic cultigens by means of artificial selection? If this is possible, we might suppose that natural selection could elicit the same response, given an available niche in which conditions are favorable for vigorous growth and high fecundity.

Two examples show that this question can be answered affirmitavely. One of them is cultivated cotton. The wild species of *Gossypium* live in semiarid or even desert climates and possess to a varying degree xeric adaptations (Hutchinson and Stephens 1947, Fryxell 1965). The cultivated varieties, on the other hand, are grown in fertile soils and require abundant moisture. These mesic adaptations are an inevitable result of selection for greater vigor and higher seed production. A second example, the cultivated tomato, has had a similar evolutionary history (Luckwill 1943). The wild species of *Lycopersicum* are nearly all adapted to the highly seasonal climate that prevails on the west coast of Peru and Chile. The cultivated species, *L. esculentum,* radiated from this environment through acquiring greater vigor and fecundity, along with the shift from self-incompatibility to predominant self-fertilization. Although the initial shift was probably under the influence of natural selection for the weedy habit, the most mesic, vigorous, and productive varieties of *L. esculentum* are products of artificial selection in relatively recent times.

These examples are instructive as analogies. They show how semixerophytes or even xerophytes might become adapted by natural selection to more favorable conditions. From the genetic and developmental point of view, two kinds of change have accompanied these adaptive shifts. First, vigor and fecundity have been achieved by quantitative increases in the amount of meristematic activity during development. Second, various xeromorphic specializations, such as elaborate trichomes, heavy cuticle, and specialized glands have been reduced or lost. The changes in meristematic activity are clearly reversible, as is evident from both selection experiments (Huether 1968) and observations of individual mutations (Gottschalk 1971). With respect to morphological specializations that appear late in ontogeny, their loss by means of any of a number of mutations that could destroy the epigenetic sequences of gene action responsible for them is to be expected. The experiences of both primitive men and modern plant breeders, therefore, indicate that selection, either artificial or natural, can guide evolutionary lines of angiosperms toward adaptation to more favorable climates, as well as to more severe conditions. Specialized xeromorphic adaptations are not an evolutionary cul-de-sac, but rather peaks of adaptation from which, under certain conditions, an evolutionary line can pass to an even higher adaptive peak represented by more efficient adaptation to a more favorable environment.

Examples of Adaptive Radiation to Xeric and Mesic Habitats

The next step toward establishing the validity of the principles and hypotheses that have been proposed is to assemble a series of examples among natural groups that are best explained on the basis of them.

Ideally, such examples should consist of sequences of fossils, in which the evolutionary succession from species found in older strata to those of younger age is clear, and in which independent paleoclimatic evidence indicates that shifts in the climate, first from more favorable conditions, have taken place. Unfortunately, however, such series are not available. Microfossils, for example pollen grains, in spite of their abundance, are of no value in this connection, since one cannot interpret the adaptive characteristics of extinct species on the basis of them. We are, therefore, forced to rely on comparisons between related groups of contemporary plants.

Four such examples have been chosen for analysis, largely because of my own familiarity with them. One of these, *Antennaria,* is a specialized and recently evolving genus of the advanced family Compositae. The second genus, *Ceanothus,* is somewhat less specialized, though still in an active state of evolution. Of the two woody families selected, the Hydrangeaceae have an intermediate phylogenetic position, whereas the Dilleniaceae are relatively primitive or archaic.

The postulated phylogeny of the four groups is presented in Figs. 9-1 to 9-4. The representation is in a form designated here an *ecophyletic cross-sectional chart.* It is an elaboration of a kind of chart that has been used by several botanists, including Babcock (1947) in his monograph on *Crepis* and me in an article on the Gramineae (Stebbins 1956*a*), and fully elaborated by Sporne (1956). This method is preferred to the conventional phylogenetic "tree," since the actual ancestors of the groups concerned are not represented by fossils and the contemporary species are nearly all characterized by some relatively advanced characteristics and other traits that are more primitive and unspecialized. The overall degree of advancement of each species or species group is indicated in a generalized fashion by its distance radially from a star, placed in the center of the diagram.

In order to emphasize adaptive radiation, the generalized habitat or habitats of each species or species group are represented on the chart by appropriate symbols. The principal character differences that have served as the basis for estimating phylogenetic advancement are represented by dividing lines of a distinctive character.

The genus Antennaria

The first genus to be discussed, *Antennaria* (Fig. 9-1), belongs to the large family Compositae, tribe Inuleae, subtribe Gnaphalinae. The largest genus of this subtribe is *Helichrysum,* containing 300 species that are adapted to the semiarid and desertic regions of Eurasia, Africa, and Australia. With respect to floret structure, *Helichrysum* is the least-specialized

Fig. 9-1. Ecophyletic chart showing the principal sexual species of the genus *Antennaria.* (From Stebbins 1972.) The species represented by the numbers are as follows:

1. Geyeri
2. Luzuloides
3. Argentea
4. Carpathica, anaphaloides, pulcherrima
5. Dioica group (dioica, virginica, microphylla, corymbosa)
6. Plantaginifolia group (plantaginifolia, parlinii, racemosa)
7. Alpina group (alaskana, media, monocephala, reflexa)
8. Suffrutescens
9. Neglecta
10. Solitaria
11. Dimorpha, flagellaris, rosulata

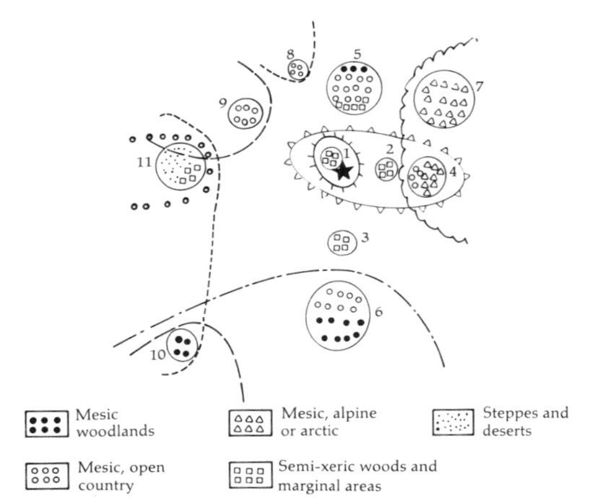

List of characters. Advanced states are listed first.

Leaves dimorphic vs monomorphic
Stolons present vs absent
Basal leaves large, several-nerved vs small, 1-nerved
Stolons long, creeping vs short, erect
Heads solitary vs several
Stems very short vs well developed
Phyllaries dark vs pale

genus of Gnaphalinae. The dense and woolly indumentum that covers the stems and leaves of species in all genera of the subtribe Gnaphalinae consists of single cells that may reach a length of 6–7 mm. Such extreme developmental specialization of cells could be expected to evolve only in response to strong selective pressure.

The distinctive characteristics of *Antennaria* compared with its immediate relatives, such as *Gnaphalium,* are the presence of stolons and the marked dimorphism of leaves that occurs in many of its species. These are, therefore, the principal specialized characters upon which the advanced condition of the species was based. There is a central group of two species in which horizontal stolons are not formed and the involucral phyllaries are pale in color. One of these is *A. Geyeri,* which most resembles other genera of Gnaphalinae in having basal leaves and stem leaves similar to each other in size and shape. Both it and *A. luzuloides* are adapted to dry, open woodlands or open country in the semiarid climate of the western United States. The other three species that lack stolons, *A. carpathica, A. anaphaloides,* and *A. pulcherrima,* all have dark-brown or black involucral phyllaries and denser tomentum. They are adapted to mesic alpine or subalpine habitats.

Among the stoloniferous species, all of which have the stem leaves and basal leaves strongly differentiated from each other, only two of those shown, *A. microphylla* and *A. argentea,* are adapted exclusively to semiarid habitats. Three others, *A. dimorpha, A. flagellaris,* and *A. rosulata,* are more xeric in their adaptation, having become adapted to *Artemisia* steppes, associated chiefly with a dwarf, cushionlike growth habit. The remaining thirteen are adapted to more mesic habitats. They include the complex of *A. alpina*, characterized by low stature and dark-brown or black phyllaries and adapted to alpine meadows or arctic tundra, and several groups that have independently become adapted to open fields or dry woods in mesic climates. With respect to both complexity of growth pattern and divergence from other species of Gnaphalinae, the most specialized species of *Antennaria* is *A. solitaria*. It has long, prostrate stolons and large basal leaves that contrast strongly with the reduced cauline leaves and are differentially tomentose, being essentially glabrous above and heavily tomentose beneath; and its inflorescence is reduced to a single capitulum. It inhabits rich woods in the mesic climate of the eastern United States.

Ceanothus, subg. Euceanothus

The second genus selected, *Ceanothus*, subg. *Euceanothus*, consists almost entirely of shrubs, as do the great majority of other genera belonging to its family, the Rhamnaceae. The great bulk of the modern species are adapted to the semiarid, Mediterranean-type climate of California. It has been well monographed by McMinn (1942). For the sake of simplicity, only the larger of its two subgenera, *Euceanothus*, is included in Fig. 9-2. Its species are all closely related to one another, and hybrids between them are frequent. In the opinion of Nobs (1963), they might all be regarded as subspecies of a single, highly polymorphic genetical species. On the

Fig. 9-2. Ecophyletic cross section of *Ceanothus*, subg. *Euceanothus*. (From Stebbins 1972.) The species represented by the numbers are as follows:

1. Ochraceus
2. Buxifolius, cordulatus, depressus, Fendleri, leucodermis
3. Incanus
4. Dentatus, foliosus, impressus, Lemmonnii, papillosus
5. Diversifolius
6. Spinosus
7. Integerrimus, Palmeri
8. Microphyllus, serpyllifolius
9. Ovatus
10. Velutinus
11. Americanus, sanguineus, Martinii
12. Arboreus, coeruleus, oliganthus, sorediatus, tomentosus
13. Parryi
14. Cyaneus, griseus, thyrsiflorus

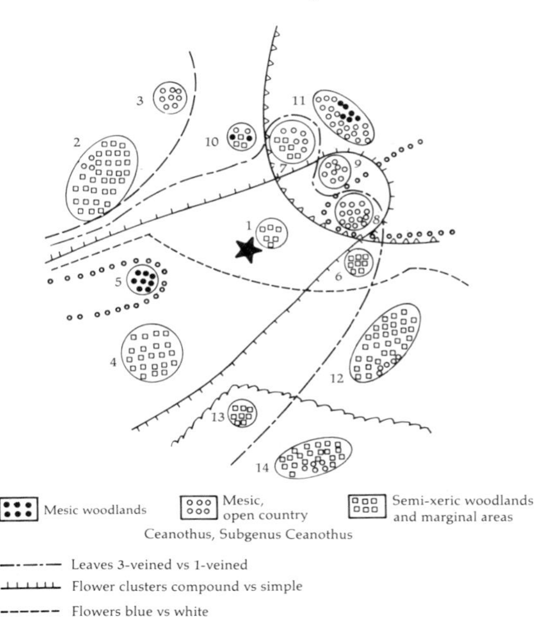

other hand, my observations in several localities indicate that most of the hybrids between sympatric species of *Euceanothus* exist for only short periods of time in disturbed habitats, and that under stable ecological conditions sympatric species may remain edaphically isolated from one another and thus retain their identity for indefinite periods of time.

I have evaluated the character states on the following basis. The prostrate growth habit is an obvious specialization that has arisen only occasionally. Rigid branches that end in spines have been evolved repeatedly in the Rhamnaceae, and are one adaptation against browsing predators in a climate where recovery from browsing is slow, owing to the rigors of periodic drought. Branchlets that are angular in cross section are peculiar to a few species of subg. *Euceanothus,* and are not found elsewhere in the Rhamnaceae. The three-veined leaf, with essentially palmate venation, is also regarded as a more specialized condition than typical pinnate venation, since it is found elsewhere in the Rhamnaceae only in relatively specialized genera, such as *Paliurus* and *Zizyphus.* Deciduous leaves are justly regarded by botanists as specialized conditions, associated with the evolution of specific physiological mechanisms involving particular hormones. The compoundly cymose inflorescences found in the larger species of subg. *Euceanothus* are unique to the subgenus, being absent from other genera of Rhamnaceae, and are not paralleled in genera of the related and perhaps more primitive family Celastraceae. They are therefore to be regarded as secondary aggregations of simple corymbs. Finally, blue flowers are unknown elsewhere in the Rhamnaceae, and must therefore be regarded as specialized in the genus *Ceanothus.*

On the basis of these considerations, the most generalized living species of *Ceanothus* is *C. ochraceus,* endemic to the semiarid mountains of northern Mexico. The great majority of the remaining species, having intermediate degrees of specialization, are endemic to the semiarid Mediterranean-type climatic region of California. Several species have, however, become adapted to more mesic climates. Each of these has acquired certain specializations. *Ceanothus incanus,* adapted to the redwood border forest, retains the spiny branches of its more xeric relatives, *C. leucodermis, C. cordulatus,* and *C. fendleri,* and differs from them chiefly in its more vigorous growth, larger leaves, and more numerous flowers. An obvious specialization of *C. diversifolius,* adapted to dense forests in the central Sierra Nevada, is its prostrate growth habit. In the same habitat occurs *C. integerrimus,* which has compound inflorescences and three-veined, deciduous or partly deciduous leaves. The most widespread of the truly deciduous species, *C. ovatus, C. americanus,* and *C. sanguineus,* are all adapted to mesic climates in the eastern United States and the Pacific Northwest.

The second subgenus of *Ceanothus, Cerastes,* contains a larger number of xeric adaptations than subg. *Euceanothus,* particularly the thick leaves provided with elaborate stomatal crypts (Nobs 1963). It has not evolved any species adapted to truly mesic habitats. If the entire genus is considered, therefore, *Ceanothus* agrees with

Antennaria and *Crepis* in having evolved both more xeric and more mesic species than the original progenitors.

In my opinion, the two examples chosen are representative of the majority of higher plant genera that occupy many different kinds of habitat. Those that have evolved in only one direction, either toward more xeric or toward more mesic adaptations, have been in the minority. These trends are most easily seen in families that are relatively specialized, and that have been evolving actively in more recent geological epochs, such as Compositae, Labiatae, Scrophulariaceae, Dipsacaceae, Cruciferae, Polygonaceae, and Gramineae. These modern families have been expanding in a climatic cycle characterized by increasing aridity, increasing extent of arid and semiarid regions, and diminishing area of mesic forests, particularly in tropical and subtropical climates. Under such conditions, one would expect that xeromorphic adaptations would in general have greater survival value than adaptations to more mesic conditions. The fact that in genera such as *Ceanothus* and *Antennaria,* as well as in many other genera of Compositae such as *Helianthus, Solidago, Achillea, Cirsium, Taraxacum, Scorzonera,* and *Tragopogon,* certain groups of species have evolved in a direction counter to the prevailing climatic trend and have produced species having more mesic adaptations is evidence suggestive of a widespread trend of evolution in angiosperms from more xeric to more mesic adaptations. The analysis by Meusel (1965) of the distribution of *Fraxinus* suggests that it has taken place in that well-known genus of woody plants. This trend is hard to detect in the more primitive families; the difficulty is probably due to the relictual nature of their genera. According to the arguments presented earlier in this chapter, we would expect to find in older genera a differential extinction of their more xeric derivatives, and preferential survival of those adapted to more mesic climates. Indications that some families of Magnoliales had ancestors that were adapted to more xeric conditions than those in which the modern forms occur are presented in Chapter 10.

Adaptive Radiations of Genera Within Families

Examples of adaptive radiation at the level of generic differentiation within a family are hard to find, for two reasons. One is that monographers have not, as a rule, paid much attention to differences between genera with respect to ecological relations. The other reason is that, as has already been mentioned repeatedly, most of the larger genera of flowering plants exhibit a wide range of ecological differentiation. Nevertheless, examples of ecological radiation at the generic level undoubtedly exist. Indications of them can be seen in most of the tribes of Compositae. In addition to the Inuleae, the Helianthieae, the Astereae, and the Cichorieae are good examples. In the Heliantheae, the genus that is least specialized with respect to reproductive characteristics is *Wyethia,* which is adapted to the semiarid climate of the western United States. Among the genera of Heliantheae having more specialized reproductive structures are mesophytes such as *Helianthus* and *Rudbeckia,* as well as extreme xerophytes such as *Encelia, Bebbia, Sanvitalia,* and *Parthenium.* In the tribe Astereae, the central genus is *Haplopappus,* which in some of its subgenera (*Pyrrocoma, Tonestes, Stenotopsis*) includes species that are at the same time the most generalized in reproductive structures of any members of the Astereae, and also show the closest connections between the Astereae and other tribes. These subgenera are all adapted to semiarid climates. More specialized genera of Astereae include *Solidago* and *Aster,* which are predominantly mesic, as well as xerophytes such as *Gutierrezia, Amphipappus, Acamptopappus,* and *Chrysothamnus.*

The situation in the tribe Cichorieae has been discussed by me in earlier publications (Stebbins 1952, 1953) and by Jeffrey (1966). The least-specialized genus of this tribe is *Tolpis,* which is adapted primarily to a semiarid climate. Predominantly mesic, more specialized genera are *Lactuca, Sonchus, Prenanthes, Hieracium,* and *Tragopogon,* whereas greater xerophytism has been associated with more specialization in *Chondrilla, Stephanomeria, Chaetadelpha,* and *Lygodesmia.*

A cursory survey of genera in several of the more advanced families of flowering plants, such as Labiatae, Scrophulariaceae, Bignoniaceae, Hydrophyllaceae, Polemoniaceae, Polygonaceae, and Gramineae, suggests that more careful study will reveal similar patterns in them also. Although in many of them the genera most often regarded as primitive occur in mesic habitats, these genera are usually small, and so distantly related to the larger genera of their families that they must be regarded as relictual.

These families have been mentioned first because their relatively advanced condition, and perhaps the more recent origin of their genera, indicates that evolutionary trends within them might be more easily traced on the basis of modern forms than they could be in the older, more primitive families, of which a high proportion of their more primitive genera are probably extinct. Nevertheless, ecophyletic cross-sectional ecological charts have been attempted for two woody families that are relatively unspecialized, the Hydrangeaceae and the Dilleniaceae.

The Family Hydrangeaceae

The Hydrangeaceae were selected because they are entirely woody, they are well-enough defined that most taxonomists interested in phylogeny agree on their limits, and they occupy a wide range of ecological habitats. Furthermore, I am reasonably familiar with all of the genera that are regarded as primitive. The character differences were obtained chiefly from the synopsis of Hutchinson (1967).

The sixteen genera of this family are shown in Fig. 9-3. The smallest number of specialized characters (one out of ten) is found in *Carpenteria,* a monotypic genus confined to a restricted portion of the semiarid foothills of the Sierra Nevada in California. The genus *Fendlera,* which is second in this respect, has three species, which occupy similar habitats in the southwestern United States and northern Mexico. The monotypic *Jamesia,* with three specialized characters out of ten, occurs in dry forests of the southwestern United States, but extends also to subalpine and alpine cliffs. The large genus

Philadelphus contains many species with deciduous leaves that inhabit mesic forests in temperate regions. A whole group of small-leaved species inhabits the arid southwestern United States, and another group, regarded by Hu (1954–1956) as the least specialized in the genus, is confined to tropical and subtropical Mexico and Central America, in both mesic and semiarid situations. The remaining twelve genera are all more specialized and occur exclusively in mesic habitats.

The Family Dilleniaceae

The Dilleniaceae (Fig. 9-4) were selected because they are the most primitive family of angiosperms that exhibits a wide range of ecological preferences. Other families of apocarpus woody angiosperms either are very small, or, as in the Magnoliaceae, Winteraceae, Annonaceae, and Monimiaceae, occur only in mesic

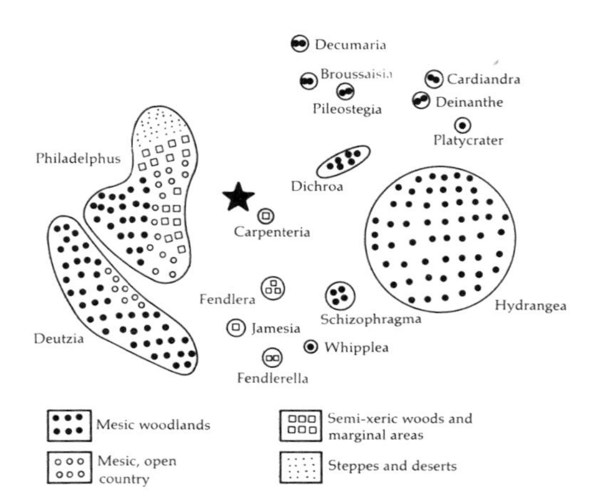

Fig. 9-3. Ecophyletic chart of the family Hydrangeaceae: list of characters. Advanced states are listed first. (From Stebbins 1972.)

1. Stems herbaceous, trailing or twining vs. shrubby or arboreal
2. Leaves deciduous vs. persistent
3. Inflorescence bracteate and cymose or paniculate vs. leafy and corymbose or flowers solitary
4. Showy or otherwise differentiated marginal sterile flowers present vs. absent
5. Perianth 4-merous or 7 to 10-merous vs. 5-merous
6. Stamen number = no. perianth parts or less, vs. stamens more numerous
7. Ovary fully inferior, vs. ovary superior or half inferior
8. Styles united vs. styles free
9. Ovary locules and/or styles ("carpels") fewer in number than petals, vs. equal in number
10. Ovary with one locule and parietal placentation vs. several locules and axial placentation
11. Ovules 1–2 per locule vs. more numerous

Advanced states found in the genera. Numbers in parentheses refer to characters found only in some of the species of the genus.

1. Carpenteria, 1 sp.: 8
2. Philadelphus, 50–60 spp.: 2, 5, (7), (8)
3. Jamesia, 1 sp.: 2, 6, 9, 10
4. Fendlera, 3 spp.: 2, 5, 6
5. Deutzia, 40 spp.: 2, 3, 7, 9
6. Fendlerella, 2 spp.: 2, 3, 6, 7, 11
7. Whipplea, 1 sp.: 1, 3, 6, 9, 11
8. Deinanthe, 2 spp.: 1, 2, 3, 4, 8
9. Cardiandra, 2 spp.: 1, 2, 3, 4, 7, 9, 10
10. Platycrater, 1 sp.: 1, 2, 3, 4, 5, 7, 9
11. Hydrangea, 80 spp.: (1), 2, 3, 4, (5), 6, 7, (9)
12. Schizophragma, 4 spp.: 1, 2, 3, 4
13. Dichroa, 9 spp.: 3, 6, (9), 11
14. Pileostegia, 2 spp.: 3, (5), 6, 7, 8
15. Broussaisia, 2 spp.: 3, 6, 7, 8
16. Decumaria, 2 spp.: 1, 2, 3, 7, 8

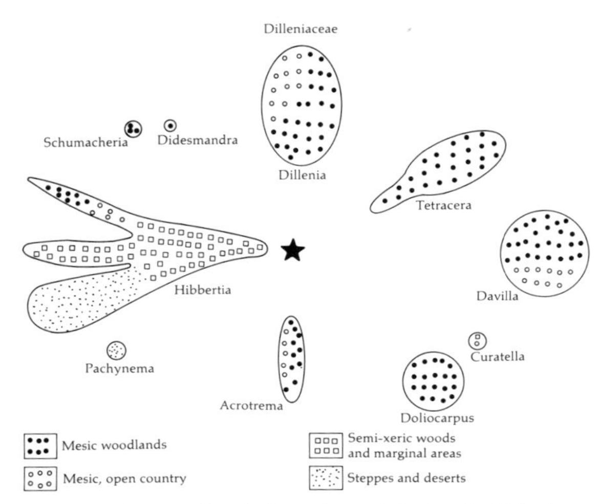

Fig. 9-4. Ecophyletic chart of the family Dilleniaceae: list of characters. Advanced states are listed first. (From Stebbins 1972.)

1. Herbaceous vs. woody
2. Plants climbing (lianas) vs. not climbing
3. Simple vessel perforations present vs. absent
4. Xylem rays of type IIA present vs. absent
5. Xylem rays uniseriate, homogeneous vs. heterogeneous
6. Multilacunar nodes present vs. absent
7. 1-lacunar nodes present vs. absent
8. Leaves scale-like vs. normal in appearance
9. Inflorescence spicate, racemose or paniculate vs. cymose and leafy bracted or flowers solitary
10. Inflorescence lateral or axillary vs. terminal
11. Flowers asymmetrical vs. actinomorphic
12. Stamens fewer or as many as perianth segments vs. stamens more numerous
13. Carpels fewer than petals vs. the same number or more numerous
14. Ovules 1–2 per carpel vs. ovules more numerous
15. Calyx enlarged and fleshy in fruit vs. not so
16. Carpels united at base vs. completely free
17. Sepals unequal (dissimilar) vs. equal and similar to each other
18. Nonmethylated flavonols present vs. absent
19. Methylated flavanols present vs. absent

habitats. Information on the Dilleniaceae has been obtained in part from the treatment by Gilg and Werdermann (1925), and in part from the studies of Dickison (1967, 1968, 1969) and Kubitzki (1968).

The most remarkable feature of the Dilleniaceae is the extraordinary diversity of its largest genus, *Hibbertia,* with respect to both growth habit and floral structure. There is probably no other genus of angiosperms that exhibits such a high degree of variation in those floral characteristics that are often regarded as "fundamental" and are usually associated with the separation of genera or even higher categories, and at the same time is equally diverse with respect to the size, growth habit, and ecological adaptations of its species. The majority of its species are, however, low-growing shrubs with slender stems and small to medium-sized, more or less xeromorphic leaves. Some of them, belonging to the section *Cyclandra* and inhabiting the western and southern portions of Australia, have a primitive combination of characters with respect to wood and nodal

Advanced states found in the genera. Numbers in parentheses refer to characters found only in some of the species of the genus.

1. Tetracera, 30 spp.: (2), 3, 4, 6, 9, (13), (14), (18), (19)
2. Davilla, 35 spp.: (2), 3, 4, 6, 9, (10), 13, 14, 15, 17, 18
3. Curatella, 2 spp.: 3, 4, 6, 9, 13, 14, 16, 18
4. Doliocarpus, 20 spp.: 3, 4, 6, 9, 10, 13, 16, 18
5. Hibbertia, 110 spp.: (2), (7), (8), (9), (10), (11), (12), (13), (14), (16), (18)
6. Pachynema, 4 spp.: 5, 7, 8, 10, 12, 13, 14, 18
7. Acrotrema, 12 spp.: 1, 5, 13, (16), (18), (19)
8. Schumacheria, 3 spp.: 2, 6, 9, 10, 11, 13, 14, 18
9. Didesmandra, 1 sp.: 6, 9, 11, 12, 13, 14, 18?
10. Dillenia, 55 spp.: 3, 5, (10), (14), 15, 16, (18), (19)

anatomy, leaf venation, inflorescence (flowers solitary or in few-flowered leafy corymbs), and floral parts (five sepals, five petals, many stamens regularly distributed, five separate carpels with two or more ovules per carpel, dehiscing ventrally). Other species are more highly specialized with respect to leaves, inflorescences, floral parts, or all of these, although all of those investigated apparently have relatively unspecialized wood and nodal anatomy (Dickison 1967, 1969). The specializations of their leaves are adaptations to the dry climate of Australia; those of their flowers will not be understood until careful investigations have been made of their pollination biology. The small genus *Pachynema* of North Australia is an extreme xeromorphic derivative of *Hibbertia.*

The isolated, relictual genus *Acrotrema,* found in Ceylon and Malaya, has a smaller number of specialized characteristics than any other genus of the family except the more unspecialized species of *Hibbertia,* in spite of its herbaceous growth habit. It is a distinct mesophyte.

The remaining seven genera consist chiefly of trees, climbing shrubs, or lianas, and nearly all of them are confined to moist tropical forests. They have intermediate to large numbers of specialized characters, the largest number being in the genus *Davilla,* which occurs chiefly in Brazil.

The Dilleniaceae provide an admirable example of ancient adaptive radiation from intermediate habitats toward both more xeric and more mesic adaptations. Other tropical and subtropical families, if they are carefully studied both morphologically and ecologically, may provide similar examples. I have seen suggestive indications of this nature in the Ochnaceae, Guttiferae (Clusiaceae and Hypericaceae), Capparaceae, Tiliaceae, Sterculiaceae, Mimosoideae, Caesalpinoideae, Rutaceae, Oleaceae, and Apocynaceae.

Reversals of Adaptation Within an Evolutionary Line

The examples just presented show that an ancestral form adapted to an intermediate habitat can give rise by adaptive radiation to descendants some of which are adapted to more xeric and others to more mesic conditions than their ancestors. This being the case, one might expect that in many evolutionary lines exposure over sufficiently long periods of time first to more xeric and later to more mesic conditions would elicit reversals in the direction of adaptive evolution. The success of such reversals could often depend on the acquisition of more efficient reproductive mechanisms during the period of adaptation to severe conditions.

Because of the long time during which they took place, detection of such sequences of reversal is very difficult in the absence of a fossil record. Nevertheless, the geographic and ecological distribution patterns of some modern groups of species suggest strongly that they have had such an evolutionary history.

Examples in the Compositae, tribe Cichorieae

One of the best examples known to me is in the Compositae, tribe Cichorieae. The two New World genera *Microseris* and *Agoseris* resemble each other in size

Fig. 9-5. Chart showing adaptive reversal in the evolution of specialized species of *Agoseris* from *Microseris*-like ancestors. (From Stebbins 1972.)

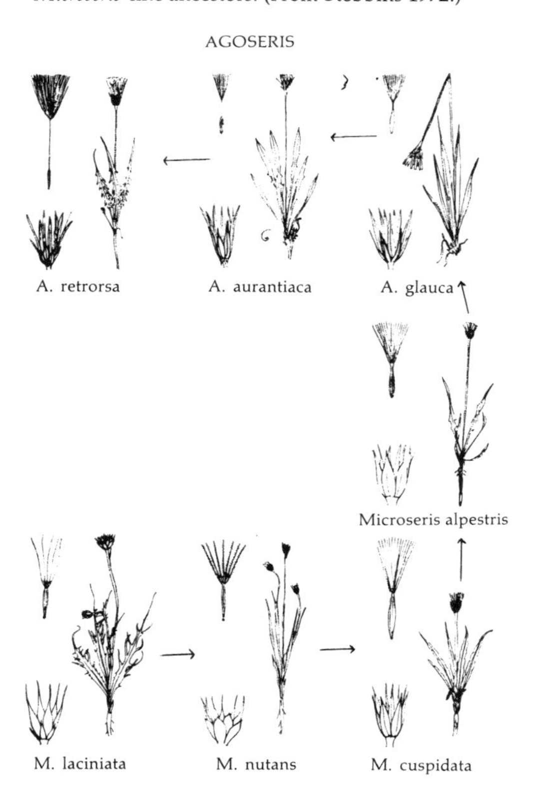

and leaf outlines, in having large capitula, yellow flower color, relatively short, broad style branches, and a peculiar orange color of their pollen, and in the number ($x = 9$), size, and morphology of their chromosomes (Stebbins 1953, Stebbins, Jenkins, and Walters 1953, Chambers 1955). They are clearly more closely related to each other than either of them is to any other of the larger genera in the tribe. For both of them, the center of distribution is Pacific North America, and, in the dry forests of northern California, species of the two genera are regularly sympatric.

The extreme forms—on the one hand *Microseris laciniata* and *M. paludosa* and on the other hand *Agoseris grandiflora* and *A. retrorsa* (the annual species *A. heterophylla* is somewhat apart)—differ principally with respect to their stems, achenes, and pappi. In all of these respects, *Agoseris* is distinctly more specialized than *Microseris* (Fig. 9-5). Nevertheless, as shown in Table 9-1, these extreme forms are connected to each other by a more or less continuous series of intermediates.

The habitats occupied by the members of this series are of particular interest. *Microseris laciniata* and *M. paludosa* have the primitive characters: leafy, branched stems, beakless achenes, and a paleaceous pappus of which the segments resemble most their homologues, calyx lobes. They occupy mesic woodlands or open meadows along the Pacific coast, in a climate that is cool in summer and mild in winter, with a long growing season. From species essentially similar to these, though probably with a different kind of pubescence (Chambers 1955), three lines of adaptive radiation can be detected. One of them led to a group of annuals inhabiting the warm, dry regions of California that developed leafless, unbranched flowering stems, but retained the paleaceous pappus. A second line became adapted to moister habitats, and also developed unbranched, leafless stems. Its termini are two rather isolated, swamp- or bog-inhabiting species: *M.* (*Apargidium*) *borealis*, which has a capillary pappus and occurs along the coast from northern California to Alaska, and the epappose *Phalacroseris Bolanderi* of the Sierra Nevada.

The third and most successful radiating line is the most interesting one for this discussion. It starts with *Microseris nutans*, which occurs at high altitudes in the mountains and differs from *M. laciniata* in its depressed stature and short lower internodes. Next come two similar species,

Table 9-1. Morphological characteristics and ecological distribution of seven species groups of the *Microseris-Agoseris* complex (Compositae, Cichorieae). The species all resemble one another in size, indumentum, involucral phyllaries, hairs on corolla, styles and stigmas, pollen mass, and chromosome number ($2n = 18$).

Species group	Life cycle	Stems	Achenes	Pappus	Habitat
Microseris laciniata, paludosa, sylvatica	Perennial	Branched, leafy	Beakless, 4–9 mm long	5–10 paleae	Dry woods, meadows, mild temperate
Nothocalais troximoides	Perennial	Unbranched, leafless	Beakless, 8–10 mm long	10–30 paleae	Dry steppes and open woods, cold temperate
Nothocalais alpestris	Perennial	Unbranched, leafless	Beakless, 5–8 mm long	30–50 bristles, paleaceous at base	Meadows and dry slopes, subalpine-alpine
Agoseris glauca	Perennial	Unbranched, leafless	With short, stout beak, body 5–8 mm	More than 50 coarse bristles	Dry meadows, rocky slopes, cold temperate
Agoseris elata, apargioides	Perennial	Unbranched, leafless	With slender beak twice length of body, body 3–8 mm	Very many fine bristles	Meadows and open slopes, cold or mild temperate
Agoseris grandiflora retrorsa	Perennial	Unbranched, leafless	Slender beak, 2–4 times body, body 4–7 mm	Very many fine bristles	Dry woods, open slopes, meadows, warm, mild, or cold temperate
Agoseris heterophylla	Annual	Unbranched, leafless	Slender beak, twice length of body, body 3–5 mm	Very many fine bristles	Meadows, open slopes, steppes, etc., warm, mild, cold temperate

M. cuspidata and *M. troximoides,* which are widespread in the cold, arid steppes of the Great Basin, the Rocky Mountain region, and the western Great Plains. Probably in response to a short growing season, they have evolved an unbranched, leafless scape bearing a single capitulum, a condition that persists in all of the remaining members of this evolutionary line. In response to the great adaptive advantage of wind dispersal in the open, windy places where they grow, they have evolved a pappus consisting of many slender, papery paleae, which offer a large amount of wind resistance. They are closely related to *M. alpestris,* which differs only in its somewhat smaller size and in having more numerous pappus parts, which appear like bristles but have broadened, palealike bases. The habitat of *M. alpestris* is more mesic, but, being alpine or subalpine, it has an even shorter growing season. All of the differences between the two species could be explained as adaptations to this different habitat; capillary pappus bristles, having fewer cells than paleae, can develop more rapidly. *Microseris alpestris* has been placed by various well-recognized authorities in any one of the three genera *Microseris* (Quentin Jones), *Nothocalais* (Chambers), or *Agoseris* (Greene and Gray, as *Troximon*).

The most primitive species of *Agoseris, A. glauca,* occupies a geographic and ecological distribution about equal to that of *Microseris troximoides* and *M. cuspidata* combined. It has three subspecies, and includes both diploid and tetraploid cytotypes. It differs from *M. alpestris* chiefly in its completely capillary pappus and in having achenes bearing a short, stout beak. Next come a group of three species, all of them more mesic in habitat than *A. glauca,* and differing from that species in the longer, more slender beaks of their achenes. Two of them, *A. aurantiaca* and *A. elata,* inhabit mountain meadows; the third, *A. apargioides,* occurs along the mild, foggy coastal strip of California, being therefore sympatric with *M. laciniata.* The ability of its ancestors to descend from the mountains and reinvade the coastal region may well have been based upon the production of a very large number of small, light, wind-borne achenes. The two remaining species groups of *Agoseris,* which probably evolved from ancestors resembling *A. apargioides,* are the perennial *A. grandiflora* and *A. retrorsa,* and the annual *A. heterophylla.* The former two species are distinctive in their very large heads, containing numerous florets, and long-beaked achenes. They have reached the ultimate degree of seed production and ease of dispersal and of establishment, since *A. grandiflora* tends to become weedy. They occur sympatrically with *M. laciniata,* though in drier habitats. *Agoseris heterophylla* is the smallest and most ephemeral of all of the species in this complex of genera. These most advanced groups have acquired a very wide distribution, both geographically and ecologically.

In summary, the most-specialized species of *Agoseris* occur sympatrically with the least-specialized species of *Microseris,* but the former are much more common and widespread than the latter. The distribution of the intermediate species groups suggests that the specialized characteristics of *Agoseris* evolved in response to the strong selective pressures for rapid development and efficient seed dispersal that were exerted in the severe steppe and arid montane habitats where these intermediate species occurred, followed by selection for fecundity under more favorable conditions.

In the Old World, the evolution of the genus *Taraxacum* from its probable ancestors in the genus *Launaea* probably took a similar course. The most primitive species of *Taraxacum,* which resemble *Agoseris glauca* in both growth habit and the morphology of their achenes, occur in semiarid mountain slopes in the Middle East (Handel-Mazetti 1907). They are linked to the most advanced species, such as the common dandelion (*T. officinale*) of western Eurasia and the circumpolar *T. ceratophorum* complex, by species such as *T. kok-saghyz,* which inhabit moist mountain meadows in central Asia and so are both morphologically and ecologically comparable to *Agoseris elata* and *A. aurantiaca.*

Polygonaceae, tribe Eriogoneae

In the subalpine timber belt of the Sierra Nevada of California, two very different representatives of the family Polygonaceae grow sympatrically in similar situations. Both of them are subshrubby perennials growing on dry soils. The first group consists of two species belonging to the genus *Polygonum, P. Davisiae* and *P. shastense.* They have leaves with well-developed leaf sheaths, in common with most other genera of Polygonaceae, and clusters of small white or greenish flowers in the axils of the leaves.

Side by side with these two species of *Polygonum* grow several subshrubby species belonging to the genus *Eriogonum,* particularly *E. marifolium* and *E. umbellatum.* In their superficial appearance, these species differ so widely from *Polygonum* that a careful examination of their small flowers and achenes is necessary to reveal the fact that they belong to the family Polygonaceae. Their elliptic or oval, long-petioled leaves emerge from nodes that lack the scarious sheaths that are so characteristic of other genera of the family. Their yellow flowers are densely clustered in bell-shaped involucres formed by "fusion" of the upper floral bracts. The involucres, in turn, are on the ends of naked peduncles, most of which emerge from the axils of a whorl of reduced foliage leaves. Further elevation of the inflorescence occurs through extensive elongation of one or more internodes between the upper whorls of leaves.

The flowers of *Eriogonum* are much like those of *Polygonum* except that they consistently have 6 sepals and 9 stamens, whereas those of *Polygonum Davisiae* and *P. shastense* have 4, 5, or 6 sepals and 7, 8, or 9 stamens, 5 sepals and 8 stamens being the modal number. In addition, *E. umbellatum, E. marifolium,* and other species of subg. *Eriogonum,* which are the ones most often found in the timber belt along with the perennial *Polygonum* species, have perianths that are specialized through "fusion" of the basal portions of the sepals.

Clearly, the species of *Eriogonum* subg. *Eriogonum* are much more specialized than the subshrubby species of *Polygonum,* and have diverged much further from the common ancestor of the two groups. Hence one naturally asks the question: Under what conditions did this greater divergence and specialization occur? Unfortunately, only a partial answer to this question can be given, since a great morphological gap exists between *Eriogonum* and its relatives on the one hand, and all other genera of Polygonaceae on the other. Nevertheless, such intermediate conditions as still exist occur in species that are adapted to much more xeric conditions than those occupied by species of either *Polygonum* or *Eriogonum* subg. *Eriogonum.* In the monotypic annual *Nemacaulis denudata,* found along the semiarid seacoast of southern California and in the Colorado Desert, the flowers and their subtending bracts form a capitate cluster, superficially similar to the involucre of *Eriogonum,* but the bracts are not at all united. In another monotypic annual, *Hollisteria lanata,* found in the arid interior valleys of south central California, the bracts are partly united, but they do not form a true involucre. These two monotypic genera may be regarded as relictual, annual derivatives of the extinct perennials that formed the link between the original stock of the Polygonaceae and the genus *Eriogonum* and its relatives.

Within the genus *Eriogonum* itself, the distribution of the species suggests that the most generalized ones occupy more xeric habitats than do those that are the most specialized (Reveal 1969). The perennial species that has the involucres least strongly developed is *E. intrafractum,* which is endemic to the desert mountains surrounding Death Valley. Of the two subgenera that contain the bulk of the perennial species, subg. *Oregonium,* in which the segments of the calyx are free from one another, is less specialized than subg. *Eriogonum,* in which they are united at its stipitate base. The species of subg. *Oregonium* occur in a wide range of habitats, including warm deserts, cold steppes, and dry chaparral. Those of subg. *Eriogonum,* which includes *E. marifolium* and *E. umbellatum,* do not occur at all in the warm deserts and only rarely in dry chaparral. The majority of them are found in the drier portions of the coniferous forests.

These facts of distribution suggest the following interpretation. The divergence of *Eriogonum* subg. *Eriogonum* from a *Polygonum*-like ancestral group of Polygonaceae occurred in two stages. The first was a series of reductions and condensations associated with adaptation to arid conditions. They included the loss of the scarious stipular sheaths that normally characterize the family, the aggregation of the leaves into whorls, the condensation of the inflorescences into capitate clusters, and the "fusion" of the floral bracts to form the involucre. These changes were basic to the evolution of the original species of *Eriogonum.* Within the genus, adaptive radiation into the semixeric forest belt was associated with the development of elongated internodes below the whorled floral leaves, the further development of the involucres so that in the most advanced species (*E. ursinum, E. marifolium*) the free portions of the bracts are reduced to teeth, and the basal "fusion" of the perianth segments.

This example resembles the previous one, since in both examples the group that has undergone the greatest amount of reduction and specialization of its floral structures has had a past evolutionary history of adaptation to a more severe climate. In the Cichorieae, these severe conditions were those of steppes and montane and alpine situations; in the Polygonaceae, aridity alone was the determining factor.

Heliantheae–Ambrosiinae

Two additional examples from the Compositae both concern the sunflower tribe (*Heliantheae*). In the eastern United States, representatives of several genera of this tribe (*Helianthus, Rudbeckia,* and others) occur sympatrically with ragweed (*Ambrosia artemisiifolia, A. trifida*) and cocklebur (*Xanthium* spp.). Both of the latter genera belong to the subtribe Ambrosiinae, the species of which are superficially so different from the true sunflowers that some botanists have placed them in a separate family. In association with their wind pollination, they have lost their conspicuous ray flowers, and the corollas that remain are greenish and inconspicuous. Moreover, the flowers have become unisexual and separated into "unisexual" capitula that are strikingly dimorphic. The staminate capitula are arranged in a terminal raceme and have flat, unarmed phyllaries, whereas the pistillate or female capitula, situated at the base of the raceme, contain phyllaries that are "fused" and bear spines or hooks.

The genera intermediate between typical Heliantheae and the most advanced Ambrosiinae (*Parthenium, Iva, Hymenoclea*) all occur in the deserts of northern Mexico and the southwestern United States. Moreover, the center of distribution of *Ambrosia* as now recognized (including the formerly separate genus *Franseria*) is in these same deserts. Only the annual ragweeds mentioned above occur in the mesic regions of eastern North America. Most probably, therefore, they are secondary mesophytes, having been derived from xerophytic ancestors. *Xanthium* is connected to *Ambrosia* via two species of the southwestern deserts, *A. ambrosioides* and *A. ilicifolia,* which have large burs covered with numerous spines. Its species, therefore, are also secondary mesophytes. Thus, this example closely resembles the one in the Polygonaceae, in that the more specialized mesophytes were derived from the ancestral stock via intermediate forms that were adapted to an arid climate.

Heliantheae, subtribe Madiinae

The final example is essentially similar. In the forests of the Sierra Nevada and the Coast Ranges of northern California various genera of Heliantheae (*Wyethia, Balsamorhiza, Helianthella,* and others) occur sympatrically with two perennial species of the genus *Madia, M. Bolanderi* and *M. madioides.* These species are definitely mesophytes. The habitat of *M. Bolanderi* is damp mountain meadows and stream banks; that of *M. madioides* is shady forests of *Sequoia* or *Pseudotsuga.* Nevertheless, the distribution of the tribe Madiinae as a whole, particularly its few perennial species, is basically xeric. The most primitive species related to *Madia* is probably the subshrubby *Adenothamnus validus,* which is localized in a desert region of Baja California. In their morphological characteristics, particularly the protection of the achenes by strongly enveloping bracts and the dense glandular pubescence that in most species covers the entire plant, the species of *Madia* reflect their xeric origin. This, then, is probably another example of specialization through initial adaptation to xeric conditions, followed by later reversion of two species to a mesic habitat.

The widespread occurrence of secondary reversions from xeric to mesic habitats

Since many botanists have almost intuitively concluded that species or genera having xerophytic specializations rarely or never give rise to descendants that are adapted to mesic conditions, and since the widespread occurrence of such reversions is essential to the hypothesis of alternating selective pressures, presentation of several more examples, taken from a variety of different plant families, appears to be desirable. In reviewing the floras with which I am familiar, as well as a few tropical groups with which I have less acquaintance, I could detect probable examples in 40 different families. Some of these have already been reviewed in the ecophyletic charts presented in Figs. 9-1 to 9-4. Three more are described here, since they represent particularly striking examples and occur in groups that are not well enough known that complete ecophyletic charts could be made of them. They are as follows.

Ericaceae. The tribe Arbutoideae of the Ericaceae clearly demonstrates adaptive radiation from submesic to more arid as well as less arid situations. The most generalized genus of the tribe is *Arbutus,* which inhabits dry forests of Mexico, the Pacific Coast of North America, and the Mediterranean region. From an *Arbutus*-like stock were derived the large genus *Arctostaphylos,* as well as the related and hardly distinct *Comarostaphylis.* In their thick, sclerophyllous leaves, their reduced inflorescences, and, in many species, the coalescence of their carpels to form a single hard stone, these genera are distinctly more xeric than *Arbutus.* Furthermore, as has been pointed out by Wells (1969), those species that have lost their ability to form sprouts from their stumps, a characteristic that exists in *Arbutus* as well as in most other genera of Ericaceae, occupy in general more xeric habitats than the stump sprouters. Nevertheless, a group of non-stump-sprouting species is characteristic of the relatively mild coastal region of central California, where they occur sympatrically with the stump sprouters. These species have relatively thin leaves with cordate bases, a characteristic that is rare or absent elsewhere in the tribe Arbutoideae and uncommon in the Ericaceae generally. Furthermore, the cordate shape does not appear in the seedlings of any of these species. It is, therefore, best interpreted as a specialization produced by late-acting genes, which has been established in connection with the secondary adaptation to mesic conditions of the species concerned.

The most mesic of these species are *Arctostaphylos Andersonii* of central California and *A. columbiana* of the Pacific Northwest. These species are in general sympatric with *Arbutus Menziesii,* but are even more mesic in their requirements. *Arbutus Menziesii* and *Arctostaphylos columbiana* represent another example of two related, sympatric species, of which the more specialized representative has been derived via an initial adaptation to more xeric conditions, followed by secondary readaptation to mesic conditions. In a completely different species, the widespread *Arctostaphylos uva-ursi,* the small, sclerophyllous leaves are retained, but increased leaf surface is acquired through evolution of the prostrate, creeping habit of growth, with great and rapid elongation of vegetative shoots and consequent increase in the number of leaves. The other widespread mesic species often placed in *Arctostaphylos, A.* (*Mairania*) *alpina,* is possibly the product of still a third line of secondary mesic radiation, but its origin is obscure.

Leguminosae (*sens. lat.*). The family Leguminosae, one of the largest in the plant kingdom, like the even larger family Compositae, contains numerous examples of secondary mesic radiation. Some of them have been mentioned in a previous discussion (Stebbins 1952). One of these is in the genus *Acacia,* whose predominant habitat is semiarid or arid regions in the tropics of both hemispheres. In one subgenus, having its center of diversification in Australia, the leaflets have disappeared, and photosynthesis is conducted by the flattened leaf rachises, termed phyllodes. A small group of phyllode-bearing acacias has radiated out of Australia through many parts of Polynesia. At the extreme limit of this radiation is the Hawaiian endemic, *A. koa.* By an increase in the size and number of its phyllodes, it has considerably increased its photosynthetic surface, and is definitely mesophytic.

Interestingly enough, another species of *Acacia, A. Farnesiana,* has become adventive in Hawaii through human introduction. This species, native to the semiarid regions of tropical America, possesses normal leaves bearing leaflets, as do all of the New World species of the genus. Nevertheless, it is adapted to drier habitats than is *A. koa.* In parts of Hawaii, therefore, we see at present the anomalous situation of two sympatric species, of which the one having apparently the most specialized leaves actually occurs in the wetter situations. *Acacia koa* is, therefore, an excellent example of increased specialization of a mesophyte through former adaptation to more xeric conditions in its ancestral evolutionary line.

Two examples in the subfamily Caesalpinoideae, *Cassia* and *Bauhinia,* have been discussed elsewhere (Stebbins 1952). With respect to *Cassia,* it is interesting to note that extensive data on chromosome numbers (Irwin and Turner 1960) have shown that the majority of the woody species, both mesic and xeric, have the somatic number $2n = 28$, which with respect to the lowest number known in the genus, $2n = 16$, appears to be a modified polyploid condition. The only perennial and woody species known to have the lower number occur in the semiarid regions of Brazil.

New World species of Galium (*Rubiaceae*)

In the western United States, the large and widespread genus *Galium* has evolved three distinct species groups, characterized by the dioecious condition. This condition is a derived specialization, and the origin of the dioecious species can be traced to monoecious species that inhabit southern California, Arizona, and northern Mexico. Two of the three species groups involved, those of *G. angustifolium* and *G. multiflorum,* are characterized by having bristly fruits. They occur predominantly in semiarid regions, and have evolved a few species that are adapted to truly desert conditions. The third group is that of the berry-fruited species, which are widespread and abundant throughout those regions of California that have a Mediterranean type of climate (Dempster and Stebbins 1968). They have radiated into a great variety of habitats, including the semiarid foothill woodland, the coastal sage formation, and dry, brush-covered slopes. Two species of this group, *G. sparsiflorum* of the Sierra Nevada and *G. muricatum,* endemic to the redwood (Sequoia) forests of the northwest coast, are distinct mesophytes.

The "species pump" hypothesis

A plausible hypothesis about the relationship between the floras of the climax forests of the tropics and those of the ecotonal regions that surround them was developed by Valentine (1967) to explain the diversity of mollusc species in tropical littoral communities (Fig. 9-6). He observed that during successive geological epochs there occurred warm periods during which tropical faunas extended far northward, alternating with cooler periods when they were much more restricted. He then pointed out that the restriction of faunas during periods of cooling would be very uneven. Isolated pockets of warm water, associated with favorable combinations of local edaphic conditions, would be expected to persist long after the surrounding waters had become much cooler and could no longer support the tropical species. The populations inhabiting these isolated pockets would be small, and would sooner or later be exposed to strong selective pressures. Divergent evolution would, therefore, be promoted in them.

With the return of a warmer, more favorable climate, some of these small populations could be expected to evolve new adaptations to tropical conditions, perhaps at a higher level of organization. Their habitats would then merge with the new tropical habitats produced by the climatic change, and the newly evolved species that proved themselves able to compete successfully with the older ones would be able to migrate southward and find vacant niches in the older tropical faunas. Valentine visualizes this migration as producing a division of the original ecological niches into smaller ones, occupied by species having progressively narrower amplitudes of ecological tolerance. This narrower tolerance of species in stable tropical communities of various kinds is equally characteristic of animals and of plants (Lowe-McConnell 1969, Cain 1969, Richards 1969, Ashton 1969). In this fashion, the more active and rapid speciation in the ecologically diversified marginal areas may from time to time give rise to populations that are "pumped" into the more stable climax communities. Since in such favorable environments extinction is likely to be slower than immigration, and many different ways of exploiting the same environment are possible, a gradual increase in the richness of these communities will take place, regardless of the amount of speciation that occurs within the communities themselves.

If Valentine's hypothesis is applied to higher plants, we must recognize the fact

Fig. 9-6. Diagrams to illustrate the "species-pump" hypothesis.

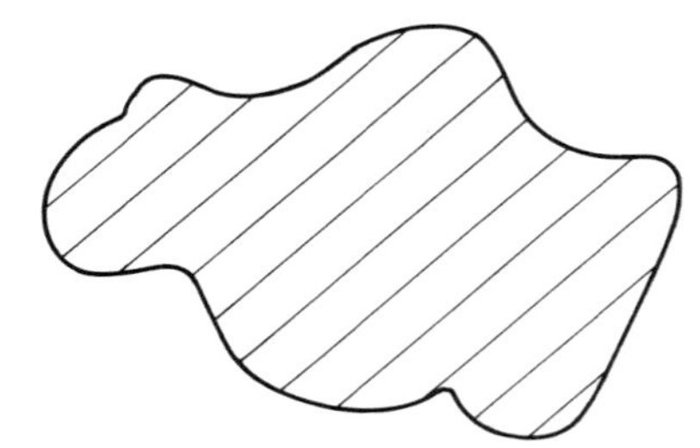

1. Widespread species adapted to favorable mesic conditions.

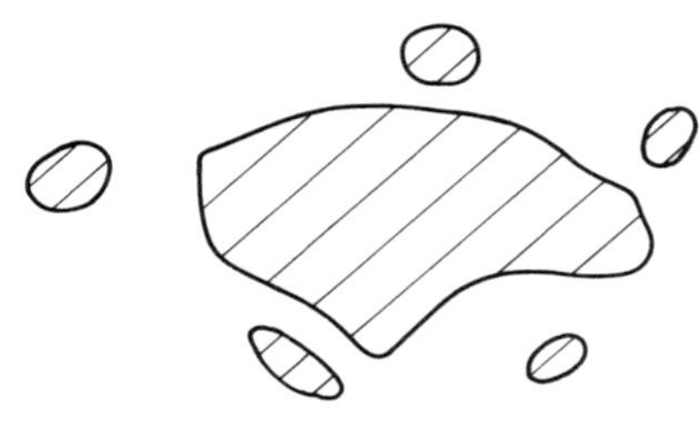

2. Contraction of species owing to onset of one or more unfavorable conditions (cold, aridity, orogenic movements, vulcanism), leaving isolated populations in isolated pockets around the periphery of the main distributional area.

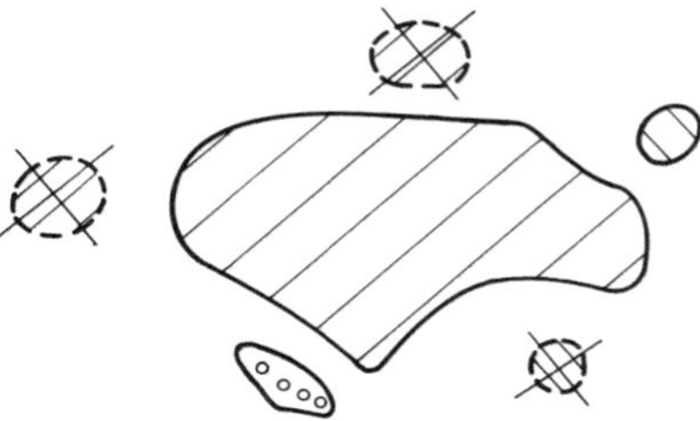

3. During persistence of unfavorable conditions, elimination of some isolated populations, evolutionary changes in others.

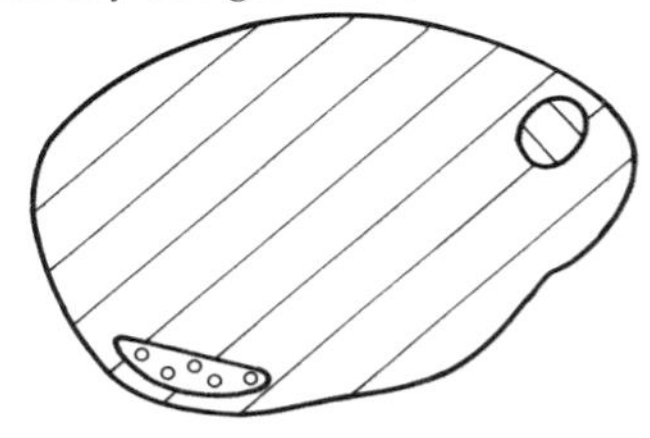

4. With return of favorable conditions, spread of the original species into regions occupied by altered populations.

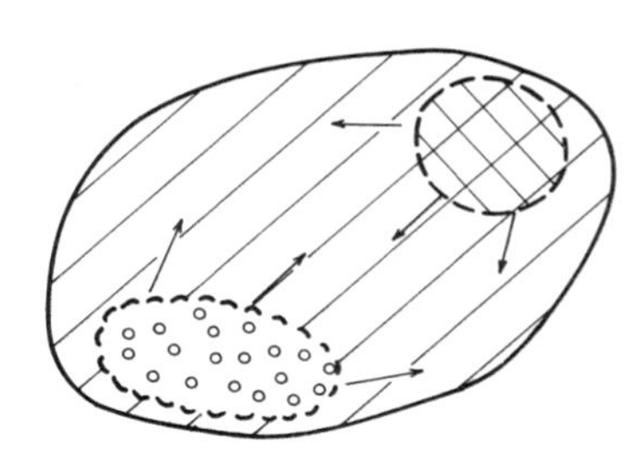

5. Mutual adjustment of the newly sympatric species, followed by spread of the formerly isolated populations into the distributional area of the widespread species.

that marginal conditions due to increasing drought surround the tropical forests to a much greater degree than do marginal conditions imposed by unfavorable temperatures. Except in eastern Asia, the tropical forests of the world are bordered not by temperate or even by subtropical plant communities, but by tropical communities adapted to drier conditions, often of a seasonal nature. Consequently, in tropical groups of higher plants the "species-pump" principle would operate more often in response to changes in available moisture than to long-time cycles of temperature change. This probability is reinforced by the fact that available moisture can be altered not only by overall climatic changes, but also, and perhaps more often, by local changes in topography, particularly the elevation of mountain ranges and their degradation through erosion. In this way, local "rain shadows" might appear and disappear over relatively short periods of time, geologically speaking.

An example from *Drosophila* which, if correctly interpreted, supports this modification of the "species-pump" hypothesis has been presented by da Cunha and Dobzhansky (1954). In the Brazilian state of Bahia there exists a race of *D. Willistonii* that is nearly uniform in its chromosomal composition. It occurs both in the lush tropical forests of the coast and in the severe environment of the caatinga desert. Da Cunha and Dobzhansky interpret this anomalous race by assuming that an "evolutionary invention" appeared among the relatively uniform inhabitants of the caatinga, and then spread secondarily into the rain forest.

In this and the previous chapters evidence has been presented which suggests that diverse trends in higher plants have reflected the diversity of organism-environment interactions, both in different regions of the earth at the same time and in the same region at different periods of the earth's history. If this is true, maximum evolution, including most of the differentiation and origin of major groups, should have taken place in those regions where and at those times when diversities of organism–environment interactions, or interactions between different kinds of organisms, have been greatest.

Summary

Sedentary plants may be exposed during geological periods of time first to increasing drought and cold, and later to more favorable climates. Adaptations to the more severe conditions involve acquisition of protective specializations, both morphological and physiological. Adaptations to more favorable conditions involve loss or reduction of xeromorphic specializations, increase in vigor and fecundity, and often diversification of flower structure and pollination mechanisms. Since loss mutations play a more significant role in adaptation to more mesic than to more xeric conditions, on the basis of principles of developmental genetics, the derivation of mesophytes from more xeric ancestors might be expected to occur more often than the reverse trend. Examples of the former trend among cultivated plants are the origin of cotton and tomatoes. Ecophyletic cross-sectional charts indicate that in the genera *Antennaria* and *Ceanothus,* as well as in the families Hydrangeaceae and Dilleniaceae, adaptive radiation has included derivation of mesophytes, extreme xerophytes, and arctic alpine forms all from semixeric inhabitants of intermediate or transitional regions. In the genera *Microseris* and *Agoseris*, as well as in other tribes of Compositae, and in certain genera of Polygonaceae, Ericaceae, Leguminosae, and Rubiaceae, evidence exists for reversals, involving first a trend from more mesic to more xeric adaptations, and later trends from more xeric to more mesic adaptations. The secondary mesophytes have greater reproductive efficiency than their original ancestors. These trends are analogous to those that form the basis of the species-pump hypothesis of Valentine, except that in these plant groups alternating availability of moisture has been the basis of the principal selection pressure, rather than alternating temperature regimes, as is true for the mollusca that formed the basis of Valentine's theory.

10 / The Nature and Origin of Primitive Angiosperms

More than a century ago, Charles Darwin declared that the origin of the angiosperms is an "abominable mystery." Unfortunately, in spite of many new fossil discoveries, a large amount of research on contemporary seed plants, and a host of new theories, the mystery still remains. Its persistence is due chiefly to one embarrassing fact: fossil remains of any plants that might provide plausible connecting links between angiosperms and any other groups of vascular plants are still completely lacking. We must, therefore, rely entirely upon deduction and inference.

Faced with this situation, a botanist might well be justified in declaring the problem to be insoluble and abandoning it altogether. Nevertheless, a new synthesis of available knowledge, based upon a different viewpoint, plus a few bits of additional information that have not been sufficiently considered in the past, might provide a basis for comparing and evaluating current theories, and for suggesting sources of additional knowledge. With this goal in mind, the present chapter has been undertaken.

History of the Problem

Until the beginning of the present century, botanists were about equally divided between two opposing groups. One of them, led by Adolf Engler, R. von Wettstein, and their followers, believed that the most primitive angiosperms were those having the simplest-appearing flowers, with the smallest numbers of parts, such as the "Amentiferae" and various aquatic monocotyledons. The alternative belief, supported by de Candolle, Bentham and Hooker, Baillon, and Bessey, was that the primitive angiosperm flower had many parts, spirally arranged, and that evolution progressed chiefly via reductions, "fusions," and various kinds of specializations. Neither group made any real attempts to establish connections between angiosperms and other groups of vascular plants, living or fossil.

Early in the present century, Arber and Parkin (1907) produced their "strobilus theory," which was the first serious effort to construct a connecting link between angiosperms and other seed plants. On the basis of the fact that the best-known modern gymnosperms, the conifers and cycads, as well as the extinct Cycadophytes or Bennettitales, have conelike or strobiloid reproductive structures, they postulated that the primitive "flower" of angiosperms also had a strobiloid structure. They constructed a hypothetical category of "Hemiangiospermae," having numerous spirally arranged, large, and branching stamens, associated with large, flat "carpels," bearing ovules on their margins, and somewhat similar to the megasporophylls of the genus *Cycas*. Since the nearest approach to such a flower in modern angiosperms occurs in the family Magnoliaceae, this family was designated as the starting point of the angiosperms themselves.

The strobilus theory has influenced the thinking of modern botanists more than any other. Bessey (1897, 1915) based his "phylogenetic system" partly upon this theory, and partly upon older systems, such as those of de Candolle and of Bentham and Hooker. Hallier (1912) postulated that the "Polycarpicae," that is, woody angiosperms having separate carpels, were derived from unknown cycads ("Cycadacées") similar to the Bennettitales, and that the more advanced angiosperms were all descended from this primitive stock. The strobilus theory has formed the basis of the systems constructed by most modern taxonomists, such as Hutchinson (1926, 1934, 1959), Takhtajan (1954, 1959, 1966, 1969, 1970), Cronquist (1968), and Thorne (1958, 1963, 1968).

An apparent advance in our understanding of angiosperm phylogeny was made with the discovery and description of the fossil order Caytoniales from Jurassic strata of Scotland (Thomas 1925). The cupule that enclosed the ovules in this order was seized upon as a possible ancestral carpel. Quickly, however, great differences between Caytonialean cupules and angiosperm carpels were pointed out, with respect to both overall structure and the position of the ovules, so that the hypothesis of homology between these two structures had to be abandoned. Because of this lack of homology, most modern authors have regarded the Caytoniales as in no way associated with the ancestors of the angiosperms. Nevertheless, a different kind of homology, between the Caytonialean cupule and the angiosperm ovule, was suggested by Gaussen (1946), and is considered in detail below. If this homology is correct, the position of the Caytoniales relative to angiosperm ancestry must be reconsidered.

The largest body of new information bearing upon the nature of primitive angiosperms and gathered during the present century was that which resulted from the morphological and anatomical research of

Bailey and his collaborators (Bailey 1944*a,b*, Bailey and Nast 1943*a,b*, 1944*a,b*, 1945*a,b*, 1948, Bailey, Nast, and Smith 1943, Bailey and Smith 1942, Bailey and Swamy 1948, 1951, Swamy 1949, Swamy and Bailey 1949, 1950). This research, chiefly on various families of angiosperms that lack vessels in their xylem, established the high probability that xylem vessels, which had previously been regarded as one of the most distinctive features of angiosperms, were absent from the most primitive members of the class. In fact, the wood of some angiosperms, such as *Tetracentron* and the Winteraceae, is actually more primitive than that of many gymnosperms. Because vesselless wood occurs in angiosperms having widely different affinities as judged from their reproductive structures, the research of Bailey and his group indicates that primitive, vesselless angiosperms were a highly diversified group, most of which are now extinct. Bailey and Nast (1945*a*) emphasize the likelihood that the modern angiosperms regarded as primitive are relictual survivors of groups that formerly were much more extensive.

The possible ecological relations of early angiosperms were discussed by Corner (1949, 1953, 1954*a,b,c*) and Axelrod (1952, 1959, 1960, 1970, 1972). Corner's "Durian theory" reaffirms the belief of Bews (1927) that angiosperms first evolved in equable tropical rain forests, and hypothesizes that the most primitive angiosperms were thick-stemmed, large-leaved trees bearing large, indehiscent fruits similar to those borne by the genus *Durio* of the Bombacaceae. Axelrod's original theory also hypothesizes that angiosperms evolved under the equable conditions found in moist tropical climates, but emphasizes his belief that the class existed long before the appearance of their abundant fossil remains in the Cretaceous Period, suggesting an origin as far back even as the end of the Permian Period. He accounts for the lack of angiosperm remains in the fossil record of Permian, Triassic, and Jurassic strata by transferring the hypothetical locale of early angiosperm evolution from tropical rain forests to upland regions having equable, tropical or subtropical climates with seasonal drought. In his most recent publication (Axelrod 1972), he agrees with my own point of view, but emphasizes the significance of poor soils derived from hard crystalline rocks, even in regions having moderate precipitation.

Current hypotheses of angiosperm origin

The hypotheses about the origin of angiosperms that are still in vogue can be roughly divided into three groups. The first group, which can now be designated as "classical," has been well expressed during the past few years by the books of Cronquist (1968), Takhtajan (1966, 1969, 1970), and Hutchinson (1969), and papers of Thorne (1963, 1968). It may be summarized as follows.

Angiosperms evolved sometime during the Mesozoic Era, in advance of the extensive appearance of their fossils in the early to middle part of the Cretaceous Period. Their evolution took place in equable tropical forests, probably in upland regions. Their immediate ancestors are unknown but may have been seed ferns similar to those described from late Paleozoic strata. The earliest angiosperms were large or small trees with thick, heavy branches, primitive vesselless wood, and relatively large, probably simple leaves. Their flowers were large, consisting of many parts spirally arranged, and bore a general resemblance to the strobili of cyads and the strobiloid flower of the Magnoliaceae. Their stamens were broad, flat, and simple, bearing elongate microsporangia on either the adaxial or the abaxial surface, or perhaps marginally. The ancestral carpel was large and leaflike, bearing many ovules, and probably already involute or conduplicate, prior to the complete enclosure of the ovules as in all modern angiosperms. It lacked a terminal stigma, but the entire margin may have been stigmatic and capable of receiving pollen. On the basis of this hypothesis, the two most primitive modern families are believed to be the Magnoliaceae and the Winteraceae.

The second group of theories is well exemplified by that of Nemejc (1956). He recognized the angiosperms as a single cohesive group, of which all the various orders and families radiated from a common ancestor. Nevertheless, he did not believe that any single modern order can be regarded as representing the living counterpart of this ancestor. Instead, he regarded several modern orders as about equally closely related to it. Their connections with one another and with the common ancestor were believed to be extinct. The Magnoliales were regarded by him as one of these radiants, which has retained a relatively high proportion of primitive or archaic characteristics. This theory is considered by Axelrod (1960) to be the most probable one, although he recognizes that some of the radiants designated by Nemejc as primary may, in fact, be secondarily derived.

On the basis of biochemical studies of flavonoids, Bate-Smith (1962, 1965, 1972) and Kubitzki (1969) have postulated a similar phylogeny. They suggest that among dicotyledons at least two major phylogenetic lines, the Ranalian and the Dillenialian-Rosalian, diverged from common ancestors that are now extinct, and did not correspond to any modern order. A similar phylogeny is implied, though not expressly stated, by Ehrendorfer *et al.* (1968).

The third group of hypotheses includes those that do not recognize the bisexual reproductive axis, bearing separate and spirally arranged parts, as the primitive condition in angiosperms. Most of them hold that the angiosperms are highly polyphyletic, and many of them seek an origin for some groups among the Gnetales. Among the latter are the hypotheses of Hagerup (1934, 1936, 1938, 1939), Fagerlind (1946), Emberger (1950, 1960), Li (1960), Maekawa (1960), Lam (1952, 1959, 1961), Melville (1960, 1962, 1963), and Meeuse (1965, 1967, 1971). Those of Hagerup and Fagerlind have followed the lead of R. von Wettstein in attempting to derive the angiosperms from the Gnetales, and in maintaining that the angiosperm flower is derived from a compound shoot or "pseudanthium."

Another hypothesis, that of Meeuse (1965, 1967), has postulated a strongly polyphyletic origin for the group. The dicotyledons are believed to have originated from the Gnetales via the Piperales, whereas the monocotyledons are derived from the fossil order Pentoxylales via the "screw pines" or Pandanales. Finally, Melville (1960, 1962, 1963) has produced his "gonophyll theory," according to which the angiosperms are not only highly polyphyletic, but in addition have come from ancestors that did not have the conventional architecture of shoots and appendages that is found in nearly all seed plants, living and fossil.

Critical comments on all of these hypotheses will be presented later in this chapter. First, however, some facts about the angiosperms and their possible relatives need to be reviewed.

Similarities and Major Differences Between Angiosperms and Other Plants

The majority of angiosperms, including all of those that are generally regarded as primitive, agree with other groups of seed plants in the following respects. They have shoots or stems bearing leaves as lateral appendages. Their leaves are flat and dorsiventral, and primitively evergreen or persistent. They have a vascular cylinder consisting of internal xylem and external phloem, which is produced by a cambial layer that forms a more or less continuous ring. They have separate microsporangiophores or microsporophylls and megasporangiophores or megasporophylls. The latter give rise after pollination to seeds having integuments and a micropyle, and at maturity an embryo consisting of hypocotyl, cotyledons, and plumule. Monocotyledons differ from this description in several respects; their position as a derived group is discussed in Chapter 13.

On the basis of these similarities, including characteristics that are very unlikely to have arisen more than once in evolution, the minimal requirements for ancestors of angiosperms are the possession of roots, shoots, and leaves as lateral appendages; primary vascular tissue produced by protoxylem and protophloem points and secondary vascular tissue produced by a continuous cambial ring; separate microsporangium- and ovule-bearing structures; and seeds equipped with an integument, endosperm, and an embryo consisting of hypocotyl, cotyledons, and plumule.

The principal characteristics by which most angiosperms differ from all gymnosperms (included here are all nonangiospermous seed plants) are the following. Their leaves have both secondary and tertiary venation, usually with isolated termination of veinlet endings. These structures are produced by "plate meristems," intercalated between the midrib and the leaf margin, that are quite different from the meristems that give rise to gymnosperm leaves, except for the anomalous genus *Gnetum.* Their microsporangium- and ovule-bearing structures, stamens and carpels, are situated on the same shoot, with the carpels uppermost. The derived status of all monoecious and dioecious angiosperms, which have stamen- and carpel-bearing flowers or inflorescences separated from one another, is a tenet of the classical theory which, in my opinion, is supported by convincing evidence (cf. Parkin 1952, Takhtajan 1959, 1969, Cronquist 1968). Bisexual reproductive structures are known in gymnosperms only in the extinct order Bennettitales and in the anomalous genus *Welwitschia,* neither of which is sufficiently similar to angiosperms that any connection between them can be postulated. This suggests that bisexual flowers arose from unisexual reproductive axes in the earliest angiosperms or their immediate ancestors, independently of their origin in any other group of seed plants.

Among the reproductive structures themselves, the most conspicuous feature of angiosperms is the closed carpel, which gives the group its name. Perhaps more fundamental, however, are the reductions in the gametophyte and the changes in the reproductive cycle. Between microspore formation and fertilization, male gametophytes of angiosperms produce only three haploid nuclei by means of two mitoses, a reduction beyond any condition found in gymnosperms. A similar reduction is the octonucleate embryo sac. The condition of double fertilization, coupled with the deferment of endosperm production until after fertilization, is of prime importance because it is the most clear-cut evidence in favor of the monophyletic origin of angiosperms. The independent origin two or more times of the complex readjustment of metabolic regulators that must have been required for this shift is hard to imagine.

Another important difference between angiosperms and nearly all other contemporary seed plants is the direct formation of the embryo from the zygote. In conifers, Taxales, Cycads, *Ginkgo,* and *Ephedra,* a stage of free nuclear division, forming a coenocytic structure of varying size, is interpolated between the zygote and the formation of the actual embryo or embryos (Schnarf 1933, Johansen 1950). Since the coenocytic "proembryo" exists in nearly all nonangiospermous seed plants of which embryo development can be studied (this is obviously impossible in extinct groups), the most reasonable assumption is that it was the original condition in them, and that the direct formation of the embryo has evolved in angiosperms by a process of reduction.

The two exceptional gymnosperms that lack a coenocytic proembryo are *Gnetum* and *Welwitschia.* In *Gnetum,* the proembryo stage consists of branched, septate filaments, from the ends of which embryos are formed. In *Welwitschia,* the embryo is formed directly from the zygote, and a proembryo stage is lacking. The sequence of cell divisions, however, differs from that found in any angiosperm (Johansen 1950). With reference to the discussion in the next section, *Welwitschia* may be regarded as "the exception that proves the rule," since it inhabits deserts and flowers only during brief periods having favorable moisture conditions. The exceptional angiosperm genus *Paeonia,* which possesses a coenocytic proembryo, is discussed below.

Hypothetical Conditions Under Which Angiosperms Arose

In considering the conditions under which angiosperms arose, major emphasis must be placed upon these distinctive characteristics of the reproductive cycle. These adjustments were too complex to have been brought about by the chance fixation of genes. They must have originated under the guidance of natural selection. The most likely selection pressure that can be imagined is that of a seasonal climate, favoring the compression of the reproductive cycle into a relatively short period of time. The hypothesis that seasonal cold was the selective agent is unlikely, since none of the primitive angiosperms have other characteristics that adapt them to a temperate, seasonal climate. A more attractive hypothesis, already suggested by Axelrod (1970), is that seasonal drought was responsible, and that the distinctive characteristics of angiosperms arose in response to a tropical or subtropical climate having distinct dry and wet seasons.

The closure of the carpel could have evolved as an adaptation to seasonal drought. Grant (1950*b*) has suggested that the closed carpel may have evolved in response to the adaptive value of protecting the ovules from coleopteran pollinators that have biting mouth parts. One difficulty with this explanation is that the hypothesis of early coleopteran pollination is disputed by entomologists (Pervukhina 1967, Malyshev 1964). Furthermore, one might suggest that, according to the hypothesis of adaptive modification along the lines of least resistance, early angiosperms or their ancestors could have acquired protection of their ovules from Coleoptera by evolving distasteful or toxic repellent substances, as the Leguminosae did at a later stage of evolution (Janzen 1969). Closure of the carpel would have had an adaptive value in a seasonally dry climate as a means of protection from both drought and potential insect predators.

The closed carpel can, in fact, be regarded as a structure that combines the advantages of protection and rapid pollination. In conifers, protection is achieved at the expense of a long interval between pollination and fertilization, since the pollen tube must digest its way through the massive tissue of the cone scale and ovular integuments. In angiosperms, on the other hand, the stylar canal that is formed by the folded margins of the conduplicate carpel provides an avenue through which the pollen tube can grow quickly and easily.

A climate having seasonal drought would exert the greatest possible selective pressure in favor of reduction in gametophytes, as well as in the development of the embryo. The evolution of double fertilization might be regarded as a compensation for the reduction of the female gametophyte. The polyploid and heterozygous condition of the endosperm nuclei that result from double fertilization is probably the principal basis of the rapid development of endosperm that follows its fertilization, and precedes the division of the zygote.

A climate having seasonal drought and relatively little wind would also favor the shift from wind to insect pollination. In such a climate, the most favorable period for flowering is after the rains have ceased and the weather conditions have become relatively quiet. This is also the most favorable time for the emergence and flight of insects, so that insect life would have a seasonal peak at this time. Under these conditions, competition between insect species for food would be most intense, and the advantage of shifting to a new food source, either pollen or nectar, would be greatest.

The selective pressure in favor of ecological flexibility would be greatest in a mountainous region having seasonal drought. Flexibility might be based upon a genetic constitution that favored the stage-specific production of growth substances in considerable quantity and the relative ease of their translocation. Such a condition would be promoted by natural selection in a climate having great but predictable shifts in temperature and precipitation, and in a topography which, because of considerable differences between localities with respect to altitude and degree of insolation, would contain a large number of different ecological niches, characterized by different variants of the overall climatic regime.

The factors of ecology and population structure that would favor the kind of rapid evolution which is envisaged for the origin and early diversification of angiosperms have been discussed in Chapter 8. In that discussion, the point was made that these factors would act most strongly in mountainous regions having seasonal drought.

Representative of the regions in which these factors have greatly promoted the diversification of the modern flora are the Cape Region of South Africa, the Ethiopian highlands, parts of the Indian Peninsula, south central Mexico, northern Venezuela, the coast of Ecuador, and the northern provinces (Salta, Jujuy) of Argentina. The families that are evolving most rapidly in these regions are the most progressive ones of the modern flora: Compositae, Gramineae, Leguminosae, Geraniaceae, Aizoaceae, Iridaceae, Cactaceae, Solanaceae, and others.

On the basis of the principles of genetic uniformitarianism, explained in Chapter 1, we would expect that the habitat and population structure of the most progressive vascular plants would have been similar in the middle of the Mesozoic Era to what it is today. At that time there were no Compositae, Gramineae, or other specialized angiosperms. The most progressive vascular plants of the Jurassic Period were probably the earliest angiosperms, which, if they existed today, would be classified as belonging to or related to the Magnoliales. A logical assumption, therefore, is that the pioneer, ecotonal and mosaic habitats found in semiarid, subtropical mountainous regions were at that time occupied by the original, ancestral Magnoliales and Dilleniales. The present situation, in which the least-modified descendants of these primitive angiosperms occupy moist, equable rain and cloud forests, is regarded as the result of differential survival. This subject is discussed more fully in Chapter 8.

Habitats similar to those found in the regions mentioned above were much less widespread during the Mesozoic era than at present, and did not exist or were poorly developed in the regions where they now occur. Hence, none of the above-mentioned regions can be regarded as the cradle of the angiosperms. They have been mentioned only to give the reader an idea of the ecological conditions under which the angiosperms are believed to have arisen and to have undergone their first diversification.

Chromosomal evolution in primitive angiosperms

The most striking fact about the basic chromosome numbers of the more primitive woody families is that they are not only high relative to those of herbaceous groups (Stebbins 1950, 1971) but also discontinuous and multimodal within individual orders and families in such a way that a polyploid origin of many woody genera and some woody families has been considered very likely by many cytologists for a long time (Raven, Kyhos, and Cave 1971). Most conspicuous in this connection is the number $n = 13$ and 43 in Winteraceae, $n = 21$ in Platanaceae, $n = 19$ in all genera of Magnoliaceae and in *Cercidiphyllum, Tetracentron,* and *Trochodendron,* and $n = 28$, 36, and 44 in Bombacaceae (Bolkovskikh *et al.* 1970). Furthermore, when lower chromosome numbers occur in Magnoliales or Dilleniales, such as $n = 7$, 8, and 9 in Annonaceae, $n = 8$ in *Hibbertia,* and $n = 5$ in *Paeonia,* they vary both between and within families. This situation caused me (Stebbins 1966) to postulate that chromosomal evolution in most woody orders of angiosperms has involved three stages. The first stage consisted of aneuploid alterations of the originally low basic number, which may have been either $n = 6$, $n = 7$, or $n = 8$. From this original number, stepwise reduction, increase, or both could have produced a continuous series of basic numbers from $n = 5$ to $n = 12$ or higher. The second stage was polyploidization. Numbers like $n = 11$, 12, 13, or 14 could have been produced either by stepwise increase or by polyploidy from numbers $n = 5$, 6, and 7. Numbers like $n = 19$, 21, or higher almost certainly are of polyploid origin, and may represent secondary cycles of polyploidy. The third stage was diversification of genera and additional speciation at these higher basic chromosome numbers, sometimes accompanied by additional cycles of polyploidy.

If we assume that evolutionary processes that took place in early angiosperms during the Mesozoic era were similar to those that take place in their contemporary descendants, then we can postulate that these changes were associated with shifts in habitat in the following manner. Modern genera that combine the life cycle of long-lived, shrubby or subshrubby perennials with extensive aneuploid series of chromosome numbers, such as *Astragalus, Astroloma, Boronia, Crepis, Haplopappus, Leucopogon, Limonium, Primula, Salvia, Saxifraga,* and *Verticordia,* are almost exclusively adapted to pioneer habitats in open alpine or semiarid situations. The chromosomal translocations that bring about their aneuploidy are probably favored in these situations because they tie together by genetic linkage adaptive combinations of genes (Stebbins 1971). Colonization of new pioneer habitats depends upon the origin of new linked combinations, and hence of new translocations. As was suggested above, the earliest angiosperms are believed to have been shrubs or subshrubs, and so may have occupied ecological sites similar to those now occupied by shrubby species of *Haplopappus* (*sens. lat.*) in North America or of *Astroloma, Boronia, Leucopogon,* and *Verticordia* in Australia. The original aneuploid series of basic numbers that they possessed may have originated under these conditions.

As habitats were opened up to them at various times and in various places, a few of the most vigorous of these early angiosperms became adapted to mesic tropical or subtropical forests. This adaptation was often accompanied by hybridization and polyploidy, a pair of processes that in modern species not only of herbaceous genera, but also of woody genera such as *Rubus, Rosa, Salix,* and *Betula,* has provided vigorous and aggressive invaders of new shrub and forest associations. In this way, the ancestors of the modern Magnoliales and other "primitive" groups acquired their present habitat. These ancestors then evolved many species and genera, and some of them spread to different continents. The same succession of aneuploidy followed by polyploidy and invasion of tropical forests took place also in more advanced families, so that from the middle of the Cretaceous through the beginning of the Tertiary Period a series of invasions of tropical forests by more and more specialized angiosperms took place. Some families, such as the Dipterocarpaceae and Annonaceae, invaded the forests with little or no polyploidy, whereas in others, such as the Bombacaceae and Tiliaceae, very high basic numbers were evolved.

Meanwhile, the angiosperms that remained in the open habitats evolved at the diploid level, giving rise to more advanced and specialized families and genera. These newly evolved groups were better able to take advantage of the succession of drastic environmental changes to which these open areas were subjected, so that they gradually replaced the ancestors of the more primitive groups.

Paleontological evidence regarding the time and place of angiosperm origin

With respect to the time when the angiosperms originated and diversified, two divergent opinions are current. A relatively recent origin is postulated by many paleobotanists who have studied principally pollen and other microfossils (Scott, Barghoorn, and Leopold 1960, Doyle 1969, Muller 1970). These authors maintain that angiosperms originated at a time shortly before the first appearance of their pollen in the fossil record, probably toward the end of the Jurassic Period. Their diversification is believed to have taken place during the Cretaceous Period, contemporaneously with the gradual increase in abundance and diversity of angiosperm pollen during this period. The principal argument in favor of this point of view is that microfossils are far more abundant in the fossil record than megafossils, and are known from a much larger number of localities and strata of different ages. The fact is recognized that much angiosperm pollen, particularly that of some Ranales, is so fragile and delicate that it is rarely preserved. This fact, however, is regarded as only a minor hindrance to the establishment of general hypotheses based upon evidence from fossil pollen. Another fact, that many insect-pollinated plants produce so little pollen that it is very unlikely to enter the record unless the plants producing it grew in the actual site of deposition, is, as a rule, overlooked altogether.

The alternative viewpoint, that angiosperms existed in localized, special habitats long before their established fossil record began, is held by a larger number of botanists (Berry 1920, Scott 1924, Wieland 1933, Arldt 1938, Camp 1947, Thomas 1947, Just 1948, Axelrod 1952, 1960, 1970, Nemejc 1956, Takhtajan 1959, 1969, 1970, Hawkes and Smith 1965, and others). The usual explanation for the almost complete lack of fossils older than early to mid-Cretaceous is that prior to this time angiosperms lived in upland areas which were sites of erosion rather than of deposition. Moreover, greater emphasis is placed by some of these authors on the fragility of much of the primitive angiosperm pollen. If, as is by no means improbable, all pre-Cretaceous angiosperms had either pollen that was so fragile that it could not be transported for more than a few miles in an intact condition or pollen that was monocolpate and so essentially indistinguishable from gymnospermous pollen (Muller 1970), the absence of recognizable angiosperm pollen in pre-Cretaceous strata could be explained.

Axelrod (1970) has pointed out an important discrepancy between the megafossil and the microfossil record of strata having the same geological age. The mid-Cretaceous Dakota formation of the central United States has long been renowned for the richness of its flora in angiosperm leaves, and more than 200 taxa of undoubted angiosperms have been described from it. Nevertheless, the microfossil assemblage of this flora includes only 5 percent of angiosperms. Similar discrepancies exist between the mega- and the microfossil records of upper Cretaceous strata, in which the evidence from megafossils indicates that angiosperms had reached almost or quite the degree of dominance that they have today. One can only conclude that the microfossil record is biased both by the fragility and paucity of the pollen produced by many species of angiosperms and by the improbability that many kinds of pollen will be transported intact more than a few hundred meters from the plants on which they developed.

Although angiosperm fossils have been reported several times from strata older than the Cretaceous Period, none of these claims have been completely verified. In some instances, the remains proved upon further examination not to be angiosperms, and in others, the original dating was found to be faulty. A few of them remain on the doubtful list. The best known of these are the leaves from the upper Triassic (Rhaetic) of Greenland described as *Furcula* (Harris 1932), and the large leaves from the Triassic of southwestern Colorado described as *Sanmiguelia* (Brown 1956). The former have reticulate venation that resembles that of angiosperms (as well as some ferns, such as *Tectaria*), but they differ from contemporary angiosperms with respect to the dichotomous forking of the midvein. The *Sanmiguelia* leaves are so poorly preserved that only the major veins can be seen. They bear a superficial resemblance to juvenile leaves of palms and to the adult leaves of some members of the family Cyclanthaceae. Their status must remain in doubt until well-preserved reproductive structures have been found associated with them.

Although the virtual absence of angiosperms from strata of Jurassic age is not easily explained by assuming that they had not yet evolved, the conclusion that during both the Triassic and the Jurassic Periods angiosperms were absent from lowland rain forests is well justified. Fossil floras of Jurassic age are plentiful in many parts of the world (Seward 1931, Arnold 1947, Delavoryas 1962). They consist chiefly of conifers, Ginkgophytes, cycads, cycadophytes, seed ferns, and various kinds of spore-bearing plants. Significantly, however, they do not include any groups which by the remotest stretch of the imagination could be regarded as ancestral to angiosperms, with the possible exception of the Caytoniales (see below). On the basis of any hypothesis about angiosperm origins, the entrance of these plants into the mesic lowland floras occurred during the Lower Cretaceous Period (Aptian-Albian). The problem of why this dramatic change in the world's flora occurred at this time is still completely unsolved. For its solution paleoclimatic and stratigraphic evidence will have to be combined with paleontological evidence regarding the Cretaceous distribution and nature of herbivorous animals as well as potential insect and bird vectors of pollination and seed dispersal. An attempt at such a synthesis by cooperation between specialists in each of these fields would be both timely and rewarding.

The Probable Nature of the Earliest Angiosperms

The following reconstruction of the extinct plants that are believed to have given rise to modern angiosperms is based upon a variety of considerations. These include (1) the morphology of various extinct seed plants, chiefly those that existed in the first half of the Mesozoic Era, (2) the kind of growth habit that in modern groups appears to show the greatest flexibility and potentiality for giving rise to a diversity of growth forms, and (3) the morphology, anatomy, and ontogeny of both vegetative and reproductive structures in all of the different groups of modern angiosperms that might be regarded as primitive in these respects. Individual characteristics of modern forms are regarded as primitive and original, and others are regarded as specialized and derived, on the basis of the following criteria:

1. Characteristics are regarded as primitive that are widespread and general in nonangiospermous seed plants but occur in angiosperms only in a few groups, and then in association with other primitive characters.
2. Characters are regarded as specialized that do not occur in nonangiospermous seed plants, or if present show evidence of independent origin, and occur in angiosperms chiefly in groups that with respect to other characteristics are regarded as more specialized.
3. Characteristics are regarded as primitive that occur in all groups of angiosperms that in other respects appear to be primitive, even if they are not found in contemporary nonangiospermous seed plants.
4. In the case of serially produced structures, such as flowers in an inflorescence, or the parts of the flower, alterations in number are believed to be possible in both an upward and a downward direction (Chapter 7); on the other hand, very high as well as very low numbers are regarded as specialized conditions, being the end points of adaptive radiation.
5. Union of parts via the development of intercalary meristems, designated as intercalary concrescence in Chapter 6, is regarded as a unidirectional trend which can only occasionally be reversed (Chapter 7), so that united or adnate structures are regarded as a derived condition.

As a basis for estimating the degree of

advancement of contemporary groups, the following assemblage of characteristics is regarded as having been present in the earliest angiosperms or their immediate predecessors.

Growth habit

The earliest angiosperms had the following growth habit: stems woody or at least with a continuous ring of secondary vascular tissue, and persistent throughout the year; stems sparingly branched but relatively slender, and able to sprout from the root crown; plants relatively low, and flowering only a few years after seed germination. The growth habit of some species of *Drimys,* subg. *Tasmannia* (for example, *D. lanceolata*), of many Australian and New Caledonian species of *Hibbertia,* and, in the Northern Hemisphere, of such species as *Daphne cneorum* and the smaller-leaved species of *Rhododendron* is visualized as similar to that of these early angiosperms.

The reasons for favoring this kind of growth habit for the original angiosperms rather than the tree habit are given elsewhere (Stebbins 1965). The principal reasons are as follows: (1) If the original angiosperms were large trees, their gymnospermous ancestors should also have been large trees. At least some of them should have entered the lowland forests of the Jurassic floras. No trace of such ancestral forms exists in these fossil floras. (2) In large trees of equable tropical forests, the adaptive value of the reductional trends in reproductive structures which led to the angiospermous condition would have been minimal, and their origin by natural selection would be very difficult to explain. (3) The semiarid pioneer habitats in which early angiosperm evolution is believed to have occurred are particularly favorable for the establishment of low shrubs of the type mentioned.

The principal divergent opinion is that the original angiosperms were thick-stemmed, unbranched or little-branched trees similar to fossil Bennettitales and modern cycads, as well as to such modern angiosperms as various Bombacaceae, the genera *Carica* and *Pandanus,* and the palms. This opinion was first expressed by Corner (1949) and has since been adopted by Takhtajan (1959, 1969, 1970). Corner based his opinion upon his belief that the Bombacaceae, and in particular the genus *Durio,* are among the most primitive living angiosperms—a belief that, so far as I am aware, is not shared by any other botanist who has studied this problem. In stating his theory, Takhtajan usually refers to the Bennettitales, and gives the impression that he regards these extinct plants as closely related to the ancestors of the angiosperms. For reasons that are given later in this chapter, I believe that the Bennettitales are not at all closely related to the ancestors of angiosperms and that all resemblances between the two groups are the result of evolutionary convergence. Since there is no other reason for postulating that the primitive angiosperms were thick-stemmed trees, this hypothesis has little to recommend it.

Takhtajan (1970) has criticized my hypothesis that the original angiosperms were shrubs on three grounds. First, he believes that angiosperms originated in tropical forests, where woody plants are nearly all trees. Second, he considers the phrase "shrubs or subshrubs" (Stebbins

1965:467) indefinite, since the latter term could mean either a small shrub or shrublet or a form transitional between a shrub and an herb. Finally, he criticizes the evidence presented because no reference was made to the most primitive living angiosperms.

The first criticism has already been answered. With respect to the second, the remarks made by Cronquist (1968:62–63) about the close similarity in stem anatomy between many herbs and shrubs are pertinent. I am not ready to specify precisely what was the growth habit of the earliest angiosperms. I merely reiterate my belief that they were low-growing plants, having a continuous ring of cambium, stems that did not die back to the ground every season, and no single well-developed trunk.

As for Takhtajan's third criticism, the modern woody Ranales or Magnoliales were purposely not discussed in connection with the theory of ancestral shrubbiness because of their relictual nature. Even though most of the primitive modern Magnoliaceae, Winteraceae, and Degeneriaceae are trees, their extinct common ancestors may not have been. In this connection it is worthy of note that the Winteraceae having the lowest chromosome number, $n = 13$ (Ehrendorfer *et al.* 1968), belong to *Drimys,* subg. *Tasmannia,* which consists chiefly of shrubs. Dickison (1967) has shown that the best-developed arboreal genus of Dilleniaceae, *Dillenia,* has more specialized wood than does the shrubby genus *Hibbertia.* The most primitive woody anatomy in the family exists in the genera *Didesmandra* and *Schumacheria,* which are, respectively, trees or shrubs and climbing shrubs. These genera, however, have a highly specialized floral structure resembling that found in the more specialized species of *Hibbertia.* The hypothesis that the original, extinct members of the Dilleniaceae were shrubs having flowers similar to the more primitive ones found in *Hibbertia* but woody anatomy resembling that of *Didesmandra* and *Schumacheria* is not unreasonable.

Meeuse (1967) has criticized me on the grounds that the primitive angiosperms may have contained plants having a great variety of different kinds of growth habit. Although Meeuse's hypothesis of a highly polyphyletic origin of angiosperms is unacceptable, I agree with the hypothesis that adaptive radiation in the very earliest stages of angiosperm evolution gave rise to a great variety of growth forms, including both herbs and trees. Nevertheless, I maintain my belief that the majority of primitive angiosperms were shrubby, and that all dicotyledonous trees are derived secondarily from either shrubs or herbs. On the other hand, many shrubs belonging to advanced families such as Fagaceae, Salicaceae, and Proteaceae may well have been derived secondarily from trees.

Takhtajan (1959), Cronquist (1968), and Carlquist (1966*b*) have mentioned the likelihood that trends in growth habit have been reversed several times during angiosperm evolution. I would like to emphasize this probability even more. In particular, I would like to respond to Cronquist's statement (1968:63) that "if any large dicot trees have herbaceous ancestors within the angiosperms, this remains to be demonstrated." Anatomical evidence suggests that the most primitive members of both the genus *Phytolacca* and the family Nyctaginaceae were herbaceous, and this viewpoint is adopted by Takhtajan (1959:38). Nevertheless, both *Phytolacca dioica* and various species of *Pisonia* (Nyctaginaceae) are certainly large trees. The genus *Paulownia,* if, as I believe, it belongs to the Scrophulariaceae, is another likely candidate for a large tree with herbaceous ancestors.

Leaves, nodal anatomy, and wood

The majority of botanists now believe that the spiral ("alternate") arrangement of leaves is more primitive than the decussate ("opposite") arrangement. Nevertheless, decussate leaves exist in some families (Austrobaileyaceae, Monimiaceae, Calycanthaceae) that are relatively primitive and in others (Chloranthaceae, Cercidipyllaceae) that have specialized flowers but primitive stem anatomy. Moreover, many relatively advanced families and even genera contain some species with spirally arranged and others with decussately arranged leaves, and in several Compositae the two conditions may exist on different parts of the same stem. Consequently, there appears to be no good reason for assuming that the earliest angiosperms had either exclusively spiral or exclusively decussate leaf arrangement.

With respect to leaf shape, both Cronquist (1968) and Takhtajan (1970) suggest that the original angiosperms had simple, pinnately veined leaves, an opinion with which I agree. Cronquist believes that their leaves were stipulate, but Takhtajan points out both that stipulelike structures are unknown in other seed plants and that within the Magnoliales they occur only in the Magnoliaceae. Since, as is pointed out later in this chapter, the primeval position of the Magnoliaceae in their order is seriously open to question, there seems to be no good reason for regarding the presence of stipules as a primitive or original condition. Instead, stipules can be regarded as early adaptations for protecting vegetative buds and young leaves, particularly from insect predation. They very probably perform this function in the Magnoliaceae. Their reduction in or absence from many of the more advanced families of woody plants is probably associated with the evolution of specialized cataphylls that serve to protect the buds, as well as of more elaborate and complex chemical defenses against predation (Levin 1971).

The type of nodal anatomy found in the primitive angiosperms has been the subject of much speculation. For forty years the hypothesis of Sinnott and Bailey (1914) that the primitive angiosperms had three leaf traces from as many gaps was accepted with little question by most anatomists. Then Bailey (1956, Marsden and Bailey 1955), as well as Canright (1955), expressed the opinion that the primitive angiosperms had unilacunar nodes, containing one or two leaf traces. Meanwhile, Ozenda (1949) had maintained that in the Magnoliaceae the multilacunar node was primitive, an opinion shared by Takhtajan (1970).

I agree with Cronquist (1968) that this question cannot be resolved on the basis of the present evidence. Most trends with respect to this character are probably reversible. Nevertheless, the multilacunar nodes of the larger species of *Magnolia* are probably derived. This hypothesis is in line with the evidence from floral anatomy, described below, that indicates that the flowers in this family have undergone secondary increase in size. Secondary increase in size may well have affected both reproductive and vegetative organs in the Magnoliaceae.

As has already been mentioned, the presence of vesselless angiosperms belonging to a variety of otherwise primitive as well as relatively specialized families suggests that the wood of the original members of the group lacked vessels. These structures probably arose several times in different evolutionary lines.

Inflorescence and overall flower structure

I am inclined to agree with both Cronquist and Takhtajan that primitive angiosperms had either solitary flowers at the ends of the branches or loosely organized cymes like those of many species of *Paeonia,* in which branches from the upper internodes bear reduced leaves and secondary flowers.

On the other hand, there are several good reasons for doubting the widely held hypothesis that the original angiosperms had strobiloid flowers, the numerous parts of which were arranged on an elongate axis. Five separate lines of evidence cast doubt upon this hypothesis: (1) comparisons with various groups of nonangiospermous seed plants, both living and fossil; (2) vascular anatomy of the Magnoliaceous flower; (3) developmental pattern of this type of flower; (4) reproductive biology of the Magnoliaceous flower; (5) relationship between strobiloid and nonstrobiloid flowers and the occurrence of other primitive characteristics.

The strobiloid hypothesis was developed originally at a time when most known gymnosperms, both living and fossil, were those having reproductive parts grouped into cones or strobili. The superficial resemblance between these strobili and the gynoecium of the Magnoliaceae is striking, and was emphasized by Arber and Parkin (1907). If, however, the structure of the various kinds of strobili is examined more carefully, and direct comparisons are made between them, the difficulty of imagining direct genetic or evolutionary connections between them becomes increasingly great. The strobili of conifers are unisexual, and the excellent evidence obtained by Florin (1944, 1950, 1951) on the evolution of this order has shown clearly that male and female strobili of the same plant are analogous rather than homologous. The strobiloid form has evolved independently and in different ways in the two sexes. The strobili of cycads are also unisexual but, with respect to both the position of the sporangia and ovules and their anatomical structure, these strobili are completely different from those of conifers.

Strobili that are bisexual, containing both microsporangia and ovules, occur in the Bennettitales, and were strongly emphasized by Arber and Parkin, since they agree with the Magnoliaceous flower in having microsporophylls below and ovules on their upper or distal regions. There, however, the resemblance ceases. The gynoecial region of Bennettitales consists of individual, orthotropus ovules having only one integument, each ovule borne on a separate stalk and alternating with bractlike interseminal scales (Darrah 1960). The transformation of such a structure into an angiospermous gynoecium, or its derivation from a strobilus of the kind found in cycads, are both very difficult to imagine. Comparative morphology, therefore, leads to the conclusion that the strobili of conifers, cycads, and Bennetitales, and the "strobiloid" flower of Magnoliaceae, are analogous, not homologous, structures. Each of them has arisen independently of the others, probably as a result of similar selection pressures.

Fig. 10-1. Diagrammatic representation of an early stage in the vascularization of floral appendages and one of the three sepallike fused bracts of a generalized magnoliaceous flower: (*at left*) one strand of the central vascular system; (*in center*) two strands of the cortical system; (*at right*) vascularization of: (*A*) bract, (*B*) tepal, (*C*) stamens, (*D*) carpels. (From Skipworth and Philipson 1966.)

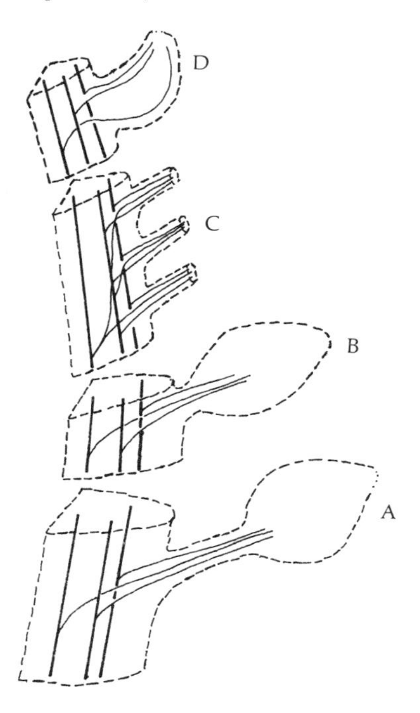

The vascular anatomy of the Magnoliaceous flower fully supports this conclusion. As has been clearly shown by the studies of Ozenda (1949), Canright (1955), Melville (1962), and Skipworth and Philipson (1966), and verified by my own observations, the anatomical structure of this flower is entirely different from that of any other floral or vegetative axis found in angiosperms. Its most striking feature is the presence of an elaborate cortical network of vascular bundles peripheral to a central ring. The ventral margins of the carpels are supplied by lateral bundles that come from the cortical network, whereas the dorsal bundle, which supplies the stigma and to which the ovular bundles are attached, emerges from the inner, central ring (Fig. 10-1). Since the two concentric cylinders of vascular tissue join each other only at the base of the flower, the vascular supply of each carpel consists of bundles that are separate from one another throughout the length of the floral axis. This condition is unique to the Magnoliaceae. Melville (1962) believes that it is so anomalous as to preclude its origin from a reproductive shoot bearing appendages and a central axis, and partly on the basis of it has developed his gonophyll theory. As is discussed below, this interpretation is unnecessary and unwarranted. Nevertheless, the vascular anatomy of the Magnoliaceous flower reveals no homologies with that of any other known strobiloid structure.

The development of the Magnoliaceous flower, particularly with respect to the differentiation of the procambial tissue that forms the basis of the vascular skeleton, has been insufficiently studied. Nevertheless, an excellent description of the overall development of the floral meristem in *Michelia* has been provided by Tucker (1960). Her evidence agrees entirely with that from vascular anatomy. *Michelia* was a particularly fortunate choice for this investigation, since it has the most strongly elongate, axislike gynoecium found anywhere among the angiosperms. If this elongate axis is a primitive condition, reflecting the affinity of the Magnoliaceae to a hypothetical ancestor that had reproductive appendages spirally arranged along an axis of the same kind that exists in the vegetative parts of the plant, then the developmental pattern of this gynoecium should resemble that of a vegetative axis more closely than is the case when floral and vegetative axes of other angiosperms are compared with one another. This conclusion has been reached not on the basis of the Haeckelian doctrine that "ontogeny recapitulates phylogeny," which I reject, but on the conviction that genes produce alterations in form by means of changing developmental patterns, and that radical changes, such as those being discussed, require interaction between many altered genes. On this basis, intermediate conditions with respect to adult form, if they are actually relictual intermediate evolutionary stages and not the products of convergent evolution, should be produced by intermediate developmental patterns.

Tucker's investigations have shown very clearly that this is not the case. In the developmental pattern of the flower of *Michelia*, the only stage that resembles even remotely the development of the vegetative axis is that of tepal differentiation. Even this stage, however, resembles that of the biseriate, whorled perianth found in the flowers of other angiosperms more than the vegetative axis of *Michelia* or any other plant. The six tepals are produced in two series, those of the inner series alternating with those of the outer. Their spiral arrangement is due to a slight displacement from one another of the members of each series or "pseudowhorl." This condition could have evolved from a normal spiral arrangement, but it could equally well have been derived from a condition of two successive whorls, through slight secondary elongation of the floral axis.

The developmental pattern of the floral axis prior to and during the formation of stamens and carpels is highly divergent (Fig. 10-2). It begins with a long period during which no appendages are differentiated. During this period, the undifferentiated meristem increases greatly in size, and its corpus becomes differentiated into three regions: the central initial zone, the peripheral zone, and the rib meristem. When finally the reproductive parts are differentiated, they are not produced either singly by individual plastochrons or by a rapid but continuous and even succession, as might be expected if an elongate axis had been greatly compressed. Instead, they are produced in tiers, each tier being the expression of a single plastochronic stage. The stamens are initiated in three tiers of 12–14 stamens each, whereas the carpels are initiated in 5 or 6 stages of 3–8 carpels each. The meristematic dome increases in size up to the

Fig. 10-2. Early development of the floral axis in *Michelia fuscata:* (*A*) at the stage of differentiation of the prefloral bract (note the similarity to the vegetative apex of *Liriodendron,* shown in Fig. 6-4); (*B*) early and (*C*) late state of carpel differentiation; (*D*) diagrams showing height and width of the undifferentiated apex, represented by triangles, and the number of appendages being differentiated, represented by vertical bars, at designated stages of development of the floral axis. Note that at vegetative stages in *Michelia* appendages are always differentiated one at a time. (From Tucker 1960.)

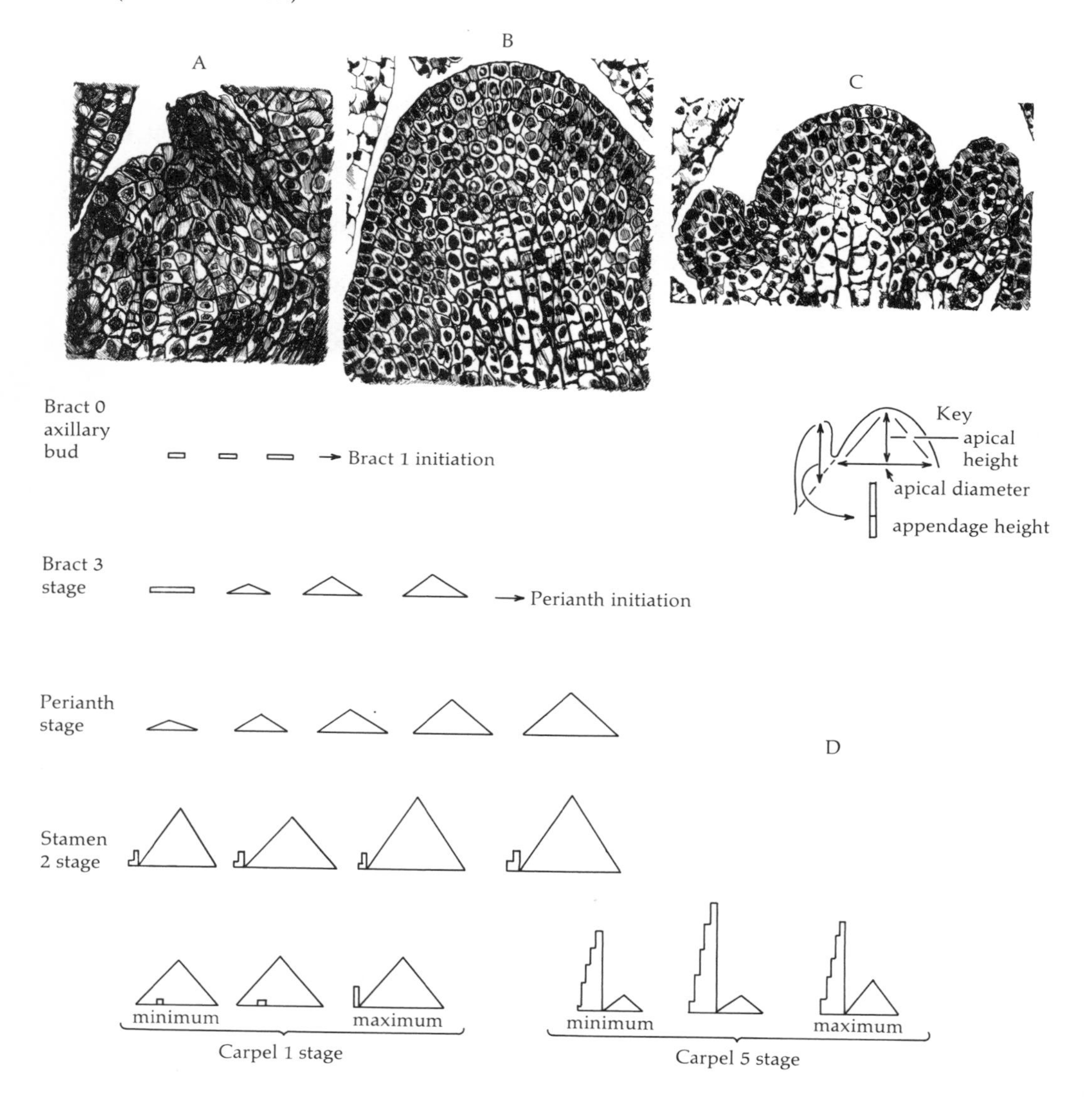

later stages of stamen initiation, after which it becomes smaller.

The only developmental pattern that resembles even remotely that of *Michelia* occurs in certain highly specialized Ranunculaceae, particularly *Ranunculus* (Tepfer 1953) and *Myosurus* (Buvat 1951, 1952). Most botanists now agree that the genera of Ranunculaceae having one-seeded indehiscent carpels, which are the only ones that resemble *Michelia* in the developmental pattern of their gynoecia, are specialized derivatives from more primitive members of their own family (Eames 1961). Consequently, no genetic or evolutionary connection can be postulated between them and the Magnoliaceae. By analogy, however, one can postulate that as in the Ranunculaceae, so in the woody Ranales, the elaborate developmental pattern of *Michelia* and presumably of other Magnoliaceae is secondarily derived, and their elongate, strobiloid or axislike gynoecium is by no means a persistence of some primitive, preangiospermous floral structure. It is a response to selection for higher seed production via an increase in number of carpels per flower (Stebbins 1967). In this connection, it is noteworthy that the gynoecium of *Myosurus,* probably the most specialized of all achene-bearing Ranunculaceae, has an axis with indefinite growth, which can produce varying numbers of carpels depending upon environmental conditions (Stone 1959). This indefinite condition is almost certainly of secondary derivation.

The evidence from reproductive biology consists chiefly of the fact that the method of carpel dehiscence in the Magnoliaceae does not agree with that expected on the basis of the classical theory of carpel structure. If carpels are derived from conduplicate megasporophylls, their normal dehiscence should be ventral or adaxial, along the suture formed by the fused margins of the sporophyll. This is the usual form of dehiscence in angiosperms, being found in *Illicium,* various Annonaceae, the follicle-bearing Ranunculaceae, and generally in the Dilleniales and Rosales. The indehiscent carpels of the Winteraceae and Degeneriaceae have well-developed ventral sutures, so that they may well have been derived from ancestors having adaxial dehiscence.

In the Magnoliaceae, on the other hand, the carpels are either indehiscent (*Liriodendron*) or abaxially (dorsally) dehiscent, by a special suture that develops beside the dorsal bundle (Fig. 10-3). The only genera having a kind of adaxial dehiscence are *Kmeria* and *Talauma,* in which the carpels split in half at maturity (Dandy 1927).

The adaptive reason for abaxial dehiscence in the Magnoliaceae becomes obvious if one follows the development of the gynoecium from anthesis up to seed maturity. During intermediate stages, the carpels are fused into a solid mass, and their adaxial sutures cannot be seen. This union is more than superficial, since it involves anastomoses between ventral bundles belonging to different carpels (Canright 1960, Melville 1962, and unpublished personal observations). Because of this fusion, the adaxial sutures are oriented inward toward the solid central portion of the carpel mass, so that ovular dehiscence along them would be impossible. This condition is hard to

Fig. 10-3. Mature gynoecia of (*A*) *Magnolia grandiflora,* showing abaxial dehiscence of carpels, and (*B*) *M. Soulangeana,* showing upper carpels not dehisced and with no obvious suture, and lower carpels dehiscing. (Original.)

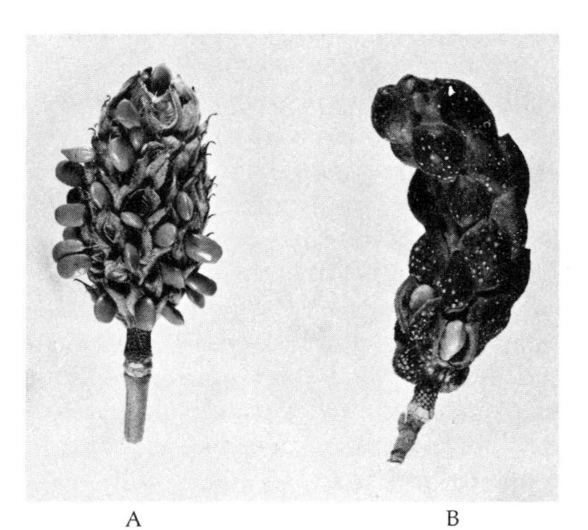

imagine as primitive. It is best explained by assuming a course of evolution in three stages, between the first two of which a change in the direction of selection occurred. During the first stage, the original gynoecium, which may well have consisted of adaxially dehiscent, pluriovulate carpels spirally arranged along an axis, became reduced by shortening of internodes, aggregation, and reduction in carpel number. At the same time, the carpels became few-ovuled, indehiscent, and achenelike, as in Himantandraceae and some other families of Magnoliales. This first stage may well have coincided with the initial diversification of the order. Selective pressures during this first stage favored reduction, in adaptation to a short favorable reproductive season. During the second stage, selection for higher seed production in response to more favorable ecological conditions gave rise to a secondary increase in the number of indehiscent carpels, producing either a mass of achenelike propagules similar to those found in *Ranunculus* or, more probably, an aggregate, indehiscent fruit similar to that of the Schisandraceae and Annonaceae such as *Annona.* The third stage was the development of secondary dehiscence along the newly differentiated abaxial suture, together with the elaborate mechanism found in the Magnoliaceae for exsertion of seeds through this suture. The second and third stages, therefore, are regarded as constituting the divergence of the Magnoliaceae from other Magnoliales and their differentiation in their own particular direction of specialization.

The final argument against the strobiloid hypothesis of the original angiosperm flower is based upon the nature of the flower in the majority of the Magnoliales. With the exception of the Magnoliaceae and Annonaceae, and perhaps of the anomalous Eupomatiaceae, the families placed in the Magnoliales by Hutchinson, Cronquist, and Takhtajan have either few carpels per gynoecium, few ovules per carpel, or both of these conditions. This is true also of the three families that have primitive vesselless wood and are placed in other orders, the Tetracentraceae, the Trochodendraceae, and *Sarcandra* in the Chloranthaceae. Consequently, without convincing evidence for the primitiveness of the Magnoliaceous flower, there is no other reason for believing that the flowers of the original angiosperms were exceptionally large and that they had very many parts. The widespread belief that reductions in numbers and sizes of parts are necessarily much more frequent than are increases in numbers and size has no basis in genetic or evolutionary principles. It can be supported, if at all, only on the basis of interpreting relationships between modern forms, the actual ancestors of which are unknown.

The hypothesis that the original angiosperms had flowers of moderate size is in harmony with the hypothesis that they were shrubs inhabiting pioneer habitats that were exposed to seasonal drought. Under these ecological conditions, rapid development of flowers and seeds would have had an adaptive advantage and would be most easily acquired by reduction in size of the reproductive shoots.

There is another reason for believing that the shortening of the floral axis that gave rise to whorled or subverticillate appendages occurred simultaneously with the origin of the bisexual flower. The adaptive advantage of having both pollen- and ovule-bearing structures on the same axis is the economy and efficiency of this condition with respect to the visits of insect pollinators. That being the case, the most efficient flower could be one in which anthers and stigmas were as near together as possible, a condition that exists in the flowers of the Winteraceae and other Magnoliales apart from the Magnoliaceae. Shortening of internodes is one of the modifications that is most easily produced from the developmental genetic viewpoint, since it involves only the suppression of the growth of certain internodes. Consequently, one might imagine that the same selection pressures that brought about the bisexual floral axis also brought about the compression of internodes that is characteristic of the angiosperm flower. I believe, therefore, that there never was a primitive, elongate, bisexual flower, having numerous spirally arranged, relatively widely spaced microsporophylls, and distal to them similarly arranged megasporophylls.

This entire discussion may be summarized as follows. In my opinion, the majority of botanists interested in angiosperm phylogeny, led by Hutchinson, Cronquist, Takhtajan, and Thorne, are correct in believing that the Magnoliales are among the most primitive angiosperms, and that the original angiosperm flower was a reproductive shoot, in which the parts were all separate and indefinite in number. Nevertheless, this does not require one to believe that this original flower was large in size and strobiloid in form. The numerous and obvious specializations that exist in the flowers of the modern Magnoliaceae render invalid the hypothesis that this type of flower is a prototype of that first evolved in angiosperms. Instead, the Magnoliaceae must be regarded as specialized offshoots of the Magnoliales, which evolved in the direction of increased size and numbers of parts of the flower. There is probably no modern group that has flowers similar to those that existed in the earliest angiosperms. Their nature can be best understood, moreover, by considering floral structure in orders other than the Magnoliales, as well as in that order.

Perianth and androecium

In the most primitive angiosperms, the perianth was not sharply differentiated into sepals and petals. The "tepals" of which it consisted were almost certainly modified leaves or bracts. In several genera, particularly *Calycanthus, Paeonia,* and a few species of *Hibbertia,* there is a gradual transition from leaves through bractlike structures to typical sepals or tepals. In these forms, the derivation of both sepals and petals from undifferentiated tepals is clear and is supported by developmental evidence (Hiepko 1965*a*). In other families, such as Ranunculaceae and Berberidaceae, on the other hand, petals are anatomically similar to stamens and staminodial nectaries (Hiepko 1965*b*). The petals also resemble sterile stamens in Loasaceae (Cronquist 1968:81) and Caryophyllaceae (Thomson 1942). This suggests that, depending upon the family, petals probably evolved in one of two different ways (Eames 1961). This dual origin of petals illustrates well the opportunistic nature of evolution. If a population is in an ecological situation in which conspicuous flowers have a potential adaptive advantage, those genetic changes that produce conspicuousness most quickly and with the least disturbance of the developmental pattern and the functioning of the flower as a whole will become established by natural selection. This path of least resistance is of necessity different from one group to another.

With respect to the androecium, recent evidence from both anatomy and morphogenesis requires that the classical theory concerning the nature of the original angiosperm stamen or microsporophyll must be revised. This evidence indicates that, in most groups having numerous stamens, the individual stamen is not serially homologous to the individual tepal or carpel. Instead, the homology is between tepals, carpels, and groups of stamens.

The hypothesis that stamens are derived from branchlets of a dichotomous branch system was put forward many years ago by Wilson (1937), but has been discounted by recent authors (Takhtajan 1959, Eames 1961) because most of the examples cited by Wilson are from relatively advanced families. Nevertheless, anatomical evidence from more primitive groups, as well as developmental evidence and comparisons with fossil forms, demands that this hypothesis be reexamined.

The anatomical evidence consists of the fact that in the majority of flowers having large numbers of stamens the vascular bundles that supply individual stamens emerge not separately from the axial cylinder of vascular tissue, but from a common or "trunk" bundle, which emerges from this cylinder and branches to produce the supply to the individual stamens. This condition has been illustrated for *Degeneria* by Swamy (1949; Fig. 10-4), for *Hibbertia* by Wilson (1965; Fig. 10-5), for *Paeonia* by Eames (1961), for *Cercidiphyllum* by Melville (1963), for both primitive and advanced members of the order Malvales by van Heel (1970), for *Hypericum* by Robson (1972), and for the Butomaceae by Kaul (1968). I have seen it in slides containing sections of the flowers of *Cananga odorata* and of *Guatteria olivaeformis* (Annonaceae) that I have received from Dr. Carl Wilson, as well as in cleared preparations of staminate flowers of *Kadsura* (Schisandraceae) and *Goniothalmus* (Annonaceae). The evidence presented by Tepfer (1953) suggesting that in *Aquilegia* and *Ranunculus* the vascular strands that supply the stamens are spirally arranged and emerge separately from the vascular cylinder is not supported by other studies of various genera of Ranunculaceae. In *Isopyrum*, which is related to *Aquilegia* and has less specialized flowers, the individual

Fig. 10-4. Diagram of procambial strands in the flower of *Degeneria* at two stages of development, showing branching of strands that supply individual stamens, which are outlined by stippled lines; (*s*) sepals; (*c*) petals; (*a*) androecium; (*g*) gynoecium. (From Swamy 1949.)

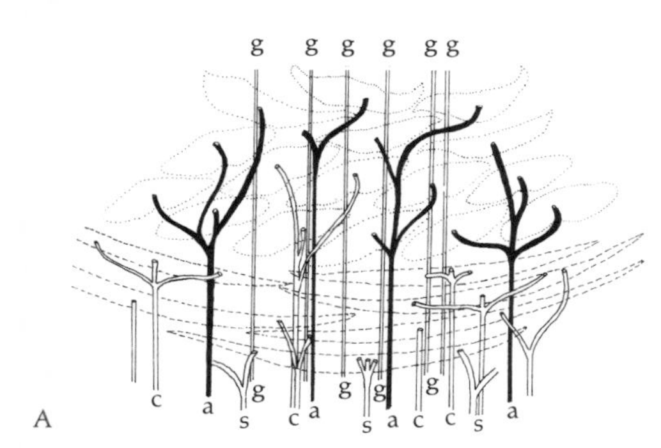

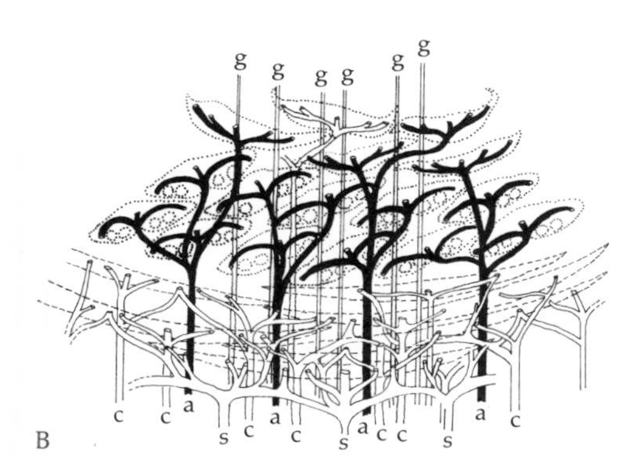

Fig. 10-5. Vascular anatomy of the androecium in four species of the genus *Hibbertia*—(*A*) *H. scandens;* (*B*) *H. vestita;* (*C*) *H. pungens;* (*D*) *H. Stirlingii* f. *squamulosa*—showing variation in stamen number associated with variation in number of staminal bundles and of stamens per bundle; each separate strand at the periphery of a diagram supplies a single stamen. (From C. L. Wilson 1965.)

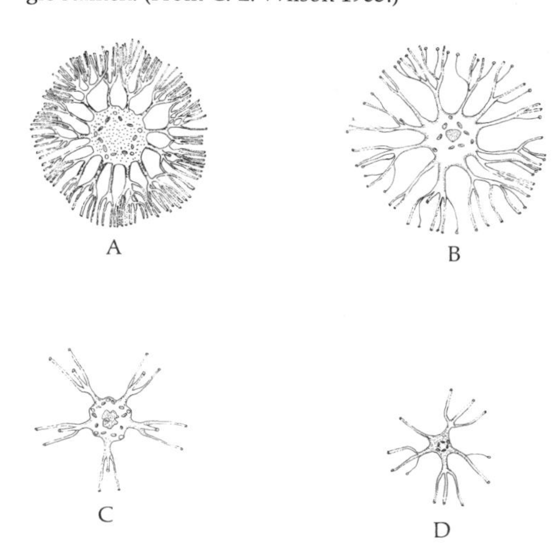

strands are parts of branching systems as in the families mentioned above (Melville 1963 and unpublished personal observations). The existence of this condition in more advanced families that have numerous stamens, such as Guttiferae, Cistaceae, Rosaceae, Papaveraceae, and Myrtaceae, has long been known (Wilson 1937, 1942, 1950, Murbeck 1912, 1941, Leins 1964*b*, Mayr 1969).

Flowers having the individual stamens and their single bundles genuinely arranged into spirals that are comparable with the phyllotactic spirals of the perianth members are found in *Illicium* (Melville 1963) and *Trochodendron* (unpublished personal observations). In these flowers, however, the stamen bundles are unusually thick, and tend to branch at the apex of the filament. On the basis of the present hypothesis, the individual stamens could be interpreted as being homologous to stamen bundles in other families, which have, during phylogeny, been reduced to single stamens. This interpretation is extended to cover all flowering plants in which flowers are oligostemonous and isomerous, that is, in which the stamens are verticillate in one or two alternating series that correspond in numbers with the perianth members, as in most Sapindales, Caryophyllales, Geraniales, and other more advanced orders.

The situation in the Magnoliaceae is complicated by the anomalous cortical vascular system that exists in their flowers. Nevertheless, the diagram presented by Skipworth and Philipson (1966) suggests strongly that in this family units comparable to tepals and carpels are not individual stamens but groups of them. My observations on cleared preparations of young flowers of *Magnolia grandifora* and *M. stellata* are that the traces to individual stamens usually emerge from trunk bundles, as in the other families mentioned above. In those flowers of the Winteraceae that have been available for study, the androecium is so condensed that the relations between the individual stamen bundles cannot be clearly seen. Nevertheless, the observations of Vink (1970) indicate that in this family the individual stamens are not produced in a spiral, acropetal succession, as would be expected on the basis of the classical theory.

A different kind of anatomical evidence against the hypothesis that stamens have evolved from flat sporophylls has been obtained by Heinsbroek and van Heel (1969) in *Victoria amazonica* (Nymphaeaceae). The large, flattened filaments of this species have vascular bundles arranged not with xylem oriented consistently toward the adaxial and the phloem toward the abaxial surface, as is true of leaves, bracts, and similar lateral appendages; instead, they are oriented in such a way as to suggest that the stamen of *Victoria* is derived from a three-dimensional structure, the homology of which is obscure.

I agree with the opinion of Carlquist (1969*c*) that, in the absence of other evidence, one must be very cautious in using evidence from the course of the vascular bundles to interpret phylogeny. With respect to the phylogeny of stamens, however, significant developmental evidence is available. This consists of a fact that was first demonstrated long ago in a general way by Payer (1857), verified by Murbeck (1912) for the Papaveraceae, by Gelius (1967) for *Philadelphus*, and by Hirmer (1917) in several families, and studied in histological detail by Hiepko (1965*a*) in *Paeonia* as well as by Leins (1964*a,b*, 1965, 1967) in *Hypericum, Melaleuca*, and *Aegle*. In these groups stamen *bundle* primordia are first differentiated from the floral meristem in positions that are related to the primordia of other parts of the flower (Fig. 10-6). In *Paeonia*, which has the floral parts spirally arranged, the bundle primordia continue the phyllotactic spiral formed by the perianth member. In verticillate flowers like those of the shrubby species of *Hypericum*, the alternate whorls consist of five sepals, five petals, five stamen bundles, and five carpels. A significant fact is that in the more advanced herbaceous species of *Hypericum*, in which the number of carpels has been reduced to three, the number of stamen bundles has been similarly reduced, apparently through fusion of two pairs of adjacent bundles (Leins 1964*a*; Fig. 10-7). Following the principle set forth by van Iterson (1907) as well as by Snow and Snow (1934, 1948), that primordia emerge from a shoot apex in positions that are determined by the largest available space not occupied by already-formed primordia,

Fig. 10-6. Diagram of a developing flower bud of *Paeonia lactiflora*, showing the spiral succession of primordia: (*8-17*) perianth; (*18-22*) stamen groups, in which the primordia of individual stamens are illustrated in only one (*22*); (*23-25*) carpel primordia. (From Hiepko 1965*a*.)

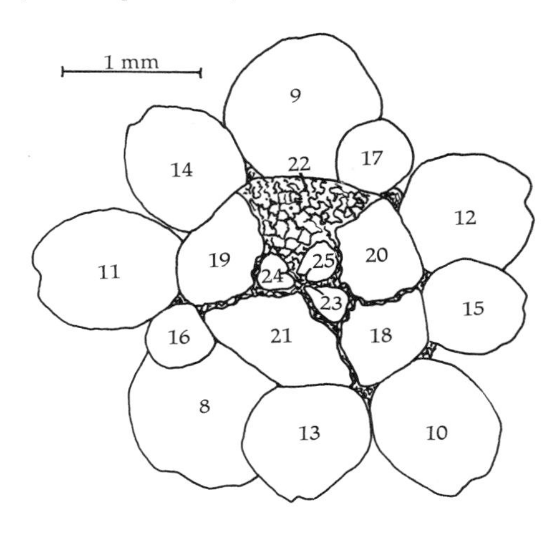

Fig. 10-7. Development of stamen bundle primordia in *Hypericum hookerianum*: (*A*) longitudinal section of half of a floral primordium, showing early differentiation of a stamen-bundle primordium (*Stb*) and a well-developed petal primordium (*P*); (*B, C*) transverse sections of a bud at the same stage; (*D-F*) transverse sections of a bud at an older stage, showing well-developed stamen-bundle primordium and early differentiation from it of a primordium for a single stamen (*Ste*); (*G*) longitudinal section of a somewhat later stage, showing well-developed stamen primordia and early differentiation of a carpel or stigma primordium (*CS*). (From Leins 1964*a*.)

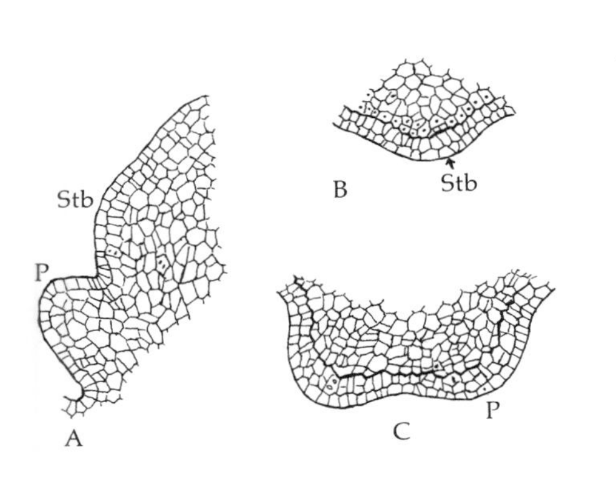

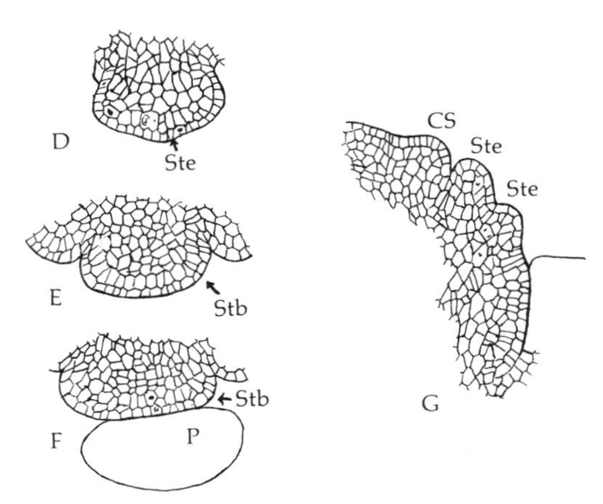

one must conclude that this principle is followed by perianth parts, carpels, and stamen bundles, but not by individual stamens in those flowers having large numbers of stamens, which are anatomically united by trunk bundles. Moreover, the differentiation of individual anthers does not follow the spiral pattern that would be expected on the assumption that they represent individual sporophylls. Instead, single-anther primordia are first differentiated almost simultaneously from each bundle primordium. From each of these initial primordia, differentiation of other primordia belonging to the same bundle complex proceeds rapidly in directions that vary according to the species. This pattern of development indicates that the developmental basis of the anatomical connections between stamens in these groups is entirely different from that of the union of petals to form a sympetalous corolla or of carpels to form a syncarpous gynoecium. In these latter examples, separate primordia are first differentiated from the floral meristem, and union results from the later development of a continuous intercalary meristem. This condition is also true of the stamen column of the Malvaceae and other families, as well as of the morphological fusion of stamens to form bundles in various genera of Myrtaceae, such as *Callistemon* (Mayr 1969). These latter examples, therefore, represent later fusions. The species concerned are descended from ancestors in which the morphological connections between separate filaments were not evident.

Cleared specimens of the remarkable stamen clusters found in the genus *Garcinia* of the Guttiferae show separate vascular bundles passing to each anther through the staminal column, rather than a single branching trunk bundle contained in it (unpublished personal observations). Consequently these structures, like the staminal bundles of the Myrtaceae, are probably of secondary derivation.

All of this evidence suggests to me that the original androecium of the angiosperms consisted of a relatively small number of branched structures, which could have been either compound, flattened microsporophylls or small lateral branchlets. The hypothesis of compound, flattened appendages or microsporophylls is part of the original strobiloid hypothesis (Arber and Parkin 1907); such structures are illustrated in their hypothetical reconstruction of hemiangiosperms.

The branchlet hypothesis would be the best interpretation of the vascular anatomy of the stamen in *Victoria*, mentioned above. However, it is not completely in accord with the evidence presented below, indicating that the angiosperm ovule is homologous to the cupule of the Caytoniales and other Mesozoic cupule-bearing Pteridosperms. In the forms in which they are preserved, the microsporangia-bearing structures may look like branchlets, as in the Caytoniales (Fig. 10-8, *A*), but they are interpreted otherwise. The nearest approach to the kind of microsporangiophore that I associate with primitive angiosperms is that found in *Pteruchus* of the Corystospermaceae (Fig. 10-8, *B*). It has been interpreted as an appendage (Townrow 1962, 1965). Such a structure could be transformed into a stamen bundle by suppression of its axis except for the vascular system, by reduction to one of the disk-like structures on which the sporangia rest, and subsequently the separation of these sporangia or anthers from the shortened axis and disk by the activation of new intercalary meristems, which gave rise to stamen filaments. Consequently, the exact homology of the stamen bundle is still uncertain.

One of the principal arguments that have been advanced in favor of the hypothesis that stamens are derived from flattened microsporophylls is that such structures exist in several groups of primitive angiosperms, such as *Magnolia, Degeneria, Austrobaileya,* and *Gabulimima* (*Himantandra*). This argument has, however, one serious weakness. These four genera, all of which are placed in the Magnoliales, differ with respect to the position of the microsporangia. In *Magnolia* and *Austrobaileya* they are on the adaxial surface, whereas in *Degeneria* and *Gabulimima* they are on the abaxial surface (Canright 1952, Eames 1961). This suggests that these four types of flattened "sporophylls" are not truly homologous with one another. Their divergence is most easily explained by assuming that they are all derived from structures which were little if at all flattened, and in which the microsporangia were borne on the margins or on the apex (Zimmermann 1959:538, Sauer and Ehrendorfer 1970). The adaxial position would then be derived by increased growth of sterile tissue on the abaxial side, and the abaxial position would be derived in the converse fashion.

Anatomical evidence that has been advanced in favor of the hypothesis that the stamens of the original angiosperms were flattened microsporophylls consists in the fact that in some primitive groups each stamen bears three vascular strands, and that in some species of *Magnolia* the three strands to a particular stamen are attached to different parts of the floral stele (Canright 1952, Eames 1961). Since this anatomical condition resembles to some degree that found in foliage leaves that are regarded as primitive, it is taken as evidence of homology between stamens and leaves.

The hazards and uncertainties in the use of such "evidence" to indicate homology have been clearly pointed out by Carlquist (1969*c*). In this particular instance, the evidence is particularly shaky. As was pointed out for *Magnolia* by Ozenda (1949), and verified by Melville (1963) as well as by my own observations on *Magnolia grandiflora* and *M. stellata,* the traces to the stamens do not arise from any vascular structure resembling a stele, but from the "cortical network." In *M. grandiflora,* Ozenda illustrates the origin of the central stamen bundle from the inner layer of the cortical network, and the lat-

Fig. 10-8. Microsporangiophores and microsporangia of (*A*) *Caytonia Nathorstii* (from Thomas 1925) and (*B*) *Pteruchus africanus* (from Thomas 1933).

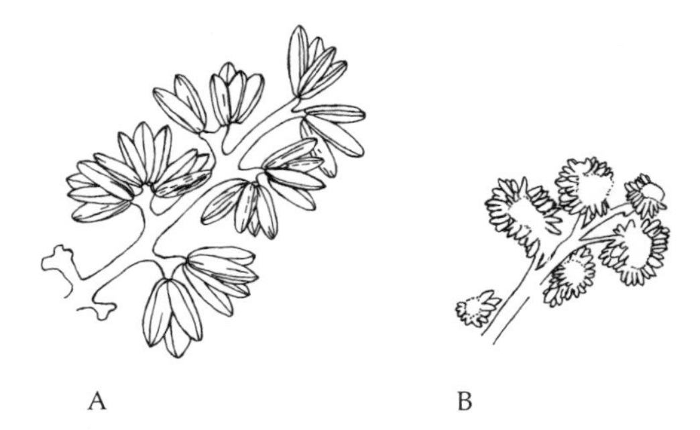

eral bundles from its outer layer. I have observed the same condition in some stamens of this species, but in other stamens of the same flowers, the three traces are united at the base of the filament. The vascular anatomy of these three-trace stamens differs in important respects from that of a three-trace foliar appendage. It is possible that the cortical network of the Magnoliaceous flower is derived by anastomosis of branches of stamen fascicles similar to those described below for the Dilleniaceae, but this requires further study.

On the basis of the assumption that modifications of microsporangium-bearing structures (microsporophylls or microsporangiophores) evolved in response to selection pressure for different methods of cross-pollination, the secondary origin of flattened stamens in early angiosperms can be explained. As is described in Chapter 4, there is direct evidence in *Magnolia* that these flowers are pollinated by Coleoptera, and the structure of the flowers in *Degeneria, Gabulimima,* and *Austrobaileya* have similar pollinators. Coleopteran-pollinated flowers are normally large, have abundant pollen, and shed their pollen in such a way that it is easily available to the insect visitors. The flowers of *Magnolia* and *Austrobaileya* are cup or bowl shaped, with the stamens spread over the inner surface of the cup, so that pollen shed from the adaxial sporangia accumulates on the bottom of the cup. In *Degeneria* and *Gabulimima,* on the other hand, the stamens are held erect, with their adaxial surfaces lying against the stigma. Given this position of the stamens, the abaxial position of the sporangia permits the pollen to be shed and to accumulate in positions that are most easily available to the insect visitors. These two kinds of flattened stamens, therefore, are regarded as alternative adaptations to coleopteran pollination. The fact that flattened stamens occur only in families of angiosperms that are regarded as relatively primitive is in accord with the much greater frequency of coleopteran pollination in these families, which probably originated and developed before the more specialized Hymenoptera, Diptera, and Lepidoptera had evolved. This kind of pollination, and the flattened stamens that are associated with it, have persisted chiefly in woody plants of tropical regions, which can tolerate relatively inefficient pollination mechanisms because of the long life of the individual plant and the equable climate, permitting each flower to remain receptive to pollinators for a long period of time.

An objection to the hypothesis that stamen bundles are primitive was raised by Corner (1946), who stated that these structures exist only in groups having centrifugal initiation of stamen primordia, a condition which he regards as derived. This objection is, however, made invalid by newer evidence already cited. Stamens arranged anatomically and developmentally into bundles are found in Degeneriaceae, Annonaceae, Rosaceae, and Myrtaceae, all of which have centripetal stamen development. The alternative conditions of centrifugal vs. centripetal stamen development apparently represent to a large extent different ways in which the floral meristem has become increased in response to selection for increased stamen number and consequently more abundant pollen. If this increase has taken place by marginal extension of receptacular meristem, centrifugal development has been the result. Distal extension of stamen-producing meristem has given rise to centripetal development (Mayr 1969). On this basis, neither of these two conditions can be regarded as more primitive or more advanced than the other.

Leins (1964*b*, 1965) has pointed out that stamen development in *Hypericum*, in which it is centrifugal, resembles the development in Myrtaceae and Rosaceae in which it is centripetal, in that in all of these groups successive stamen primordia are produced basipetally on the flanks of the already differentiated bundle primordia. In *Hypericum*, which has a convex floral meristem, the basipetal direction is of necessity centrifugal, whereas in Rosaceae and Myrtaceae, which have a concave, cup-shaped floral meristem, basipetal development is centripetal, toward the center of the "cup."

Leins believes that the families which he has studied have secondarily acquired a "false polyandry," and are derived from families having an isomerous androecium in one or two whorls, through a process of pinnation ("Fiederung") or "dédoublement." The other morphological characteristics of the families in question make this hypothesis highly unlikely. I agree in general with the phylogenetic arrangement of Takhtajan, placing the Dilleniales, Theales (including Guttiferae and *Hypericum*), and Rosales near one another and regarding them as probably descended from a common stock. Since the less specialized families of these orders all have numerous stamens, this condition appears to be original in them, and probably also in the Myrtales. Moreover, the developmental pattern of the staminal bundles in *Hypericum* is apparently very similar to that in *Hibbertia* (Payer 1857), which is very unlikely to have been derived from an isomerous, oligostemonous ancestor.

Additional developmental studies are needed in families such as the Annonaceae, in which anatomical union of stamen traces is combined with a convex floral meristem and centripetal development. In such groups, centripetal stamen development is not basipetal, as in Rosales, so that a different developmental pattern of their bundles might be expected.

Form of the pollen grains

The great majority of dicotyledonous angiosperms have pollen grains with either three apertures situated in as many furrows or grooves, or a larger number of pores or apertures. In gymnosperms, on the other hand, the pollen grains have a single aperture, usually situated on a single furrow. Uniaperturate pollen grains are found also in Magnoliales and in monocotyledons. Since pollen grains of this kind are associated with other primitive characteristics, they have been generally regarded as more primitive than triaperturate grains (Wodehouse 1935, Bailey and Nast 1943*a,b,* Eames 1961, Cronquist 1968, Takhtajan 1959, 1969). The transition from uniaperturate to triaperturate grains cannot be traced in any living group, but a possible intermediate stage has been seen in the family Canellaceae, of the Magnoliales (Wilson 1964).

Homologies of the Gynoecium

The classical concept of the carpel as a modified phyllome or leaflike structure is now so well supported by a variety of lines of evidence that, at least in apocarpous groups, such as the Magnoliales and the Ranunculales, its evolution from such structures is reasonably clear. The evolutionary process involved was chiefly infolding or conduplication, apparently associated with asymmetrical growth, so that during development the abaxial region grows more rapidly than the adaxial region. In addition, many carpels have during development a "peltate" character, which has been regarded as analogous to peltate leaves, such as those of Nymphaeaceae and *Tropeolum* (Troll 1932, Baum 1952, 1953, Leinfellner 1950, Rohweder 1967). This form of growth apparently results from the activity of an intercalary meristem at the base of the carpel. Such a meristem is most probably of secondary origin, and may have been a response to selection for greater seed production via increase in the number of seeds per carpel.

Once the phyllome character of the carpel has been accepted, four problems still remain: (1) the nature of the stigma; (2) the question of marginal vs. laminal placentation; (3) the homology of ovules and their integuments; and (4) the question whether phylogenetic union or fusion of carpels, followed by various modifications of the resulting compound ovary, can account for the numerous and diverse gynoecial structures that, according to the classical concept, are regarded as syncarpous. The first three of these problems will be considered in turn. The fourth is taken up in Chapter 11.

Homologies and evolution of the stigma

Morphologists have had two hypotheses concerning the origin of the stigma (Thomas 1934). One is that the stigma was terminal in the most primitive carpels, as it is in most modern carpels, and that the lateral stigma found in some Magnoliales, which sometimes consists only of the expanded margins of the conduplicate carpel, is derived by reduction. The second is that the lateral, marginal stigma is primitive and that the terminal stigma commonly found in angiosperms has been derived by reduction of the stigmatic surface to the upper part of the carpel (Bailey and Swamy 1951).

The following evidence suggests to me that at least in some primitive carpels the stigma was terminal and vascular, and that the terminal, vascular stigma of most modern angiosperms is, consequently, not always a derived condition. In the genus *Paeonia,* the separate carpels terminate in a broad stigma containing a large number of repeatedly dichotomizing vascular bundles (Fig. 10-9). A somewhat similar though more reduced vascular stigma exists in *Crossosoma* and some genera of Annonaceae such as *Uvaria,* both of which are apocarpous and relatively primitive compared with angiosperms as a whole. It is hard to see how this rich vascular supply could have arisen secondarily

A

Fig. 10-9. (*A*) Carpels of *Paeonia corallina,* showing short, broad, nearly sessile stigmas. (*B, C*) Cleared stigmas of (*B*) *P. lutea* (entire) and (*C*) *P. lactiflora* (split, only one half shown), showing extensively branched, dichotomous vascular bundles. (Original.)

B

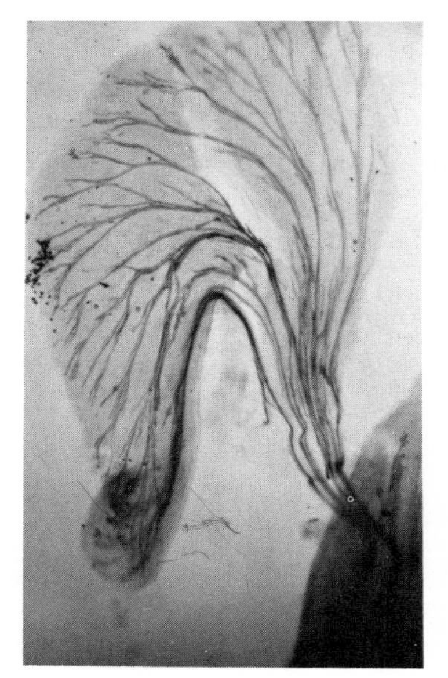

C

after the stigma became terminal, as would be required by the theory of Bailey and Swamy. Extensive vasculation of a large stigma is not required for its function to be efficiently performed. In types that are generally recognized to be relatively advanced, such as *Choisya* (Rutaceae) and *Garrya*, large stigmas are found which are superficially similar to those of *Paeonia* and the Annonaceae, but these are completely nonvascular. In my opinion, the vasculation of the carpel of *Paeonia* and similar forms is due to a homology between this structure and the sterile tip of the primitive sporophyll.

There are some reasons for suggesting that the absence of a terminal stigma in the carpels of *Degeneria* and the Winteraceae is a result of secondary reduction. These plants are primitive in many features, particularly their vesselless wood, but the gynoecium shows some decidedly specialized features. On the assumption that the primitive dicotyledons had at least one full spiral of five carpels, the

Fig. 10-10. Cleared carpels of (*A*) *Drimys piperita* (from Bailey and Swamy 1951) and (*B*) *Degeneria vitiensis*, showing laminar-lateral placentation of ovules (from Swamy 1949).

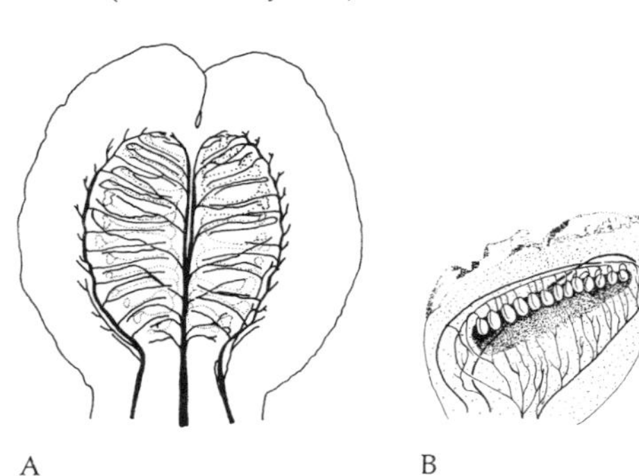

presence of two or three carpels in many genera of Winteraceae indicates that their gynoecium has suffered reduction, and on any assumption the single carpel of *Degeneria* is a derived condition. In many of the Winteraceae the size of the carpels is also reduced as compared with other woody Ranales. Furthermore, the carpels become fleshy at maturity, a characteristic that in other families of angiosperms is generally regarded as a derived condition. The reduction and disappearance of a massive vascular terminal stigma would have a positive adaptive value in association with the evolution of fleshiness, since the elimination of the woody stigmatic tissue would increase this trend, and would probably have made the fruit more palatable to the animals that normally ate it and dispersed the seed.

Homologies of the angiosperm ovule

Before 1951, proponents of the classical theory of carpel structure all were agreed that the original placement of the ovules on the phyllome was marginal or submarginal (Eames 1931). As a result of their intensive studies of the anatomy of the Magnoliales, particularly the newly discovered genus *Degeneria*, Bailey and Swamy (1951) developed the alternative hypothesis, that in the primitive carpel the ovules were arranged on its adaxial surface in two rows, situated between the midrib and the margins, a position designated by Takhtajan (1959) as *laminar-lateral* (Fig. 10-10). Bailey and Swamy regard the marginal position as secondarily derived through suppression of meristematic activity of the marginal portion of the phyllome. This hypothesis has been accepted by Cronquist (1968) and Takhtajan (1959, 1970), but other anatomists (Puri 1951, 1961, Rohweder 1967) have reiterated their belief that the submarginal position was primitive.

Evidence in favor of the primitive nature of the laminar-lateral position consists not only in the fact that the position of the ovules in such relatively primitive genera as *Degeneria* is not marginal or submarginal. In addition, in the carpels of several apocarpous genera, such as *Magnolia* (Canright 1960), and various genera of Rosaceae (Sterling 1964, 1966*a,b*), the vascular supply to the ovules emerges not from the marginal carpellary bundles, as in carpels that have typical marginal or submarginal placentation, but from special ovular bundles, which may even be more closely associated with the dorsal than the ventral or marginal bundles. It is true, as Rohweder points out, that in carpels such as those of the Ranunculaceae the marginal position is evident as soon as ovule primordia become differentiated. This condition, however, can easily be explained on the assumption that in the evolution of these carpels marginal growth has been suppressed at such an early stage that it preceded ovule differentiation.

The hypothesis that laminal placentation was primitive and characteristic of the ancestral sporophyll of angiosperms and their forerunners encounters considerable difficulties for establishing homologies between the carpel and the ovule-bearing structures of any other seed plants. In the various orders of gymnosperms, ovules were borne originally either on branched megasporangiophores, as in conifers (Florin 1944), Ginkgophytes, and Caytoniales, on individual stalks, as in Bennettitales and Gnetales (the stalks being much reduced in the latter order), or on the margins of flattened megasporophylls, as in seed ferns and cycads. Flattened megasporophylls bearing ovules on their adaxial surfaces are unknown in seed plants, living or fossil. Eames (1961) was so much impressed by this difficulty that he suggested (p. 284) that the ovules of angiosperms are emergences that are not homologous with the ovules of other seed plants. A possible solution of this dilemma is presented below.

The principal problem connected with the homology of the ovules and their integuments lies in the fact that, as most botanists now recognize (Cronquist 1968, Takhtajan 1959, 1969, 1970), the ovules of the most primitive angiosperms were anatropous, with the ovular portion bent back so that the micropyle is near the stalk or funiculus (Fig. 10-12); and that they had two well-developed integuments. The problems are: (1) Why should the primitive orientation be anatropous rather than orthotropous, as is true of the ovules of all other seed plants?, and (2) What are the homologies of the two integuments, in view of the fact that ovules of gymnosperms characteristically possess only one?

A hypothesis that may provide an answer to these problems was suggested originally by Gaussen (1946), and was conceived independently by me before I saw Gaussen's work. This is that the ovule of the angiosperms is homologous not to the unitegmic, orthotropous ovule of other seed plants, but to the cupule of the more

Fig. 10-11. Megasporangiophores and cupules of (*A–C*) *Caytonia Nathorstii* (from Thomas 1925) and (*D, E*) *Umkomasia MacLeanii* (from Thomas 1933).

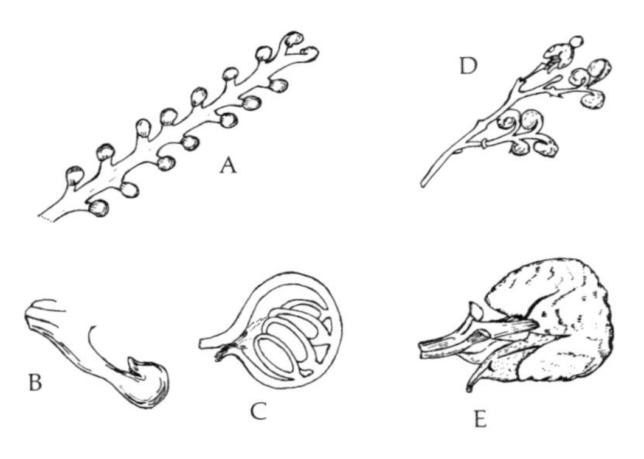

advanced Pteridosperms, particularly the Caytoniales, including both the Corystospermaceae and the Caytoniaceae (Thomas 1933, 1955, Harris 1951). The position of these cupules relative to their stalk was strikingly similar to that of an anatropous ovule having a well-developed funiculus (Fig. 10-11). In the Caytoniaceae, the cupules were multiovulate and dehiscent, but in the Corystospermaceae they were uniovulate and indehiscent. If one regards the cupule of this latter family as homologous with the angiosperm ovule, then the outer integument of the latter becomes homologous with the cupule wall, and the inner integument with the single integument of the Caytonialian ovule, as well as with the ovules of other gymnosperms. Within the "cupule," or anatropous outer integument, the "ovule" of angiosperms (inner integument plus nucellus) has an orthotropous arrangement, corresponding to other gymnosperms. The ovular stalk is much reduced, as is characteristic of both the Caytoniales and many other groups of living and fossil gymnosperms. Hence, the anatropous ovule of angiosperms is here regarded as homologous to an indehiscent uniovulate cupule.

Two strong arguments can be advanced in favor of this hypothesis. First, its acceptance provides a clue to the relationships of the extinct ancestors of the angiosperms, which otherwise are completely obscure. Second, in many families of relatively primitive angiosperms, as well as in some that are more advanced but can be related to relatively primitive groups that have the same condition, the outer and inner integuments of the ovule differ greatly from each other in their morphology and their histological structure (Brandza 1891). In these forms, the outer integument is thicker than the inner and has specialized epidermal cells, in some cases including stomata (Fig. 10-12). Moreover, the micropyle may be differ-

Fig. 10-12. Anatropous ovule at anthesis, in longitudinal section, of *Hibbertia stricta*, showing "zigzag micropyle" and strong histological differentiation of the two ovular integuments. (From Sastri 1958.)

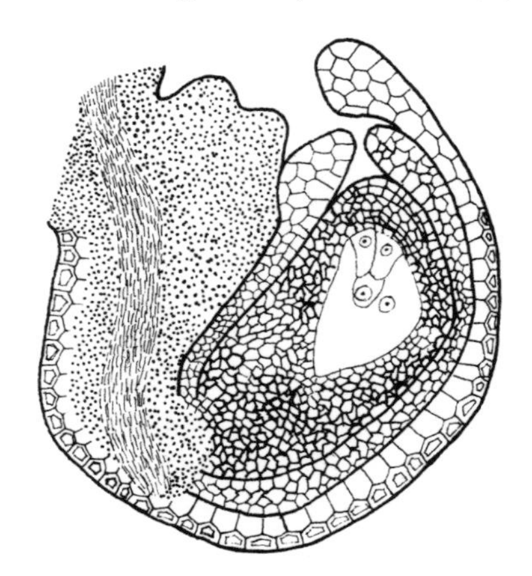

ently shaped in the two integuments, and in a few genera the distal portion of the outer integument is lobed (van Heel 1970).

A striking aspect of this difference is that in many groups the position of the opening or micropyle of the outer integument is different from that of the inner integument. This is the condition characterized by Davis (1966) as "zigzag micropyle." Table 10-1 shows the distribution of this condition among the families of angiosperms that have been investigated. In this connection our lack of knowledge of this characteristic in many of the smaller, more primitive families, such as Austrobaileyliaceae, Himantandraceae, Schisandraceae, Eupomatiaceae, Amborellaceae, Trimeniaceae, Gomortegaceae, Cercidiphyllaceae, Tetracentraceae, and Trochodendraceae, must be noted. When these families have been investigated, the distribution of ovules with dissimilar integuments may turn out to be even more widespread among primitive groups than we now realize.

Another bit of evidence in favor of this interpretation of the angiosperm ovules is

Table 10-1. Distribution of "zigzag micropyle" in families of angiosperms. Data from Davis (1966) except where otherwise indicated.

Order	Family	References
Magnoliales	Canellaceae	
	Magnoliaceae (*Magnolia, Liriodendron*)	Maneval 1914, Kaeiser and Boyce 1962, unpublished personal observations
Ranunculales	Berberidaceae (*Vancouveria*)	Berg 1972
Papaverales	Papaveraceae (*Argemone*)	Sachar 1955
Hamamelidales	Hamamelidaceae (*Hamamelis*)	
Dilleniales	Dilleniaceae	
Malvales	Bombacaceae	
	Elaeocarpaceae	
	Malvaceae	
	Sterculiaceae	
	Tiliaceae	
Violales	Bixaceae (*Cochlospermum*)	
	Caricales	Kratzer 1918
	Flacourtiaceae	
	Passifloraceae	Kratzer 1918
Capparales	Capparaceae	
	Moringaceae	
	Tovariaceae	
Rosales	Cunoniaceae	
Fabales	Leguminosae	
Myrtales	Combretaceae	
Liliales	Liliaceae (*Curculigo, Hypoxis*)	
	Dioscoreaceae (*Trichopus*)	
Cyclanthales	Cyclanthaceae	

the situation that has been reported in two genera of Capparaceae, *Capparis* (Rao 1938, Narayama 1962) and *Isomeris* (Khan 1950). Occasional ovules in these genera contain two nucelli, each with its own inner integument but surrounded by a common outer integument. On the basis of the present interpretation, these structures could be regarded as "cupules" containing two "ovules," rather than a single "ovule," as is generally true in angiosperms. Although I distrust in general evidence from teratological phenomena such as these, nevertheless, such examples serve to strengthen somewhat the hypothesis that was developed on other grounds.

If this hypothesis is accepted, then the fact must be considered that in the Caytoniales the cupules are borne not on flattened sporophylls but on branched axes.

Fig. 10-13. Bracts and cupules of *Lidgettonia africana*. (From Thomas 1958.)

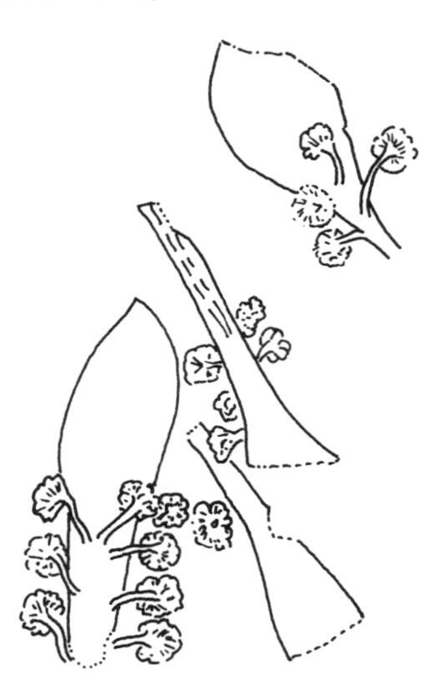

Two suggestions are offered for resolving the discrepancy that exists between this situation and the fact that in primitive angiosperms the "cupules" (that is, the ovules) are situated on the adaxial surface of a flattened sporophyll. The first is that the flattened sporophyll of angiospermous ancestors, which later became conduplicate to produce the angiosperm carpel, was derived from a cupule-bearing rachis like that of *Caytonia Nathorstii* (Fig. 10-11, *A*) via the development of an active marginal meristem.

The second hypothesis is that a branched cupule-bearing rachis, similar to that of *Umkomasia MacLeanii* (Fig. 10-11, *B*), which had an axillary position on its main axis, became adnate to its subtending bract. The carpel is thus regarded as a compound structure, analogous to the ovule-bearing cone scale of conifers (Florin 1944). This hypothesis was largely suggested by the fossil *Lidgettonia* (Thomas 1958), which was found in Permian strata of South Africa, in association with leaves of the *Glossopteris* type. In this fossil the elliptic "sporophyll" apparently bears on its surface two rows of dehiscent cupules (Fig. 10-13). Since the fossil is only an impression, of which no anatomical or histological features are preserved, the nature of the attachment of the cupular stalks to the sporophyll cannot be determined. Such a structure could, however, arise if a dichotomously branched megasporangiophore became adnate to a bract, except for the side branches, to which the cupules would have been attached. The transformation of such a structure into a primitive angiosperm carpel could occur in the following manner. First, the number of ovules per cupule could become reduced to one, and the cupule could assume the anatropous orientation found in *Caytonia* and *Umkomasia.* Second, the sporophyll could become conduplicate, as is postulated for the original angiosperm carpel, placing the cupules or "ovules" on its inner surface.

On the basis of the present rather scanty evidence, I am not ready to accept unequivocally either of these alternative hypotheses. Both of them suggest that the ancestors of the angiosperms were advanced seed ferns belonging to the Glossopteridalean-Caytonialean alliance that flourished during the Permian, Triassic, and Jurassic Periods. A preference for the second hypothesis is based upon the following considerations. In the first place, the modification of a *Lidgettonia*-like sporophyll in the manner described above is somewhat easier to imagine than the transformation of a secondary axis into a flattened phyllome by means of marginal growth. Secondly, the separate vascular bundles of the ovules in the carpels of *Degeneria* and *Magnolia* can be explained on this hypothesis by assuming that they are homologous to the bundles of the original megasporangiophore that became adnate to its subtending bract during the evolution of preangiospermous ancestors. Finally, adoption of this hypothesis resolves in part the question of "stachyospory" vs. "phyllospory" that has bothered some anatomists concerned with angiosperm evolution (Lam 1959). If the carpel is a compound structure, as this second hypothesis implies, then both it and the compound ovaries that are derived from fusion of carpels are at the same time both "stachyosporous" and "phyllosporous." The predominance of histological and developmental characteristics that have led to the hypothesis of stachyospory would then be due to preferential growth during development of placental tissues, which are homologous to the ancestral sporangiophore. The "phyllosporous" condition would result from preferential development of carpel wall tissue that is homologous to the subtending bract in the ancestral form. Under this interpretation, every possible intermediate stage between "stachyospory" and "phyllospory" would be expected in different angiosperms, and this is exactly what is found.

The course of evolution in the female reproductive structures in the evolutionary line that led to the angiosperms, as postulated by the present theory, may be summarized as follows:

1. Existence in primitive vascular plants of dichotomous sporangiophores bearing terminal sporangia;
2. Differentiation of micro- and megasporangiophores from one another, with drastic reduction of spore number in the latter;
3. Surrounding of the megasporangium with an envelope derived by modification of flattened sterile branches, which thus become the original ovule integument, as in *Lyginopteris* and many other Paleozoic seeds; the sporangium wall thus becomes the nucellus;
4. Grouping of these primitive seeds into a bowllike cupule, as in the Carboniferous genus *Calathospermum* (Walton 1953);
5. Reduction in size of cupule and ar-

rangement of cupules on short, probably lateral sporangiophores, as in Caytoniales and their relatives;

6. Adnation of these sporangiophores to subtending bracts, as in several Glossopteridean genera;

7. Reduction of the primary ovules to one per cupule, transforming the cupule wall into the outer ovular integument;

8. Infolding of the bract to form the closed carpel, with the sterile tip of the bract becoming transformed into the stigma;

9. Adnation of the bundles of the sporangiophore to the ventral carpel bundles, to give the typical three-bundle carpel of primitive angiosperms.

The Ancestral Form of the Embryo Sac and Embryo Development

The eight-celled embryo sac, containing an egg cell, two synergids, two polar nuclei, and three antipodals, designated as the *Polygonum* type by embryologists (Maheshwari 1950, Davis 1966), is so widespread in angiosperms that it must certainly have existed in their common ancestors. The various modifications of it, such as the *Peperomia, Allium,* and *Fritillaria* types, represent specializations that have occurred sporadically in isolated groups. Equally universal is double fertilization and the triploid endosperm. The extent to which these common characteristics indicate a monophyletic origin for the angiosperms is considered below.

Until recently, all known examples of embryo development in angiosperms differed in an important respect from nearly all orders of nonangiospermous seed plants. The embryo develops directly from the egg cell, starting with a single cell division, accompanied by wall formation that separates the embryo proper from the suspensor. This form of development is so simple, direct, and widespread that a natural and logical conclusion is that it is primitive and ancestral. Nevertheless, in nearly all living "gymnosperms," including cycads, *Ginkgo,* conifers, and *Ephedra* in the Gnetales (Johansen 1950), a coenocytic proembryo stage is intercalated between the zygote and the embryo proper. The exceptional genera that lack a coenocytic stage, *Gnetum* and *Welwitschia,* are highly specialized with respect to other aspects of embryo development. The coenocytic stage differs greatly from one group to another. It may or may not involve the multiplication of embryos by cleavage.

Because of these facts, I believe that the coenocytic proembryo stage that has recently been found in the genus *Paeonia* (Yakovlev and Yoffe 1957, Cave, Arnott, and Cook 1961, Mathiessen 1962) has more than ordinary significance. To be sure, embryogeny in *Paeonia* is radically different from that in any living gymnosperm. This fact might be expected on the basis of any hypothesis that postulates, as the ancestors of angiosperms, a group that is completely extinct and that was only distantly related to the ancestors of any of the modern orders of gymnosperms. The peculiar situation in *Paeonia* is very difficult to imagine as a recent specialization. Any functional or adaptational significance for it is inconceivable. Moreover, *Paeonia* is adapted to regions having a highly seasonal climate in which selection pressures would be expected to favor acceleration of embryo development. The evolution *de novo* of an intercalated stage that prolongs development is, therefore, particularly difficult to imagine. *Paeonia* displays many other primitive characteristics (see Table 10-2), and its low chromosome number renders highly unlikely its descent from any modern family that is equally primitive in all of these characteristics. Consequently, the coenocytic proembryo in this genus is here regarded as a vestigial character, which reflects the course of embryo development in the extinct Glossopteridalean or Caytonialian alliance from which the angiosperms are believed to have evolved.

The original nature of endosperm development in angiosperms is problematical. Both Takhtajan (1959) and Cronquist (1968) postulate that the nuclear pattern of development, in which the initial formation of endosperm is by nuclear division without cell-wall formation, is original and both the cellular pattern, characterized by *ab initio* cell-wall formation, and the intermediate helobial pattern are derived. Nevertheless, according to Davis (1966) and several of the original references cited by her that I have checked, endosperm development is *ab initio* cellular in all families of Magnoliales that have been investigated, except for Myristicaceae and most Lauraceae. It is also cellular in Piperales, except for the genus *Piper*, in Lardizabalaceae of the Ranunculales, in Trochodendraceae, and in most of the Nymphaeaceae. On the other hand, it is nuclear in other orders that are regarded as relatively primitive, such as Dilleniales, Theales, Malvales, and most of the Ranunculales. It is either nuclear or helobial in all of the monocotyledons. Finally, cellular endosperm reappears in the most advanced orders of dicotyledons, particularly the Lamiales, Scrophulariales, Campanulales, Dipsacales, and Asterales.

Wunderlich (1959), after a careful review of the relation between endosperm development and other characteristics of the nucellus and embryo sac in a large number of angiosperm families, has concluded that cellular endosperm is correlated with a small nucellar cavity or space, and nuclear endosperm with a large cavity. She regards the presence of a small cavity as primitive, largely because both this condition and cellular endosperm are found in the woody Magnoliidae.

If a coenocytic proembryo is regarded as an ancestral condition of angiosperms, then the coenocytic or nuclear development should be regarded as less specialized than cellular endosperm development. Since endosperm and embryo development occur simultaneously and in nearly the same site, the formation of cell walls in the endosperm while the proembryo is undergoing nuclear division without wall formation would require a marked differentiation between adjacent tissues of similar origin. Moreover, on the basis of any phylogenetic scheme that has been proposed, the sympetalous orders that have cellular endosperm must be descended from ancestors that had nuclear endosperm. On the other hand, the apparently consistent presence of cellular endosperm in Magnoliales indicates that, if this form of development is derived, it must have appeared for the first time very early in angiosperm evolution. The presence of nuclear endosperm in such families as Lauraceae, which are probably derived from a common ancestor with other Magnoliales, suggests that a reversion from the cellular to the nuclear condition is not impossible. Consequently, the character nuclear vs. cellular endosperm must be regarded as sufficiently reversible that its phylogenetic value is dubious.

I agree completely with other authors in believing that primitive angiosperm seeds had a small embryo surrounded by copious endosperm. In fact, the reduction of the amount of endosperm and the increase in the size of the embryo can be traced in many groups of trees, both temperate and tropical, and is one indication of the specialized nature of many tropical arboreal genera.

The Probable Form and Structure of the Earliest Angiosperms

The argument of the previous sections can be summarized by the following description of the hypothetical earliest angiosperms. They inhabited pioneer habitats in regions having a tropical or subtropical climate, equable as to temperature but semiarid, in which seasons of adequate moisture alternated with seasonal drought. They were low-growing shrubs, with slender or moderately thick stems that bore many leaves in a spiral phyllotaxy and were unbranched or little branched. Their leaves were small or moderate in size, undivided, with entire or dentate margins, short petioles, and no stipules. Their wood was in a continuous cylinder, vesselless, and with primitive, heterogeneous parenchyma rays. Their nodes were either unilacunar with two leaf traces or trilacunar with four traces. The guard cells of the stomata were about the same size as the neighboring epidermal cells, and were flanked by an indefinite number of subsidiary cells.

Their flowers were solitary at the ends of the stems, or also on short leafy branches, forming a leafy cyme. The transition from leaves to sepals was gradual. The parts of the flower were spirally arranged or subverticillate. They were indefinite in number, but not necessarily very numerous. Colored petals were either absent or similar to and continuous with the sepals. Pollination was by insects, either Coleoptera or primitive Hymenoptera. The stamens were in clusters, or were inserted on short lateral emergences or branches. Their filaments were short or nonexistent. Their anthers dehisced by lateral slits. The pollen was monocolpate.

The gynoecium consisted of a moderate number of subverticillate or verticillate carpels that were separate from one another and dehisced by adaxial sutures. The stigmatic surface extended from the adaxial suture to the terminal, sterile part of the carpel, which may well have been extended into a broad, sessile terminal stigma.

The ovules were borne on the adaxial surface of the carpel or were submarginal. They were indefinite in number, but not very numerous. They were anatropous, bitegmic, crassinucellar, and with a "zigzag micropyle," the two integuments being well differentiated from each other. The embryo sac was of the *Polygonum* tupe, octonuclear before fusion of the polar nuclei. They had double fertilization and a triploid endosperm, the development of which was initially nuclear, without cell-wall formation. The initial embryo formation was indirect, with a coenocytic pro-embryo preceding the formation of the true embryo. The mature seed had a small, dicotyledonous embryo embedded in a copious endosperm.

No modern angiosperms possess all of these characteristics. Table 10-2 lists the advanced conditions with respect to these characters that exist in the most generalized members of those families that most nearly approach the hypothetical primitive condition. This table shows that five families—Winteraceae (7), Dilleniaceae (6), Illiciaceae (6), Amborellaceae (5–7), and Magnoliaceae (8)—diverge from this hypothetical primitive condition with respect to approximately the same number of characters, but in different directions. Two of these, Magnoliaceae and Winteraceae, are placed in the Magnoliales by all authors. The Illiciaceae are placed in the Magnoliales by Hutchinson, Cronquist, and Thorne but in a separate order, Illiciales, by Takhtajan. *Amborella* is placed in the Magnoliales by Cronquist and Thorne, but in the Laurales by Takhtajan and Hutchinson. The Dilleniaceae are placed in a separate order by all modern authors. Consequently, these least-specialized families represent at least two and possibly four different orders. As Bailey and Nast (1945*a*) remarked, even the Winteraceae and Magnoliaceae, which are almost always associated with each other, possess in common only those characters that represent generalized states. With respect to their more specialized characteristics, including nodal anatomy, development of stipules, inflorescence, floral anatomy, number and dehiscence of carpels, structures of ovules, and basic chromosome numbers, these two families diverge widely from each other. The Illiciaceae, which because of their centripetal stamen development are usually placed closer to the Magnoliales than to the Dilleniales, nevertheless resemble the latter order in their tricolpate pollen and adaxially dehiscent carpels.

This divergence in morphological characteristics that is displayed by modern relatively unspecialized orders is reinforced by a biochemical characteristic that has been intensively studied: the nature of the most generalized phenolic compounds (Bate-Smith 1962, 1965, Kubitzki 1968, 1969). The Illiciaceae and Dilleniaceae, along with most of the orders that are associated with the Dilleniales as possible derivatives of a common stock (that is, the Dillenidae and Rosidae of Cronquist; see Fig. 11-1) nearly all possess the generalized phenolic compound, ellagic acid. This compound is absent from the other families that are generally included in the Magnoliales, as well as from the herbaceous Ranunculales and the monocotyledons. Bateman and Kubitzki agree in postulating that various widely distributed phenolic compounds found in more specialized families are probably derived from ellagic acid via more complex biosynthetic pathways.

For all of these reasons, the selection of any single modern family or order as closer than any of the others to the original primitive angiosperms requires so

Table 10-2. Distribution of advanced conditions with respect to characters in 17 generalized families of angiosperms.

Characters chosen largely from those listed by Davis and Heywood (1963), the anatomical characteristics from Metcalf and Chalk (1950) and additional ones from Takhtajan (1966). Characters not included by any of these authors are marked with an asterisk (*), and explained in the text. Families are assigned a + (advanced condition) for a character only if all members of the family possess this advanced condition.

Character[a]	Family																
	Austrobaileyaceae	Magnoliaceae	Winteraceae	Degeneriaceae	Himantandraceae	Annonaceae	Myristicaceae	Canellaceae	Illiciaceae	Schisandraceae	Eupomatiaceae	Amborellaceae	Monimiaceae	Tetracentraceae	Trochodendraceae	Dilleniaceae	Paeoniaceae
1. Climber vs. shrub or tree	+	0	0	0	0	0	0	0	0	+	0	0	0	0	0	0	0
2. Vessels present vs. absent	+	+	0	+	0	+	+	+	+	+	+	0	+	0	0	+	+
3. Vessel perforations simple vs. scalariform	0	0	0	0	0	+	0	0	0	0	0	0	0	0	0	0	0
4. Intervascular pitting not scalariform vs. scalariform	+	0	0	0	+	0	+	0	0	0	0	0	0	0	0	0	+
5. Intervascular pitting alternate vs. not alternate	0	0	0	0	+	0	0	0	0	0	0	0	0	0	0	0	0
6. Nodes multilacunar vs. uni- or trilacunar	0	+	0	+	0	0	0	0	0	0	+	0	0	0	+	0	0
7. Leaves compound vs. simple	0	0	0	0	0	0	0	0	0	0	0	0	0	0	0	0	+
8. Phyllotaxy decussate vs. spiral	+	0	0	0	0	0	0	0	0	0	0	0	+	0	0	0	0
9. Stipules present vs. absent	0	+	0	0	0	0	0	0	0	0	0	0	0	+	0	0	0
10. Transition to reproductive shoot abrupt or separate vs. gradual*	+	+	+	+	+	+	+	+	+	+	+	+	0	+	+	0	0
11. Inflorescence well developed vs. flowers solitary or in a leafy cyme[b]	0	0	+	0	0	0	+	0	0	0	0	0	0	+	+	0	0
12. Flowers secondarily unisexual vs. bisexual	0	0	0	0	0	0	+	0	0	+	0	+	+	0	0	0	0
13. Receptacle cup shaped vs. flat or convex	0	0	0	0	0	0	0	0	0	0	0	+	+	+	0	0	0

15. Perianth strongly differentiated into sepals and petals vs. poorly differentiated or absent	0	0	0	+	+	+	+	+	0	0	+	0	0	0	0	+	+
16. Perianth much reduced or absent vs. perianth present	0	0	0	0	0	0	0	0	0	0	0	0	0	+	+	0	0
17. Stamens 10 or fewer vs. numerous	0	0	0	0	0	0	0	+	0	0	0	0	0	+	0	0	0
18. Stamen filaments narrow, elongate, or united into a column, vs. filaments short, stout, and separate	0	0	0	0	0	0	+	+	0	+	0	0	0	+	0	0	+
19. Stamen development centrifugal vs. centripetal	0	0	0	0	0	0	0	0	0	0	0	0	0	0	0	+	+
20. Pollen grains tricolpate or otherwise modified vs. monocolpate with 1 aperture	0	0	+	0	0	0	0	0	+	+	+	0	+	+	+	+	+
21. Pollen grains in tetrads vs. separate	0	0	+	0	0	0	0	0	0	0	0	0	0	0	0	0	0
22. Carpels partly or wholly united vs. carpels free	0	+	0	0	+	0	0	+	0	0	+	0	0	+	+	0	0
23. Carpels indehiscent or dorsally dehiscent, vs. carpels ventrally dehiscent	0	+	+	+	+	0	+	+	0	+	+	+	+	0	0	0	0
24. Carpels reduced in number vs. carpels 5 or more	0	0	+	+	0	0	+	+	0	0	0	0	0	+	0	0	+
25. Style well developed vs. stigma sessile or lateral	+	0	0	0	+	0	0	0	+	0	0	0	0	0	0	+	0
26. Ovules 1–2 per carpel vs. ovules more numerous	0	0	0	0	+	0	+	0	0	+	0	+	+	0	0	0	0
27. Ovule integuments little differentiated or the outer poorly developed vs. integuments well differentiated and with usually "zigzag micropyle"*	?	0	+	+	?	+	+	0	+	+	?	?	+	?	?	0	0
28. Embryo development direct vs. coenocytic proembryo*	?	+	+	+	?	+	+	+	+	+	?	?	+	?	?	+	0
Total of advanced states	7–9	8	7	9	7–9	9	12	11	6	10	8–10	5–7	10	11–13	8–10	6	8

[a] Advanced condition is listed first, and is indicated by a + sign. If the specialized structure (for example, a vessel) is absent, the family is assigned a 0.

[b] Flowers in some genera of Winteraceae solitary through secondary reduction.

much distortion or overlooking of relevant facts that it must be regarded as a misrepresentation. The common ancestors of modern angiosperms were families and orders that are now completely extinct. This conclusion should not be regarded as surprising or revolutionary. This condition is known to be true of all major groups of animals and plants for which an adequate fossil record is available.

Possibilities for weighting individual characters

A possible source of disagreement with the conclusion reached in the last paragraph could be based on the fact that none of the characters listed in Table 10-2 are given any preferential weighting relative to the others. Most botanists would probably agree in believing that not all characteristics are of equal importance in determining the degree of phylogenetic advancement of a group. On the other hand, considerable disagreement might exist with respect to which characters are most important and which are less. Consider, for instance, the differences between the four families that have the smallest numbers of advanced states: Magnoliaceae, Winteraceae, Illiciaceae, and Dilleniaceae (*Hibbertia*). The characters involved are presence vs. absence of xylem vessels; numbers of lacunae and vascular strands at the stem-leaf nodes; distinctness of the reproductive parts and development of the inflorescence; differentiation of the perianth; centripetal vs. centrifugal development of the individual stamens; structure of the pollen; presence vs. absence of a distinct style; dehiscence vs. indehiscence of the carpels; development of the outer ovule integuments; and cellular vs. nuclear endosperm.

These characters affect very different parts of the plant, and so cannot be compared directly. Indirect criteria for comparison might be based on one or another of the following considerations.

(1) Those characters for which differences between states are known to exist only between families and never between different genera of the same family might be regarded as more important than those in which, in at least some families, differences between states exist at the level of genera. Of all the 29 characters listed in Table 10-2, only three differ, so far as is known, exclusively at the family level: stamens centrifugal vs. centripetal; pollen grains tricolpate vs. monocolpate; outer ovule integument reduced vs. well developed. With respect to these three characters the distribution of advanced vs. primitive states in the four families mentioned above are: Magnoliaceae, 0–3; Winteraceae 1–2; Illiciaceae 1–2; Dilleniaceae 2–1. Since in the Magnoliaceae the centripetal stamen development is associated with a unique anatomical specialization of the androecium, the development of a complex cortical vascular system, the androecium as a whole must be regarded as just as specialized in this family as in the other three. Consequently, weighting characters on the basis of the distribution of their states leaves unaltered the conclusion that these four families are all about equally primitive or specialized.

(2) A second basis of weighting would be to follow Sporne (1948, 1959, 1960), and give greater weight to those characters that show the highest correlation with

the rest. An attempt was made to determine the degree of correspondence between advanced states with respect to 12 of the 29 characters listed in Table 10-2, selecting those that differentiate the four families having the smallest number of advanced states. The result of this attempt was that all of the 12 characters are correlated with one another to about the same degree. With respect to primitiveness or advancement, each character has a mean concordance with the other 11 characters of 4.6 (42 percent) to 5.5 (50 percent) with the exception of a single character, cyclic vs. spiral arrangement of the perianth segments, for which the value is 6.0 (55 percent). With respect to this character, the Magnoliaceae, Winteraceae, Illiciaceae, and Dilleniaceae all have the unspecialized state, so that for them it is nondiscriminatory.

(3) The third, and undoubtedly the most significant, method of weighting would be on the basis of the number of genes controlling each character difference and the degree of their developmental correlation with one another. Since, however, nothing at all is known about the genetic basis of any of the characters concerned, this criterion cannot be used. Consequently, no usable system of weighting characters, other than purely subjective and speculative opinion, gives results that differ from those based upon equal weight for all characters.

Monophylesis or Polyphylesis?

The foregoing discussion should have made clear my belief that the angiosperms are not strictly monophyletic, to the extent that they are descended from a single identifiable common ancestor. On the other hand, they are almost certainly not completely polyphyletic, in the sense that different orders of the class are descended from entirely different groups of preangiospermous plants, and have become similar to one another by virtue of convergent evolution, as has been postulated by Lam (1952, 1961), Emberger (1950, 1960), Melville (1960, 1963), and Meeuse (1965). The strongest arguments for rejecting the latter point of view are (1) the lack of any generally recognizable discontinuities by means of which dicotyledonous angiosperms can be separated into two or three clearly defined groups, and (2) the common possession by both dicotyledons and monocotyledons of the octonucleate embryo sac, double fertilization, and triploid endosperm.

Nevertheless, rejection of this point of view does not require acceptance of the hypothesis that all of the advanced characteristics found in most angiosperms—xylem vessels, bisexual flowers, bi- or trinucleate pollen grains, closed carpels or their derivatives, bitegmic anatropous ovules, octonucleate embryo sacs, double fertilization, triploid endosperm, and direct embryo development—originated at the same time in a single ancestral group. Vessels almost certainly originated separately in dicotyledons and monocotyledons (Cheadle 1953), and they may have originated separately in different orders of dicotyledons. Certainly, the evolution of vessels took place after the angiosperm flower had evolved its present general plan. The incomplete closure of carpels in several distantly related groups, such as

Winteraceae and Butomaceae (Eames 1961), might lead one to speculate that the closed carpel originated independently in more than one evolutionary line. In view of the facts discussed in the next chapter, indicating that the union of carpels to produce a compound ovary took place independently in many different lines, this hypothesis is by no means unlikely. It is not contradicted by the presence of the octonucleate embryo sac and double fertilization in all generalized angiosperms. Although this condition does not exist in any modern gymnosperms, one cannot conclude from this fact that it evolved simultaneously with the evolution of bisexual flowers and closed carpels. The extinct Glossopteridales and Caytoniales, which I believe to be near to the actual ancestors of angiosperms, are only distantly related to living orders of gymnosperms. Consequently, we can never know what their gametophyte and embryo development was like. However, we do know that, compared with modern gymnosperms, they had relatively small seeds (Harris 1951, 1964, Thomas 1955). Furthermore, some of them lived in harsh, seasonal climates in which morphological reduction associated with acceleration of developmental stages would have had a high adaptive value. Consequently, the hypothesis that octonucleate embryo sacs and double fertilization existed already in the extinct gymnosperms that were the immediate ancestors of angiosperms is by no means unreasonable. If this were so, the independent origin of other angiospermous characteristics by parallel evolution in several independent lines that originated from a common stock is not hard to postulate.

This argument is presented largely to show that the strict dichotomy, monophylesis vs. polyphylesis, is a taxonomic rather than an evolutionary concept. It assumes that the combinations of characters existing in modern groups also existed in all of their extinct ancestors. The evolutionist recognizes, however, that extinct groups may have possessed character combinations that were quite different from any combinations possessed by modern groups. This belief is supported by the situation in all groups of which an adequate fossil record is available.

Summary

Current hypotheses about the origin of angiosperms are of three kinds: (1) "classical" hypotheses, which follow Arber and Parkin in regarding the strobiloid form of flower as the most primitive condition and the Magnoliaceae as similar to the earliest angiosperms; (2) an intermediate group, which regards the angiosperms as basically monophyletic but does not attempt to identify any modern group as the most primitive angiosperms; (3) theories put forward by Melville, Meeuse, Emberger, and others, which regard the angiosperms as polyphyletic, and suggest affinities with Gnetales, Pentoxylales, and various other groups of plants.

As a basis for deciding between these hypotheses, the distinctive features of angiosperms are summarized. The theory is advanced that they originated in a climate having a marked seasonal drought and a short season favorable for the formation of flowers and seeds. Under such conditions, strong selective pressures would exist for speeding up the reproductive cycle, thereby favoring morphological reductions as well as the rapid endosperm and embryo development that results from double fertilization. Highly seasonal climates tend to favor the most advanced forms of plants that exist at any period of time. When angiosperms were first evolving, only those similar to modern Magnoliidae existed, and they were the most advanced vascular plants in existence. In modern times, these original angiosperms have been displaced by their still more advanced descendants, and the only surviving Magnoliidae are those that radiated into more favorable and stable habitats, where they have been preserved as relictual forms. The time of angiosperm origin is uncertain, but the most likely geological periods are the Triassic or the Jurassic.

Evidence from reproductive ecology, morphology, and ontogeny favors the hypothesis that the strobiloid flower of the Magnoliaceae represents an early specialization rather than a primitive or ancestral condition. The primitive and ancestral angiosperms are believed to be completely extinct and not represented by any single order, living or fossil. They are postulated as low-growing shrubs, having spirally arranged simple leaves lacking stipules. Their wood formed a single vascular cylinder and lacked specialized vessels or parenchyma cells, other than wood rays. Their flowers were borne at the ends of branches or, more probably, in a loose, leafy cymose inflorescence. They were bisexual, with the floral axis already short and compressed. The floral envelopes were undifferentiated. The stamens were arranged in bundles that represented compound sporophylls, each bundle being formed in a phyllotactic spiral continuing that of the floral envelopes. The stamen filaments were short. The pollen was monocolpate. The gynoecium consisted of carpels that were infolded megasporophylls bearing terminal stigmas, no stylar elongation, and ovules in a laminar or submarginal position. The ovules were bitegmentary and anatropous, being homologous to uniovulate indehiscent cupules such as are found in some Caytoniales. The earliest development of both endosperm and embryo was by nuclear divisions without cell-wall formation. The seeds had copious endosperm and a dicotyledonous embryo.

11 / Trends of Specialization within the Angiosperms

In this and the two following chapters, an attempt is made to explain some of the principal trends of morphological specialization in flowering plants on the basis of the principles and factors discussed in previous chapters. Admittedly, many of the hypotheses offered are speculative. They are, however, presented in the hope that botanists who are interested in phylogeny will come to realize that the explanations which have proved to be highly satisfactory for many groups of animals (Rensch 1960, Mayr 1963) can also be applied to flowering plants, provided that the right kinds of factual knowledge are available. In the last chapter of this book, suggestions are offered about the kind of research that is needed in order to provide this knowledge.

A scheme of relationships between the orders and families mentioned in the text and tables of this chapter is presented in Fig. 11-1. Its format was suggested by Sporne (1956), Rodriguez (1956), Eckardt (1964), and others, and used by me (Stebbins 1956*a*) for the family Gramineae. This format is used rather than any modification of the conventional phylogenetic tree because the common ancestor of any two or more advanced orders was probably so different from any of them that, if it were known, it would be placed in a separate extinct order. The diagram is a series of cross sections of the "branches" of the "evolutionary tree" that represent as nearly as possible their affinities to one another, as well as their degree of specialization, which is represented by their radial distance from the center of the diagram, the position of the extinct ancestral complex of primitive angiosperms. The

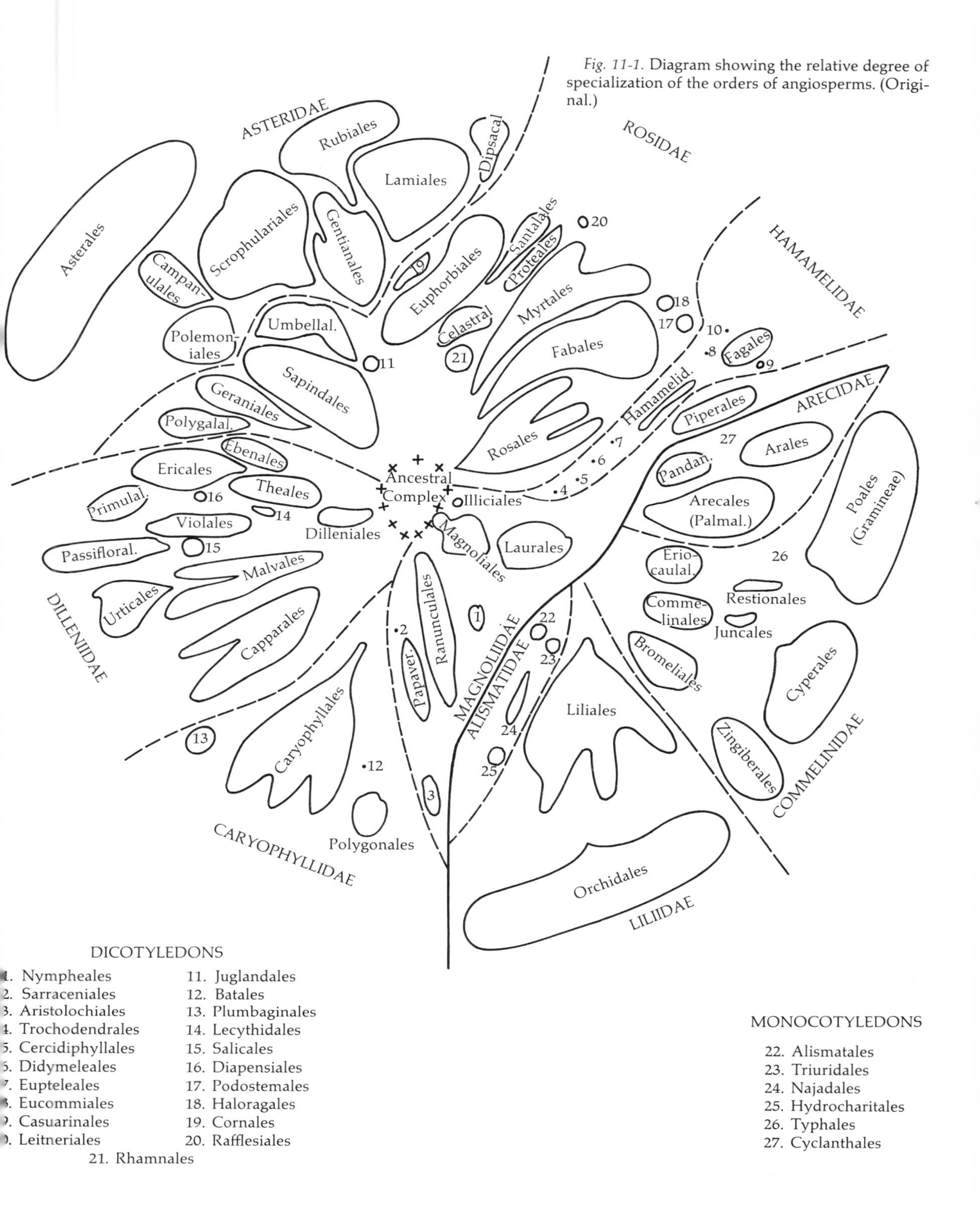

Fig. 11-1. Diagram showing the relative degree of specialization of the orders of angiosperms. (Original.)

characters that formed the basis for estimating these distances are chiefly the eight that appear most frequently in the taxonomic keys to families (Stebbins 1951): apetaly, sympetaly, zygomorphy, reduction in stamen number, syncarpy, reduction in ovule number, parietal, basal, or free central placentation, and epigyny. Additional characters were used for particular groups, especially the herbaceous condition, saprophytism, parasitism, and other vegetative specializations, unisexual flowers, union of stamen filaments, specialized inflorescences, and so on. The size of its "cross section" reflects as nearly as possible the number of species in the order. The radial length expresses the amount of specialization found within the order. In some instances, markedly divergent lines of specialization within an order are expressed by centrifugal lobes of its "cross section."

After careful study of various modern systems, that of Cronquist (1968) was selected as the principal basis of the diagram. The solid, heavy line separates monocotyledons from dicotyledons. Within each of these two classes, the subclasses recognized by Cronquist are bounded by broken lines. Deviations from Cronquist's system were made chiefly in accord with some of Takhtajan's suggestions; the separation of the Illiciales, Laurales, Cercidiphyllales, Didymeleales, Eupteleales, and Fabales as separate orders, and a few other minor changes. In addition, the proposals of Thorne (1968) were followed with respect to the position of the Urticales and Juglandales. Where possible, the subclasses and orders were placed in positions such that alternative hypotheses concerning phylogenetic connections could be represented by drawing lines in any one of two or more possible directions. Thus, the Hamamelidae are placed between the Magnoliidae and the Rosidae, so that appropriate lines between these subclasses could be drawn according to either the hypothesis of Takhtajan and Cronquist, that the Hamamelidae are specialized derivatives of Magnoliidae, or that of Endress (1967), which derives them from the Rosidae.

Trends in Growth Habit

With respect to growth habit, the widely accepted classification of Raunkiaer (1934) can serve as a basis for discussion. Botanists are generally agreed that the members of his first class, that of phanerophytes, are the most generalized, as he believed them to be. Nevertheless, this is a very heterogeneous group, having in common only the characteristic that their buds and apical shoots are far above ground and do not die back during unfavorable seasons of cold or drought. Within this group of life forms, one can distinguish between trees, woody vines, and shrubs, which, respectively, have a single main stem or trunk that is self-supporting, a main stem that is slender, relying on other plants for support and usually with a rather complex anatomy, and no main stem. In the case of both trees and shrubs, one can distinguish between those that are able to regenerate new stems or trunks from the root crown and those that are not. A fourth group of phanerophytes, those having herbaceous stems, occurs in tropical floras.

In Chapter 10 and in a previous publication (Stebbins 1965), reasons have been given for my belief that, although the original angiosperms were probably woody, they were more likely small, slender-branched shrubs than massive, tall trees. This does not imply, however, that all shrubs are believed to be more primitive or older than trees. With respect to growth habit, perhaps more than any other single characteristic, reversals in the direction of evolution might be expected to occur. This is certainly true with respect to trees and shrubs. Although the considerations mentioned above suggest that most if not all modern dicotyledonous trees have evolved from shrubby ancestors, the reverse direction of evolution, from trees to shrubs, has undoubtedly occurred many times. Probable examples can be found in the genera *Magnolia, Quercus, Acer,* and *Salix.* These trends have taken place in response to the general climatic trends that occurred during the Tertiary Period, from moist equable conditions to increasing prevalence of drought and cold.

Under some conditions, the generally recognized trend from shrubs to herbaceous phanerophytes, or to subshrubs and herbs that are characterized by seasonal dying back followed by regeneration (chamaephytes, hemicryptophytes, cryptophytes), has been reversed. Shrubs or even trees can be descended from herbaceous ancestors. The most striking examples of such reversals are monocotyledons such as palms and bamboos, which are discussed in Chapter 13. Some clearly documented examples in dicotyledons are certain genera of oceanic islands, such as *Dendroseris* on Juan Fernandez, *Sonchus* and *Echium* on the Canary islands, *Scaevola* on various islands of the Pacific, and Campanulaceae-Lobelioideae on Hawaii (Carlquist 1966*b*, 1967, 1969*a,b*, 1970). Carlquist gives both an adaptation-based reason why this reversal of the usual trend might be expected to occur on oceanic islands and a reasonable hypothesis, based upon developmental considerations, of the evolutionary mechanism by which it could be accomplished. Following Darwin, he points out that on oceanic islands, particularly those of volcanic origin, the plants that would be most likely to become established upon their raw soils following accidental long-distance dispersal would be hardy, drought-resistant herbs or subshrubs. Plants having these growth forms would, therefore, be the initial colonizers of such islands. Later, as the soil weathered and became more favorable for plant growth, habitats that could support shrubs or trees would develop. Because of the difficulties associated with the transport of seeds belonging to already evolved shrubs and trees from distant continental regions, reduced competition would permit some of the herbs and subshrubs to evolve into shrubs or trees *in situ.*

The evolutionary mechanism by which this could happen involves the principle of paedomorphosis. Carlquist points out that the specialized wood characteristic of the more advanced angiosperms is localized in the secondary vascular tissue. The procambial initials of primary vascular tissue have the long narrow shape that is characteristic of primitive xylem and phloem cells. In the reverse evolution, from herbaceous species to shrubs, the specialized secondary wood that is characteristic of the herbaceous condition may be retained in some instances, but in other evolutionary lines the later developmental steps that give rise to such secondary wood may be suppressed, and wood having a more primitive appearance may develop directly from elongated cambial initials.

I believe that these trends of evolution have not been restricted to remote oceanic islands, but have occurred also on larger islands, and, under special circumstances, even in continental floras. An example from a relatively large insular area of a group of shrubs that probably evolved from herbs is the genus *Veronica* (including *Hebe*) in New Zealand. The numerous species of this genus are one of the commonest and most widespread shrubby elements of the New Zealand vegetation. All of them are of polyploid origin, having $n = 21$, 20, or multiples of these numbers, whereas the herbaceous *Veronica* species of the Northern Hemisphere have chiefly the basic number $x = 7$ (Frankel and Hair 1937). The derivation of the shrubby New Zealand species of this genus from herbaceous or subshrubby ancestors that now occur in the Northern Hemisphere is, therefore, highly probable.

An example of a similar evolutionary trend in a region that is now continental exists in California, and also involves the family Scrophulariaceae. The genus *Mimulus* belongs to a tribe that consists almost entirely of herbs, most of them extreme herbs of wet habitats. Within the genus itself, those species that have the least specialized flowers, such as *M. moschatus,* in which the corollas are almost regular, belong in this category. Nevertheless, there exists in California a group of shrubby *Mimulus* species, which are sometimes placed in the separate genus *Diplacus.* The probable origin of these shrubs from herbaceous ancestors was called to my attention many years ago by the late F. W. Pennell (*in litt.*), a well-known monographer of the family. This example, however, may not contradict the principle developed by Carlquist, that the origin of shrubs from herbs is most likely to take place on islands. The shrubby species of *Mimulus* are most abundant and diversified in the coast ranges of California and the neighboring Channel Islands. These regions were continental islands during the middle part of the Tertiary Period, which was most probably the time when the genus *Mimulus* began its active evolution. Although they are not volcanic, they were subject to great orogenic or mountain-building activity, and owing to the gradual desiccation of the climate, with the disappearance of summer rains (Axelrod 1958), small pockets of semiarid climate and desiccated soil must have developed quickly and frequently on these Tertiary islands. In a plastic and genetically rich genus like *Mimulus,* the evolution of species best suited to these conditions, in a direction counter to the usual trend, would be a not unexpected occurrence.

The example of shrubby *Mimulus* species suggests that the trend from herbs to shrubs may exist also in continental regions. In this connection, two of the principal groups of shrubby species found in arid regions of the Northern Hemisphere deserve particular attention. These are the genus *Artemisia*, the sagebrushes, and the various shrubby genera of the Chenopodiaceae, such as *Atriplex, Eurotia, Grayia*, and *Sarcobatus.*

The genus *Artemisia* is one of the most specialized genera of the family Compositae. Belonging to a family that has normally insect-pollinated flowers, it has become specialized for wind pollination. In this connection, its heads or capitulae have become much reduced and aggregated into compound inflorescences. The tribe Anthemidae, to which *Artemisia* belongs, consists principally of herbaceous species. This is true not only of the genera *Chrysanthemum* and *Tanacetum*, which are the nearest, less-specialized relatives of *Artemisia*, but also of three genera, *Hymenopappus, Hymenothrix*, and *Leucampyx*, which apparently provide connecting links between the Anthemidae and other tribes of the family, and so must be regarded as among the most primitive members of the tribe. Hence, regardless of whether one believes, following Cronquist (1955), that the entire family Compositae was basically and originally herbaceous or agrees with Carlquist (1966) that the earliest Compositae were shrubs, the prevailing evidence regarding *Artemisia* suggests that its shrubby species are secondary. Hall and Clements (1923) point out that the section *Seriphidium*, which includes most of the shrubby species such as the North American sagebrushes, is relatively specialized with respect to floral characteristics. They suggest that in *Artemisia* the evolution from the herbaceous to the shrubby habit of growth may have taken place in three separate evolutionary lines.

This hypothesis is supported by studies of stem anatomy. Diettert (1938) has shown that the young stem of *Artemisia tridentata*, even when it is well enough developed that abundant secondary xylem has been formed, still contains a circle of separate bundles, separated by parenchyma, rather than a continuous ring of vascular tissue. An anomalous feature of this species is the formation of layers of cork between the growth rings of xylem (Diettert 1938, Moss 1940). This condition is an adaptation to severe conditions, either cold or drought, which is found chiefly in herbaceous species of various genera. Moss records it in both herbaceous and woody species of *Artemisia*, and points out that only the more specialized species of both herbs and shrubs possess this characteristic.

The woody genera of the Chenopodiaceae share with the herbaceous genera of this family an anatomical characteristic that is best explained by assuming that the entire family, together with several related families, originated from a common herbaceous stock, which has repeatedly given rise to evolution in the direction of shrubs and even trees (Takhtajan 1959: 38). This is the feature of anomalous secondary thickening (Metcalfe and Chalk 1950:1075). The first-formed vascular bundles may either form a circle of widely separated units or be scattered through the stem, giving a superficial resemblance to monocotyledons. Later, concentric zones of collateral vascular bundles arise from a succession of arcs or rings of cambium, which are situated in the pericycle or, occasionally, the phloem. Consequently, the woody stems of such shrubs as saltbush (*Atriplex*), winter fat (*Eurotia*), and greasewood (*Sarcobatus*) do not have the usual dicotyledonous wood structure, consisting of an internal zone of xylem separated by a single ring of cambium from an outer zone of phloem. Instead, they have a series of concentric rings of cambium, each of which has formed its own xylem and phloem. This mode of growth can give rise not only to woody shrubs, but even to thick-stemmed plants that look like trees, such as the saxaul tree (*Haloxylon*), found in the sandy deserts of central Asia.

One might justly ask the question, What selective pressures or other conditions made possible the establishment of this unusual anatomical condition, which with respect to growth form has caused an entire order of flowering plants to evolve in a direction opposite to that of the usual trend? In the examples previously mentioned, the reversal was associated with the colonization of island habitats. Could similar conditions have been responsible for the secondary origin of shrubs from herbs in xeromorphic Compositae and Chemopodiaceae?

An affirmative answer to this question is made probable by the evidence presented by Axelrod (1958) and others with respect to the climate and plant associations that existed in the early part of the Tertiary Period, when the groups concerned were probably beginning the most active phase of their evolution. At that time, the prevailing climate throughout the Northern Hemisphere was mesic, with ample rainfall throughout the year, so that continuous forests existed across most of North America and Eurasia. Nevertheless, there were probably "ecological islands" of more xeric climate, resulting from the rain shadows formed by isolated mountain ranges, or from other unusual combinations of ecological conditions. As the overall climate became more arid during the middle of the Tertiary Period, these xeric islands increased in size, and new ecological islands probably appeared.

The forest trees and mesic shrubs that formed the "sea" of forest surrounding these islands were already too specialized and rigid in their requirements to be able to give rise to xeric shrubs optimally adapted to the conditions on them. In this connection, it is noteworthy that few if any of the dominant shrubs found in the arid portions of North America appear to be derived from immediate common ancestors with the woody elements of the neighboring mesic forests. Given this restriction, plus the ease of seed transport and establishment that both now and formerly have been characteristic of families such as the Compositae and the Chenopodiaceae, one can imagine without much difficulty that the evolution of xeric shrubs from semixeric herbs belonging to these families could take place more quickly and easily than could the evolution of xeromorphic adaptations in trees and shrubs belonging to genera such as *Acer, Alnus, Betula, Carya, Cornus, Fagus, Hamamelis, Liquidambar, Lithocarpus,* deciduous species of *Quercus, Salix,* and other inhabitants of the surrounding forests.

The evolution of these secondary shrubs is, therefore, a good example of the general principle of evolutionary opportunism.

The Evolution of the Deciduous Condition

An evolutionary trend that has occurred repeatedly in many groups of woody plants, both shrubs and trees, is the acquisition of deciduous foliage. Since deciduous trees are most common in high latitudes and are usually associated with long, cold winters, most botanists have assumed that the deciduous habit has been acquired in response to cold. Axelrod (1966), however, has pointed out that fossil floras of early Tertiary and Cretaceous age, when deciduousness most probably evolved, contain a mixture of deciduous and broad-leaved evergreen species, many of the latter having tropical affinities. Such associations could not have tolerated the severe winters now prevailing in regions like the northeastern and central United States, where the greatest concentration of deciduous species among contemporary floras occurs. Axelrod has, therefore, suggested that drought rather than cold has been the principal selective agent in promoting the evolution of the deciduous habit.

The evidence from shrubs that inhabit temperate arid regions supports Axelrod's hypothesis that cold climates by themselves cannot be the only environmental factor that has favored the deciduous condition. At the same time, this evidence is also against the idea that drought was the only factor. Sagebrush (*Artemisia tridentata*) and related species of *Artemisia* that inhabit central Asia illustrate both of these points very well. They inhabit regions which, with respect to the severity of climate imposed by both severe drought and extreme cold, are not equaled anywhere else in the world in regions that support woody plants. Nevertheless, they have persistent leaves. The same is true of several other genera and species of arid-land, cold-resistant shrubs, such as *Chrysothamnus* and *Atriplex hymenelytra,* although perhaps the majority of them are deciduous.

A modification of the hypothesis that postulates simply cold or drought is the hypothesis that seasonal alternation between favorable and unfavorable conditions has been the principal selective pressure. In tropical regions, the forests that exist in regions having an alternation between wet and dry seasons are characterized by a high proportion of trees and shrubs having deciduous leaves (Walter 1962). In subtropical or warm-temperate climates, the proportion of woody species having deciduous foliage is very high in regions having summer rainfall, but much lower in those that have a summer drought and winter rainfall, as in the Mediterranean regions and California. Moreover, at comparable latitudes in the Southern Hemisphere the proportion of deciduous species is much lower than it is in the Northern Hemisphere. This is due to the fact that Southern Hemisphere climates are, in general, more equable, with fewer temperature extremes, than are climates of comparable regions in the

Northern Hemisphere (Axelrod 1966, 1970).

To botanists living in the north temperate zone, the origin of mesophytic trees and shrubs that are deciduous in winter presents the most interesting and relevant problem. I have arrived at a hypothesis about the origin of the deciduous condition that postulates a combination of factors: seasonal cold, enough to cause mild but not heavy frosts; summer moisture, enough to make possible a high rate of photosynthesis and transpiration; and preference for stream banks or other habitats in which the soil is saturated at the time when frosts begin.

This hypothesis is based upon observations made in California, a region that has a relatively equable climate, and supports many evergreen as well as deciduous species of angiosperms which may grow side by side in the same or in different plant associations. Of the 137 species of angiospermous trees and shrubs found in the Coast Ranges of central California, 76 are evergreen and 61 are deciduous. This region is climatically marginal. Its coastal strip is essentially frost free, and in its wettest portions adequate moisture is present throughout the year. Its colder interior portions have mild frosts and a growing season of 210 to 240 days, and in its drier sections a deficiency of moisture exists during five months of summer.

If drought were the principal environmental factor promoting the deciduous condition, one would expect to find the highest proportion of deciduous species on the hotter, drier slopes of the interior, and the lowest proportion of them in the moister regions, regardless of temperature. Such is not the case. Sixteen species of deciduous trees are found in this region. Of them, five (*Populus trichocarpa, Salix lasiandra, S. laevigata, Aesculus californica, Acer negundo* var. *californica*) are found throughout the region; six (*Populus fremontii, Alnus rhombifolia, Juglans hindsii, Quercus douglasii, Q. lobata, Fraxinus latifolia*) are absent from the coastal section, but are more or less abundant throughout the interior; two (*Acer macrophyllum, Quercus garryana*) are coastal and northern; one (*Quercus kelloggii*) is interior and northern; and one (*Platanus racemosa*) is confined to the southern part of the region. The patterns of distribution of the 45 deciduous shrubby species are somewhat similar.

Two facts should be noted about these species. One is that the largest group of deciduous species that are restricted to a portion of the region are those confined to the interior portion. These species are better adapted to the relatively marked differences in temperature between summer and winter that prevail in the interior than to the equable climate found along the coast. In other words, the relatively more continental climate of the interior supports a higher proportion of deciduous species than the more insular climate of the coast. This condition is probably widespread. In the eastern United States, which has a much more continental climate than the west coast, many deciduous species of trees extend southward to regions having a growing season of up to 240 days.

An even more striking fact is the degree to which the deciduous species are concentrated along stream banks (Table 11-1). Ten of the sixteen species of deciduous trees occur entirely or primarily in this single rather restricted habitat. Of the remaining six species, two (*Quercus garryana, Q. kelloggii*) reach their southern limit in central coastal California; a third (*Juglans hindsii*) is doubtfully native to the area. The concentration of deciduous shrubs along stream banks is equally striking. In addition to several species of *Salix*, one can mention *Calycanthus occidentalis, Cercis occidentalis, Clematis ligusticifolia, Physocarpus capitatus, Euonymus occidentalis, Cornus glabrata* and *C. californica, Rhododendron occidentale, Cephalanthus occidentalis, Sambucus callicarpa*, and *Baccharis viminea*.

Of equal significance is the fact that among the 76 evergreen species of angiosperms, none is confined to stream banks, and only a few, such as *Umbellularia californica*, occur commonly in this habitat. This fact suggests that in a climate having mild winter frosts and relatively dry summers, the most favorable habitat for the origin of the deciduous condition is that of stream banks.

These observations should be extended further. In particular, similar observations should be made in the less arid portions of southern Arizona, southern Texas, and northern Mexico, where light winter frosts occur but where the dry season is during the cold rather than the warm months. Furthermore, the situation should be investigated in wooded swamps, marshes, and deltas, habitats that are absent from central California. At present, however, I feel justified in presenting the following working hypothesis. The deciduous habit, including particularly the elaboration of genetically conditioned abscission tissue at the base of leaf petioles, originated very often in trees or shrubs inhabiting stream banks or other habitats where the soil is perpetually saturated with moisture. This was particularly likely to happen in a climate having a dry atmosphere and sharp fluctuations in temperature. Such climates would have had some winter frosts but would still have been mild enough to permit the survival of many species of broad-leaved evergreen angiosperms.

The physiological reason why this habitat and climate would be more favorable for the origin of the deciduous condition will be apparent to anyone who has experience in cultivating tender evergreens in a climate having mild frosts. If such plants are heavily watered during the autumn, they are much more likely to be damaged by frost than if they are allowed to harden by withholding water. Moreover, if the first frost is followed immediately by hot, dry weather, as is likely in a semiarid climate, the damage is more severe than if the frost is followed by a cool, moist spell.

We can imagine, therefore, that in the

Table 11-1. Comparison of habitats of deciduous and evergreen species of woody angiosperms in the central coast ranges of California.

Plants	Deciduous		Evergreen	
	Chiefly stream banks	Chiefly hill slopes	Chiefly stream banks	Chiefly hill slopes
Trees	10	6	0	7
Shrubs	22	23	0	69
	—	—	—	—
Total	32	29	0	76

marginal type of climate described above, woody plants inhabiting stream banks or swamps, with their roots kept perpetually moist, would be much more susceptible to frost damage than those growing on the drier hillsides. This damage would be particularly severe in a semiarid, continental climate, in which daily fluctuations of temperature would be great and the dry atmosphere would be poor insulation against the heat. Since the stream-bank habitat is otherwise a very favorable one for woody plants, we might expect in this habitat a maximum selection pressure in favor of genetically controlled mechanisms such as leaf abscission, which would protect them against this type of damage.

The hypothesis postulates, therefore, that the chief basis for the origin of the deciduous condition in mesic trees and shrubs of temperate regions was the selective advantage that this condition provided in hardening the plants prior to the onset of frosts. This selection pressure would have been exerted most strongly on plants inhabiting stream banks and other sites having moisture-saturated soils, since they would not have been subject to hardening by drought.

The trend from persistent to deciduous foliage can probably be reversed from time to time. A good example is the tribe Prunoideae of the Rosaceae. Its largest genus, *Prunus*, consists mostly of deciduous species, evergreen species being found only in those species of the subgenus *Padus* that occur in the Mediterranean region and in tropical and subtropical America. Moreover, the two small genera that link the Prunoideae with the less-specialized Spiraeoideae, *Exochorda* and *Osmaronia*, are both deciduous. One way of explaining this situation would be to assume that a whole series of forms having persistent foliage once connected the Spiraeoideae with the Prunoideae, but that these are now extinct. An alternative hypothesis is that the original ancestors of the subfamily had deciduous foliage, and that as the highly successful genus *Prunus* entered regions having warmer climates, selection favored the evolution of thick, leathery leaves and the loss of the abscission mechanism. The fact that the single Californian species, *P. ilicifolia*, does not resemble the Mediterranean group in any characteristics except the persistent foliage plus characteristics that are common to the entire subgenus *Padus* would favor the latter hypothesis. If it is correct, then a secondary origin of persistent leaves in the tropical genera *Maddenia* and *Prinsepia*, which resemble *Prunus* but are more specialized in their floral characteristics, seems probable.

From the physiologic and genetic point of view, reversion from deciduous to persistent foliage is not improbable. The deciduous condition owes its existence to the formation of a specialized abscission layer, which is activated by the hormone abscissin (Wareing and Ryback 1970). Any mutation that interfered with the biosynthetic pathway leading to this hormone would prevent its action, and would eliminate the abscission process.

The Origin and Spread of Geophytes

A well-documented and generally accepted series of trends are those that were first clearly outlined by Raunkiaer, from phanerophytes through chamaephytes and hemicryptophytes to cryptophytes. In response to cold, drought, or both, genotypes are selected that are dormant during the unfavorable season. The buds or shoot apices that are destined to regenerate new growth when favorable conditions return are situated nearer and nearer to the ground level, and finally, in cryptophytes, entirely underground. Raunkiaer recognizes three forms of cryptophytes: geophytes, helophytes, and hydrophytes. They are distinguished only by the places in which they grow: land, swamps or other saturated soil, and water. Helophytes are particularly hard to distinguish, and Raunkiaer admits that the same species can be either a rhizome geophyte or a helophyte. There appears to be no good reason, therefore, for recognizing them as a separate group. Hydrophytes are a very diverse lot, and have probably evolved via many different pathways. They will not be discussed further.

Four kinds of geophytes are recognized: rhizome geophytes, stem-tuber geophytes, bulb geophytes, and root-tuber geophytes. The first is the most generalized, since rhizomes are merely underground stems. Rhizome geophytes are adapted to alternation between a dormant period and a long period of vegetative activity, so that there is no particular adaptive value in storing large amounts of food. They are best adapted to mesic or even wet situations, and in dry regions occupy sandy, porous soils, having roots that penetrate to great depths. The trend from rhizome geophytes to stem-tuber geophytes takes place in response to shortening of the growing season. It can come about either through shortening of the rhizome to form a corm, as in *Crocus* and other Iridaceae, including species of *Iris,* or by the formation of specialized tubers at the ends of slender rhizomes, as in the potato (*Solanum tuberosum*).

Bulb geophytes arise from stem-tuber geophytes through the modification of bracts or bases of leaves that arise from the shortened stem. These structures may become much thickened and filled with stored food, or they may become hard, scaly, or fibrous, so that they form a protective covering around the bulb. In the bulbs that are best adapted to arid or seasonally dry regions, as in species of *Allium,* the inner scales are modified for food storage and the outer scales for protection. In some bulb geophytes, such as *Tydaea* and *Isoloma* of the Gesneriaceae, which occur in species inhabiting mesic tropical habitats, the bulbs are elongate, and are produced at the ends of scaly rhizomes. Such bulbs apparently serve the purpose of vegetative reproduction as much as protection against an unfavorable season.

The simplest form of root-tuber geophyte has swollen taproots, as in many familiar root vegetables (turnip, beet, radish). In them, new roots are regenerated from the lower part of the taproot, and shoots from buds on its upper portion. In monocotyledons, particularly Orchidaceae, a swollen storage root is formed at the beginning of the unfavorable season, but this root ceases to function after regeneration is complete. Functional roots arise from the lower nodes of the new shoot (Fig. 11-2).

The diversity of geophytes with respect to both their morphology and the habitats that they occupy illustrates very well two principles that are discussed in earlier chapters of this book: adaptive modification along the lines of least resistance and conservation of organization. Given a habitat that favors the geophyte adaptation, species that already have a well-developed branching stem system, with a plentiful supply of hormones that are responsible for elongation of internodes, can become modified relatively easily into rhizome geophytes. Response to increasing severity of the habitat, to produce stem-tuber geophytes, will depend upon the nature of the rhizomes that are present before the modification begins. If they are thick and tough, and contain some stored material, as in *Iris,* modification along the lines of least resistance is shortening to produce a corm. If, on the other hand, they are relatively slender and soft, as in nontuberous species of *Solanum* and other Solanaceae, the path of least resistance is the development of terminal swellings, leading to root tubers.

Fig. 11-2. (*A*) Young seedling (from Goebel 1931) and (*B*) mature plant (from Hegi 1906) of *Cyclamen persicum,* showing the development of a tuber as the swollen base of the hypocotyl.

If a nonrhizomatous species is responding to a habitat that favors long dormancy and a short growing season, the path of least resistance is either the formation of a stem tuber via swelling of the basal portion of the stem, as in the tuberous species of *Begonia, Pelargonium flavum,* and the genus *Cyclamen,* or the production of various kinds of root tubers. Three species of the genus *Lewisia* (Portulacaceae) native to the Sierra Nevada of California illustrate very well the producton of increasingly distinctive root tubers, correlated with occupation of habitats having an increasingly long dry season. In *Lewisia pygmaea,* a species of moist alpine situations in which ample moisture is always available to plants having long roots, the roots are long but otherwise unspecialized (Fig. 11-3, *A*). *Lewisia nevadensis,* found chiefly in the edges of meadows that are wet in spring and early summer but dry in late summer and autumn, has a short, swollen turnip-shaped taproot (Fig. 11-3, *B*). Finally, *L. triphylla,* a small, slender species inhabiting gravelly flats that are moistened only by melting snow in spring and early autumn, has a round root tuber that is the only part of the plant to remain alive for about 10 months of the year (Fig. 11-3, *C*). The tuber produces a shoot that flowers and matures its seeds in about 2 months after the snow has melted.

Certain geophytes illustrate very well the principles of selective inertia and conservation of organization. The strong selective pressures that are most likely to produce the profound modifications characteristic of stem-tuber, bulb, and root-tuber geophytes are found in climates having seasonal drought as well as cold. Nevertheless, the mesic deciduous forests of temperate, continental climates provide a number of suitable niches for plants having these growth forms. This is because the heavy canopy of leaves during the summer is highly unfavorable for light-loving plants, so that they are most suc-

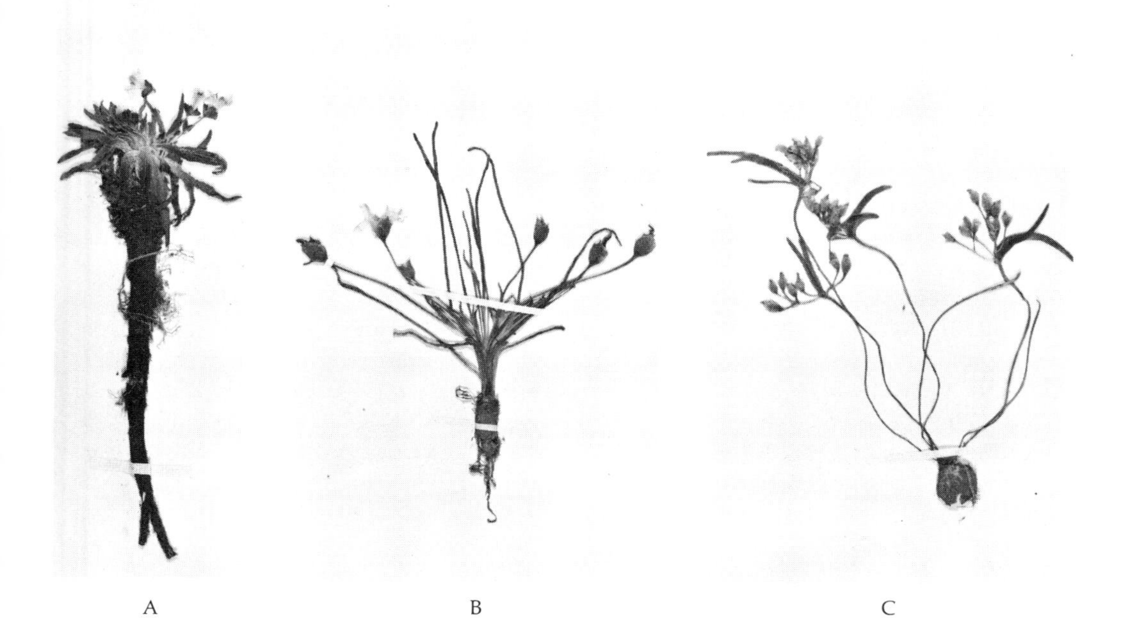

Fig. 11-3. Roots of three species of the genus *Lewisia* from the Sierra Nevada, California: (*A*) *L. pygmaea;* (*B*) *L. nevadensis;* (*C*) *L. triphylla.* (From specimens in the herbarium, University of California at Davis.)

cessful if genetically adjusted to spring growth, during the period between the onset of favorable temperatures and the full development of the leafy canopy. Consequently, a number of genera that have evolved chiefly in montane, semiarid, or alpine regions have produced species that have entered the deciduous forests and form conspicuous members of its spring flora. In the deciduous forests of the eastern United States, outlying species belonging to genera or species groups that are more characteristically western and montane are as follows:

Stem-tuber geophytes

Dicentra canadensis and *D. cucullaria.* The genus *Dicentra* has a diversity of montane species in both western North America and Asia, but its greatest diversity of growth forms is in the western Cordillera, plus the Mediterranean climatic region of California.

Claytonia virginica and *C. caroliniana.* These two species differ in little besides the width of their leaves, although within each species there is a great diversity of chromosome numbers (Lewis, Oliver, and Suda 1967). *Claytonia caroliniana* is probably also a western species, since the entity usually recognized as *C. lanceolata,* and common throughout the western Cordillera, is probably not specifically distinct from it. If, as Swanson (1964) has suggested, *Claytonia* and *Montia* should be merged into a single genus, then the center of diversity of this genus, particularly with reference to growth form, is in the western Cordillera.

Dentaria laciniata and *D. heterophylla.* The genus *Dentaria* is very close to *Cardamine,* and some species of both genera are rhizome geophytes. In the two species of eastern North America mentioned above, the rhizome is replaced by a series of stem tubers. These species are high polyploids ($2n = 128$ and 256). Since diploids ($2n = 16$) and tetraploids ($2n = 32$) exist in the complex of *D. californica* and *D. integrifolia,* which also are root-tuber geophytes, the origin of this kind of life form in species of California woodlands, where long, dry summers prevail, appears likely. At least one and perhaps more genomes of western root-tuber geophytes figured in the ancestry of the highly polyploid eastern species.

Bulb geophytes

Erythronium americanum and other woodland species of eastern North America. The great center of species diversity in the genus *Erythronium* is in western North America, where, in addition to species adapted to mesic woodlands, there are a number of "glacier lilies" or "avalanche lilies" that carpet slopes above timberline. Lower down, species can be found on cliffs that are wet in spring and dry in summer, and even in dry woodlands of the Sierra Nevada foothills and coast ranges, where the rainless summer may last for 5 months. Consequently, the adaptation of *Erythronium* to a very short season of active growth and a long period of dormancy probably originated under the more rigorous conditions of the western United States, and has persisted in species whose ancestors entered the mesic deciduous forests.

Allium canadense, A. cernuum, and *A. tricoccum.* The genus *Allium* is highly developed and diversified throughout the semiarid portions of western North America, and has only a few outliers in the mesic forests of the eastern region. The first two occur chiefly in rocky woods, in habitats that approach as nearly to those of the semiarid west as any that can be found. *Allium tricoccum* has become much modified for life in the deciduous forest, since it forms broad, thin leaves early in the season, and produces flower stalks much later, after the leaves have dried up.

Root geophytes

Delphinium tricorne. This species belongs to a group of larkspurs having short, thick, tuberous roots. Eight to ten species of this kind are common throughout the western Cordillera, where they occur mostly in dry areas, flower in spring or early summer, and then become dormant. The eastern species, which is also characterized by early flowering and summer dormancy, is an outlier that has become adapted to the vernal flora of the deciduous forest.

Trends in the Inflorescence

Trends in the inflorescence are particularly relevant to the present discussion. Admittedly, they are not of great value in delimiting major categories, although some large families, such as Cruciferae, Leguminosae, Umbelliferae, and Compositae, are remarkably homogeneous with respect to the basic architecture of the inflorescence. This disadvantage is compensated by a number of advantages. Morphological trends can be relatively easily recognized, have been much discussed, and can be compared with trends in other characteristics as a means of understanding better the basis of the trends. In addition, the development of the reproductive shoot and flower has been intensively studied by many botanists in a great variety of families, so that information is available for the formulation of reasonable hypotheses about the relation between adult form and the developmental processes that bring it into being. Finally, the extensive parallelism that exists with respect to trends in the inflorescence, which is a major reason for their relatively low value in delimiting major categories, makes possible repeated comparisons between the overall nature, reproductive biology, and habitat preferences of groups that differ from one another with respect to similar inflorescence characters. Such comparisons provide a clue to the adaptive significance of these characters.

The present discussion of inflorescences will be in three parts: adult morphology, developmental pattern at the histological level, and ecology with reference to possible adaptive values. Morphological differences and their classification have been reviewed by a number of authors. The first comprehensive review that is based upon modern concepts of phylogeny is that of Parkin (1914). Rickett (1944) has provided an extensive review of the literature, including that of the 18th and 19th centuries, and has emphasized some of the difficulties inherent in any precise classification of inflorescence types. In recent years, a typology of inflorescences, based upon examination of about 10,000 species belonging to a great variety of families, has been developed by Troll (1964), and has been summarized by Weberling (1964, 1965). Comparative studies of many related species belonging to the same family, which form the best basis for drawing conclusions, are relatively few in number, one of the best being that of Woodson (1935) on the Apocynaceae.

Discussion has centered about the following points: (1) the nature of the inflorescence in the original angiosperms; (2) the relation between determinate and indeterminate inflorescences; (3) amplifica-

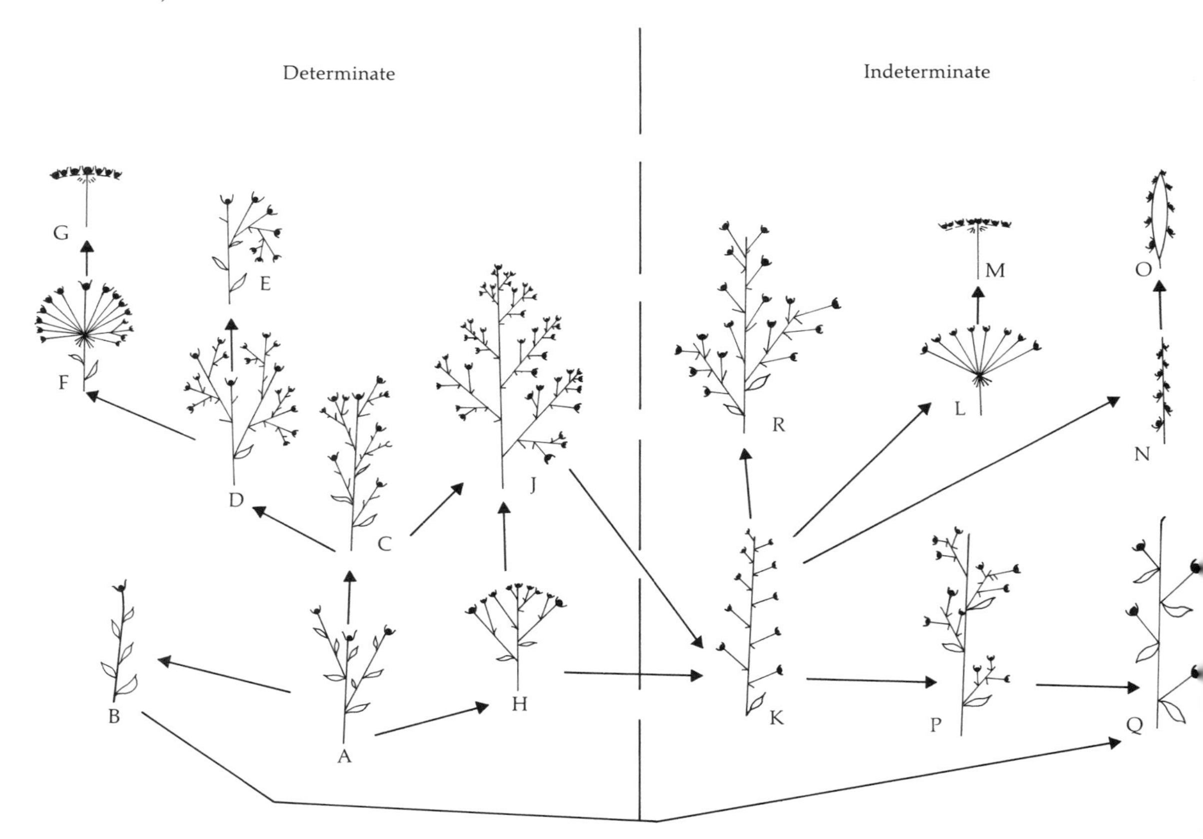

Fig. 11-4. Diagrams showing the principal trends of evolution in the inflorescence: (*A*) leafy cyme; (*B*) solitary terminal flower; (*C*) pleiochasium; (*D*) compound dichasium; (*E*) cincinnus or helicoid cyme; (*F*) umbelliform dichasium; (*G*) capitate dichasium; (*H,J*) panicles; (*K*) raceme; (*L*) umbel; (*M*) capitulum; (*N*) spike; (*O*) spadix; (*P*) axillary racemes; (*Q*) solitary axillary flowers; (*R*) compound raceme. (Original.)

tion of the inflorescence through additional branching and its reduction through suppression of branches; (4) shifts in the order of maturation of the flowers; (5) elongation of the pedicel of the flower by accentuation of intercalary growth and its shortening through suppression of this growth. These will be taken up in turn.

The three dominant hypotheses concerning the inflorescence of the original angiosperms are those of Parkin (1914), who believed that they bore a single terminal flower at the end of a leafy branch; of Rickett (1944), who postulated as the original unit a simple dichasium, consisting of a terminal flower and two lateral flowers, each of the latter being subtended by a reduced leaf or bract (Fig. 11-4, *A*); and of Takhtajan (1970), who postulated a leafy cyme as the original condition (Fig. 11-4, *B*).

As is mentioned in Chapter 10, the original angiosperms are almost certainly extinct and not represented by any single modern group. Consequently, their characteristics can be reconstructed only by inference. With respect to the inflorescence, the best basis for inference would be extrapolation from the most frequent trends that can be recognized in relatively primitive modern groups. These are summarized below, by families. Since vegetative anatomy has evolved to a considerable extent at rates independent of floral evolution, so that a number of genera combine primitive anatomical structure with relatively specialized flowers (*Sarcandra, Tetracentron, Trochodendron*), only floral characteristics have been considered in selecting families for review.

Magnoliaceae. Perhaps the strongest argument for regarding the single terminal flower as primitive is its universal presence in the Magnoliaceae, which are often regarded as the most primitive plant family. Nevertheless, the specialization of this flower with respect to both vascular anatomy and carpel dehiscence is mentioned in Chapter 10. In addition, the fact must be noted that the transition from strictly vegetative development to floral development is not smooth and gradual, as it is in some of the groups discussed below. The Magnoliaceae always contain, between the leaves and the flower, three specialized bracts that form a pseudocalyx. In view of the fact that peduncular bracts in other groups are often a sign of reduction from an ancestral inflorescence that contained more flowers, such an interpretation of the calycoid bracts in the Magnoliaceae cannot be ignored. Whether the inflorescence in this family is primitively one-flowered, or is derived from an ancestral inflorescence that had several flowers, is an open question.

Winteraceae. Most of the genera of this family have flowers in axillary cymes or dichasia, thus corresponding to the condition postulated by Rickett as primitive. Solitary flowers, such as those of *Zygogynum*, are probably derived by reduction (Bailey and Nast 1945*a*). Two considerations, however, weaken the evidence from Winteraceae as a support for the hypothesis that their type of inflorescence is primitive. First, their inflorescences are always axillary, and are produced by specialized reproductive shoots, whereas in many other primitive families transitions from the vegetative to the reproductive condition of a main axis or branch are found. Second, compared with other primitive angiosperms, the flowers of the Winteraceae show a certain amount of reduction, particularly the crowding of the stamens and the small number of carpels. In some of the families discussed below, such reductions are associated with the shift from leafy cymes to bracteate cymes or dichasia. Consequently, one cannot exclude the possibility that the ancestors of the Winteraceae had leafy cymes.

Annonaceae. According to the descriptions of Fries (1919, 1959), inflorescences of this family resemble those of the Winteraceae in being bracteate cymes or variously modified dichasia, and the one-flowered condition, as in *Rollinia, Oxymitra,* and *Hornschuchia,* is derived. An important difference, however, is that the inflorescences in several genera of Annonaceae are basically terminal, although they often appear to be axial because of the development of an axillary branch from one of the upper leaf or bract axils, resulting in a sympodial condition. the Annonaceae, therefore, support Rickett's hypothesis.

Dilleniaceae. In this family, the Australian section *Cyclandra* of *Hibbertia* contains the species that are most generalized with respect to floral structure. Many of these species have flowers solitary and sessile at the ends of leafy branches, with the upper leaves passing gradually into the sepals (Gilg and Werdermann 1925). In other species, the flowers are on axillary peduncles, but dichasia are not found in *Hibbertia.* In three sections found in New Caledonia, *Trisema, Polystiche,* and *Spicatae,* the inflorescence is either a simple or a compound spike, as also in the Australian section *Hemistemma,* which is specialized with respect to both the reduction of carpel number to two and the lateral placement of stamens and staminodes. Within the family, dichasia occur chiefly in the subfamily Tetraceroideae, which is regarded as more specialized than the Dillenoideae, to which *Hibbertia* belongs (Hoogland 1952, 1953, Dickison 1967, Kubitzki 1968). In the herbaceous genus *Acrotrema,* the flowers are solitary, on the ends of long scapes, or racemose. In the Dilleniaceae, therefore, the most generalized condition is that of solitary flowers at the ends of leafy branches, the leaves passing gradually into the sepals. The only difference between this condition and the leafy cyme is the absence of branches from the axils of the upper leaves.

Paeoniaceae. Parkin (1914) used *Paeonia* as one support for his hypothesis that the solitary, terminal flower is primitive, noting the fact that solitary flowers exist in the shrubby species *P. suffrutescens,* whereas the herbaceous garden peony, *P. albiflora,* has a leafy cyme. He neglected, however, the second shrubby species, *P. Delavayi,* which has a leafy cyme. With respect to leaf, calyx, and staminodial morphology, *P. Delavayi* is less specialized than *P. suffruticosa* (Stebbins 1939). Within the herbaceous subgenus *Paeon,* solitary flowers are associated with a relatively abrupt transition from the vegetative to the flowering condition, and they appear in relatively distantly related species, *P. obovata* of eastern Asia and the *P. daurica-corallina* complex of the Mediterranean regions, both of which appear to represent end lines of evolution. Finally, the two species of the New World, *P. Brownii* and *P. californica*, both have leafy cymes, so that this kind of inflorescence is the only one found in all three subgenera. In *Paeonia*, therefore, the original inflorescence was almost certainly a leafy cyme, and the solitary terminal flower is derived by reduction.

Ranunculaceae. The most widespread form of inflorescence in this family is the leafy cyme. It is found in the least-specialized genera of the tribe Helleboreae, such as *Helleborus, Trollius, Caltha,* and *Isopyrum.* In *Ranunculus,* the trend from a leafy cyme through a bracteate cyme to a solitary flower can be traced. Most species of the genus have a leafy cyme, but in some, such as *R. flammula* and *R. alsimaefolius,* the cyme is bracteate. In *Myosurus,* a derived, rosette-forming annual, the flowers are solitary at the ends of leafless scapes. Comparison between *Isopyrum* and the more specialized genus *Thalictrum* suggests that the bracteate cyme in *Thalictrum* is derived from a leafy cyme similar to that in *Isopyrum.* The genus *Clematis,* which is specialized with respect to its achenes bearing long plumose styles, has bracteate cymes in most of its species. Solitary flowers are associated with vegetative specialization in *Eranthis, Coptis, Hepatica,* and the subalpine species of *Caltha.* In the Ranunculaceae, therefore, the original condition of a leafy cyme appears to have given rise in various lines to more specialized conditions, both bracteate cymes and solitary flowers.

Papaveraceae. Leafy cymes occur in a number of species of *Papaver,* such as *P. somniferum,* as well as in the genera *Argemone, Dendromecon, Eschscholtzia, Glaucidium, Hypecoum, Hunnemania, Meconella, Platystemon,* and *Romneya.* Species of *Papaver* (*P. alpinum*), *Eschscholtzia* (*E. Lobbii*), *Meconella,* and *Canbya* having solitary flowers on long scapes are relatively specialized. The bracteate cyme that appears in the genus *Chelidonium* and in members of the related family Fumariaceae is definitely associated with floral specialization. Evolutionary trends in the Papaveraceae, therefore, have resembled those in the Ranunculaceae.

Rosaceae. In the tribe Spiraeae, which is the most generalized in the family, the inflorescence in the best-known genera, *Spiraea, Physcocarpus, Aruncus, Sorbaria,* and *Holodiscus,* is cymiform, bracteate, and highly compound. These genera are all deciduous shrubs or herbs. Leafy cymes exist or are approached in the genera *Kageneckia* and *Quillaja,* which have persistent foliage, and the monotypic genus *Lindleya* of Mexico has solitary flowers at the ends of leafy branches. Consequently, leafy cymes may well have been the original kind of inflorescence in the Spiraeoideae.

The evolutionary trend from leafy cymes to bracteate dichasia and compound cymes, as well as the reduction to solitary flowers, can be followed in the tribes *Dryadeae* and *Potentilleae.* Leafy cymes occur in *Fallugia, Cowania,* and *Purshia,* the most generalized genera of Dryadeae, as well as in *Geum* and in *Potentilla fruticosa* and other generalized species of that genus. In the prostrate arctic-alpine genus *Dryas* and in the creeping species *Potentilla anserina,* the inflorescence is reduced to a single scapose flower. In other species of *Potentilla*, such as *P. tridentata*, the inflorescence is a bracteate cyme, and the same structure is found in all species of the more specialized genus *Fragaria*.

The evidence summarized above supports the hypothesis of Takhtajan that the leafy cyme is the most primitive form of inflorescence but that both the dichasium and the solitary flower have been derived from it repeatedly by relatively simple changes: in the former instance by the reduction in branch length and the modification of leaves to bracts, and in the latter by the suppression of branches and nodes, usually accompanied by the elongation of the scape.

A noteworthy fact is that the leafy cyme is most common in shrubs and herbs, and is not found in any group of trees, including the woody Magnoliales. This absence cannot be taken as evidence that the leafy cyme is a derived condition, since it is not found in shrubs or herbs having relatively specialized flowers, and the trends of specialization away from this condition can be traced in every family in which it occurs. A more plausible explanation is that the adaptive value of shifts toward more specialized inflorescences is higher in trees and large shrubs than it is in small shrubs and herbs, so that such shifts occurred during the earliest evolution of angiospermous trees from the ancestral shrubs. This hypothesis is supported by the presence of conditions that are generally recognized to be derived in such otherwise primitive woody families as Winteraceae (intercalary axillary inflorescences), Degeneriaceae, Illiciaceae, Himantandraceae, and Austrobaileyaceae (solitary flowers in leaf axils). In addition, some of the most highly specialized inflorescences occur in trees that are known to belong to ancient families, such as Platanaceae, Piperaceae, Hamamelidaceae, and Fagaceae.

The relation between determinate and indeterminate inflorescences

Most classifications of inflorescences have recognized two principal categories, determinate and indeterminate. Rickett (1944) has, however, pointed out that these categories are not sharply separated from each other. Two kinds of intermediate situations exist. In one, the main axis ends in a flower, but the indeterminate condition is produced by repeated branching from the axils of the uppermost floral bracts, in either a symmetrical or an asymmetrical pattern (Fig. 11-4, *C*). This gives rise to the "scorpioid cyme" or "cincinnus," as in the Boraginaceae. In the second and truly intermediate situation, the terminal flower may be produced, but instead of becoming differentiated and flowering first, as in a typical determinate inflorescence, it may be delayed in development so that it flowers later than many or most of the lateral flowers. Examples are given below in which this latter condition coexists in the same individual as a truly indeterminate axis.

The different approaches to the problem of determinate and indeterminate inflorescences illustrate very well the contrast that often exists between the taxonomic and the evolutionary approach to variation. Rickett minimizes this difference because of the numerous intermediate situations; except in certain restricted groups it is not a good diagnostic character. On the other hand, an evolutionist who is familiar with angiosperms as a whole must recognize the fact that inflorescences with a determinate main axis and those with an indeterminate one represent two large modes. Inflorescences that belong to one or the other of these categories far outnumber those that are intermediate. Morever, in very many plant families, probably the majority of them, inflorescences are either exclusively determinate or indeterminate as to their main axis. In all Leguminosae (Fabaceae), Cruciferae, Onagraceae, Ericaceae, Scrophulariaceae, Labiatae, and nearly all families of monocotyledons the axis is indeterminate, and in Umbelliferae (Apiaceae) and Compositae (Asteraceae) the compressed inflorescence is clearly derived from an indeterminate condition. Families that have exclusively inflorescences with a determinate main axis are less common, but the Winteraceae, Rutaceae, Geraniaceae, Linaceae, Caryophyllaceae, Gentianaceae, Polemoniaceae, and Rubiaceae are examples, whereas the great majority of Ranunculaceae, Papaveraceae, Lauraceae, and Celastraceae have determinate inflorescence axes. Thus the indeterminate axis, which, as is demonstrated below, is clearly the derived condition, has evolved independently many times, and, in general, has been remarkably stable once it has evolved. The question of how it can evolve is therefore, a significant one for the evolutionist.

The derived nature of the indeterminate axis is supported by evidence from a number of examples in which both conditions exist within the same family or closely related families. Representative ones are as follows:

Ranunculaceae. In the majority of the genera of this family the axis is clearly determinate. The indeterminate condition, usually a raceme, exists in *Delphinium* and *Aconitum,* both of which have specialized, zygomorphic flowers, and in *Actaea* and *Zanthorhiza,* which have flowers much reduced in size.

Papaverales. The great majority of genera of Papaveraceae have determinate axes. In the Fumariaceae, which have specialized zygomorphic flowers, the predominant condition is intermediate, so that the inflorescence is thyrsoid or racemose in appearance (Troll 1964). Truly racemose inflorescences are found in *Dicentra spectabilis* and most species of *Corydalis.*

Rosaceae. Most genera of this family have determinate floral axes. The indeterminate condition (raceme) is well developed in species of *Prunus* and *Agrimonia,* both of which have reduced flowers that produce only one seed per gynoecium.

Saxifragaceae. In the majority of herbaceous genera, including the large genus *Saxifraga* itself, the axis is determinate. The indeterminate condition (raceme or thyrse) occurs in *Tiarella, Mitella, Tellima, Tolmiea,* and *Heuchera,* where it is associated with specialized gynoecia having uniloculate ovaries with parietal placentation.

Apocynaceae. Indeterminate axes occur rarely in this family, but exist in the relatively specialized genus *Mandevilla* and a few others (Woodson 1935).

Campanulales. In most genera of Campanulaceae the axis is determinate, but intermediate conditions or indeterminate axes occur in a few species of *Campanula,* such as *C. rapunculoides.* In the related Lobeliaceae, which are not always recognized as a separate family, the axis is usually indeterminate. This is associated with more specialized zygomorphic flowers.

Violaceae. In the genus *Rinorea* both determinate and indeterminate axes exist (Troll 1964). Comparison with other genera indicates that the species with complex, determinate inflorescences, such as *R. dentata* and *R. brachypetala,* are the least specialized, and the indeterminate, racemose condition in such species as *R. racemosa* and *R. passoura* is derived.

The derived position of indeterminate inflorescences is also supported by the phylogenetic position of the families in which they occur. In the Magnoliidae, racemose inflorescences are not found in any of the more primitive woody families, but only in Menispermaceae, Trimeniaceae, and others that are relatively specialized. In the Dilleniidae they are most characteristic of relatively advanced orders, such as Violales, Ericales, and Passiflorales. In the Rosales, the Leguminosae, in which the inflorescence is consistently either a raceme or one of its derivatives, such as a spike or head, are more specialized in both woody anatomy and floral structure than the Rosaceae and Cunoniaceae, in which cymose inflorescences are common. Finally, the raceme or its derivative spike, umbel, corymb, or capitulum is by far the most widespread form of inflorescence in the most advanced class of dicotyledons, the Asteridae. The situation in the monocotyledons is discussed in Chapter 13.

The pathways of evolution from determinate to indeterminate axes are discussed in a later section.

Amplification and reduction through alteration of branching patterns

The simple dichasial unit rarely forms an entire inflorescence. More often, it is part of an elaborate branching system. Moreover, as Woodson (1935) has pointed out, it appears in its typical form only in plants having decussate phyllotaxy, at least in the inflorescence. If the branchlets of the inflorescence are spirally arranged, which is usually the case following spiral arrangements of leaves and vegetative branches, the lateral flowers immediately below the terminal one are likely to develop at unequal rates, and if the inflorescence is highly compound, side branches from adjacent nodes may have different numbers of flowers (Fig. 11-4, *D*).

Amplification of the inflorescence through increase in number of branches and its reduction through suppression of branches occur with almost equal frequency. Evidence for both trends can be found within a number of families, as follows:

Dilleniaceae. Amplification can be seen in *Tetracera,* a genus of woody vines belonging to the more specialized subfamily Tetraceroideae, and reduction in the one-flowered inflorescences found in species of *Dillenia,* and in the scapose flowers of *Acrotema.*

Ranunculaceae. Amplification is evident in *Clematis, Thalictrum, Cimicifuga,* and *Zan-*

thorhiza, reduction in many species have solitary, scapose flowers.

Papaveraceae. The elaborate inflorescences of *Bocconia* and *Macleaya* are the end points of amplification trends; the scapose flowers are the end points of reduction.

Rosaceae. Trends of amplification have culminated in species of *Spiraea, Aruncus, Adenostoma, Sorbus, Rubus,* and many other genera; *Dryas, Fragaria, Dalibarda,* and a few others are the end points of reduction.

Rutaceae. Inflorescences of *Ruta* and of most genera of the subfamily Zanthoxyloideae are the products of amplification; those of *Cneoridium,* many genera of *Boronieae* and *Diosmeae,* as well as *Citrus* and other genera of Aurantoideae are the products of reduction.

Oleaceae. The elaborate inflorescences of *Ligustrum, Syringa,* and *Fraxinus* (Troll 1964) have resulted from amplification; those of *Adelia* and *Forsythia,* from reduction.

These examples could be multiplied almost indefinitely. In nearly every large family of angiosperms, both kinds of trends can be detected.

Amplification has occurred in four different ways. In some instances, the number of primary branches has greatly increased, but the number of secondary branches has remained small, so that an elongate, thyrsoid inflorescence results (Fig. 11-4, C). In other instances, the number of secondary branches on each primary branch equals or exceeds the total number of primary branches, so that the inflorescence is rounded or flattened (Fig. 11-4, *H*). A third form of amplification is the production of tertiary and quaternary branchlets so that the inflorescence becomes increasingly compound (Fig. 11-4, *J*). Finally, the primary branches immediately below the terminal flower may produce a short secondary branch bearing a single bract and a terminal flower; a tertiary branch of the same kind then forms in the axil of the bract, a quaternary branch forms in the axil of the bract of the tertiary branch, and so on indefinitely (Fig. 11-4, *D*). This sympodial branching gives rise to the wandlike side branches found in species of *Ruta, Linum, Sedum,* and *Sempervivum* (Troll 1964), as well as to the "scorpioid" inflorescences of Boraginaceae and Hydrophyllaceae. In more advanced members of these latter families, reduction of internode length, suppression of bracts, and crowding of flowers render the morphological nature and phylogenetic origin of their inflorescences obscure.

Shifts in the order of maturation of the flowers. In both the leafy cyme and the simple dichasium, the terminal flower is differentiated some time before any of the lateral ones, and comes into flower first, usually followed by those at the ends of the primary branches, in descending order. On the other hand, the order of flowering in a highly complex inflorescence is often very irregular (Parkin 1914:532, 543). In inflorescences that have developed a large number of similar primary branches, regularity may be found again. The terminal flowers of the lateral branches appear in a regular acropetal sequence. The terminal flower of the main axis of the inflorescence may open before any of the laterals, as in *Clematis Jackmanii* (Parkin 1914:553), and in *Aesculus californica* (unpublished personal observations); it may be intermediate in the sequence, as in *Clematis viticella,* or it may be the last to bloom, as in *Convulvulus mauretanicus* (Parkin 1914:553). This series significantly illustrates the most probable kind of transition from a determinate to an axially indeterminate inflorescence.

Elongation and suppression of pedicels. These changes are relatively simple from the developmental point of view, but they may alter radically the appearance of the inflorescence. If suppression of internode development occurs in the main axis, it converts a raceme to an umbel or a spike to a head or capitulum (Fig. 11-4, *L,M*). Suppression of internode development in the floral pedicel converts a raceme to a spike or spadix (Fig. 11-4, *N,O*). Extension of internode development is found in many solitary axillary or basal flowers. Its occurrence in the lower flowers of an abbreviated raceme or umbel produces a flattened umbel or corymb (Fig. 11-4, *L*).

Inflorescences that contain two different superimposed patterns

In several of the more advanced groups of angiosperms, the inflorescence consists of two kinds of elements, which differ from each other radically with respect to both their structure and their development. The two best-studied examples are the Compositae (Asteraceae) and the Gramineae (Poaceae). In the Compositae, the basic unit is the capitulum, in which the flowers are sessile on the involucral receptacle, are surrounded by an involucre of specialized phyllaries, and develop in an acropetal fashion. This structure has evolved by reduction from a raceme, via the intermediate condition of a corymb (Fig. 11-4, *M*).

In many Compositae, including the most generalized species of the tribe Heliantheae, the capitula are either solitary or arranged in leafy cymes. Bracteate cymes, with the capitulum as the floral units, have evolved repeatedly, as in *Bidens, Chrysothamnus, Aster, Gnaphalium, Senecio, Eupatorium, Vernonia,* and *Lactuca.* Less common has been the evolution of indeterminate axes bearing capitula, but this has occurred in *Ambrosia, Artemisia,* and a few other genera (Payne 1963). Finally, in a small number of genera, particularly *Elephantopus* and *Echinops,* the much-reduced primary capitula have been aggregated into capitula of a second order.

In the Gramineae, the basic unit is the spikelet, in which the florets are situated on a vertical axis and develop in acropetal succession. Each floret is subtended by either one or two bracts (depending upon whether one interprets the palea as a bract or as a modified calyx), and the spikelet is enveloped by two or more sterile bracts. In many genera of this family, the dual nature of the inflorescence is accentuated by the fact that although the basic units or spikelets develop acropetally, the larger units, containing aggregates of spikelets, often develop basipetally. This is true of the "panicles" of all genera of the subfamilies Arundoideae and Festucoideae (Weber 1939, Evans and Grover 1940, Bonnett 1937, 1961). On the other hand, in more specialized species of Agrostidae, such as *Alopecurus,* and the subfamily Panicoideae, the development of the spikelets along the principal branches of the inflorescence is often acropetal (Weber 1938), and this axis is, therefore, indeterminate.

In some ways, therefore, the "panicle" of the more primitive tribes of Gramineae resembles the compound cyme found in many families of dicotyledons having an intermediate stage of advancement. If one accepts the spikelet as a basic unit that existed in the ancestors of the Gramineae, one can postulate an initial condition in which the development of the inflorescence was chiefly or entirely acropetal, and it consisted of a series of spikelets arranged in a racemose fashion, as in the modern genus *Brachypodium.* Amplification occurred through the development of single branches or clusters of branches at the lower nodes and the elongation of these branches through the activity of newly developed intercalary meristems. In the tribe Hordeae, the spike evolved through reduction of the peduncles of the spikelets. The acropetal development of spikelets in the Panicoideae evolved secondarily via a shift similar to that which in dicotyledons converted determinate to indeterminate inflorescences.

This hypothesis is supported by the fact that the "panicle" of the grasses is a unique form of inflorescence, which both in its manner of branching and in its developmental pattern is different from anything known in other families of monocotyledons.

If one adopts this interpretation of the evolution of the grass panicle, one must logically interpret in a similar fashion the complex inflorescences found in the larger species belonging to the Juncaceae and the Cyperaceae. In them, also, a pattern of basipetal development of the inflorescence as a whole is superimposed upon a pattern of acropetal development for the individual units, or spikelets.

Developmental Patterns of the Inflorescence

The histological and biochemical nature of the shoot apex in both the vegetative and the flowering condition, as well as the events that accompany the transition between them, have been the objects of extensive research and a large volume of publication during the past 25 years. In addition to providing a great body of facts, the workers in this field have advanced a number of theories, many of which are highly controversial. An extensive and excellent review of this material has recently been made by Gifford and Corson (1971), and a shorter digest is presented by Cutter (1971).

For the most part, the workers in this field have attempted to prove or disprove certain general theories, particularly the concept of foliar helices advanced by Plantefol (1948), and beliefs about the nature of the shoot apex that have been advanced by his students, particularly Buvat (1952) and Lance-Nougarède (Lance 1957, Lance-Nourgarède 1961, (Nourgarède 1965). They have been less concerned with attempts to discover relationships between species and genera on the basis of comparisons between their developmental patterns. Nevertheless, comparative morphogenesis based on modern techniques is potentially a valuable tool for increasing our understanding of evolutionary trends. In the present section, the literature will be surveyed with this objective in mind.

The most important fact is that in many plants great differences exist between the vegetative- and the reproductive-shoot apex, with respect to histology, physiology, and biochemical syntheses. In any flowering-plant species, therefore, a radical transformation takes place in its shoots at a particular stage of development. Furthermore, the timing of this transformation relative to the formation of lateral appendages, such as leaves, bracts, and floral parts, varies greatly from one group of plants to another, but usually is highly constant for a particular species, and is often constant for genera and even entire families. The complexity of the inflorescence depends largely upon the number of parts that are differentiated between the appearance of the reproductive condition of the apex and the differentiation of the final flowers. Moreover, species can differ from one another with respect to the gradualness or abruptness of the transition from the vegetative to the flowering condition, as well as the degree and kinds of differences that exist between the two states. All of these differences affect the appearance of the mature inflorescence.

Fig. 11-5. Diagram showing the layering and zonation of a typical shoot apex. (From Gifford and Corson 1971.)

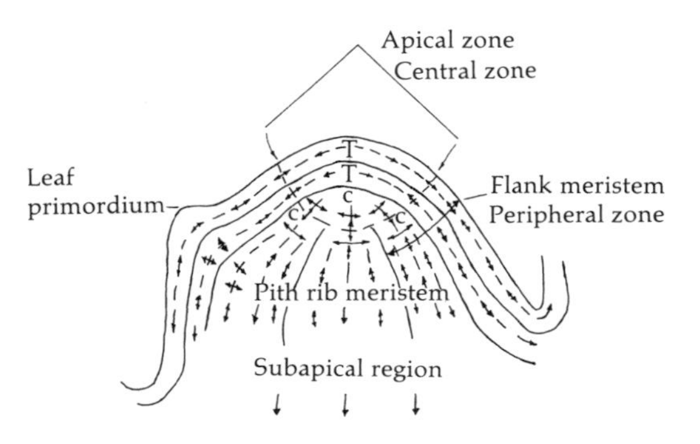

The histological features of the transformation of the apex from the vegetative to the reproductive condition can best be understood by comparisons with the active vegetative apex. After their extensive review of the literature as well as their own research. Gifford and Corson have concluded that the best concept of the apex is the holistic one developed by Wardlaw, (1965), according to which regions of differential activity are recognized, but are not sharply delimited, and no one region contains exclusively the determinants for structures that appear later in ontogeny, as some authors have proposed. The differentiation is of a dual nature, layering and zonation (Fig. 11-5). Two layers can be recognized: a superficial tunica, which overlaps the more profound corpus. Cells in the tunica later divide anticlinally, with their spindles formed parallel to the surface of the shoot, and the new cross walls at right angles to it, except during the initiation of primordia. Those in the corpus divide in various directions, periclinal divisions in the outer part of the corpus being the usual prelude to the initiation of appendages. The zones are a central or apical zone, in which cells are active metabolically but divide infrequently, a peripheral zone, which is chiefly responsible for the initiation of appendages, and an internal zone of rib meristem. The size and distribution of these zones vary in a regular, cyclic fashion, accompanying the differentiation of successive leaves. Each cycle of leaf differentiation, known as a plastochron, involves a characteristic succession of histological conditions, so that apices can be compared only when they are at the same plastochronic stage. Differences between species at comparable stages of the plastochron cycle often determine differences in the nature and arrangement of the leaves. For instance, in *Linum* and other species having many leaves that are produced in quick succession, the peripheral zone exists continuously and is altered very little during the plastochron cycle, whereas in *Magnolia, Piper,* and *Peperomia,* which have large leaves or leaf bases and a low (1/2 or 1/3) phyllotactic spiral, the peripheral zone is almost entirely used up for the formations of each leaf, and is reconstituted later in the plastochron cycle (Hagemann 1960).

The histological changes that accompany the transformation of the shoot apex to the flowering condition vary considerably from one species to another, but the following changes are widespread. The size of the undifferentiated apex often increases, and plastochron cycles may become shorter (cf. *Michelia,* Fig. 10-2). Mitotic activity increases in all parts of the apex, except for the central core, the cells of which become larger and more vacuolated. The increase in mitotic index is greatest in the apical part of the central zone, and in many species is such that the differentiation of this zone disappears entirely. Accompanying this increase in mitotic activity is an overall reduction in cell size, and in many instances an increase in the frequency of periclinal divisions, so that the layering of the apex becomes much less evident. The individual cells become characterized by weakly staining nuclei and large, heavily staining nucleoli. With respect to fine structure, the number and extent of cellular organelles, particularly dictyosomes and endoplasmic reticulum, are greatly increased (Gifford and Stewart 1965).

These changes at the cellular level are in part the result of altered biochemical activity. Histochemical tests have demonstrated marked increases in the amount of RNA, protein, and sulfhydryl (—SS—) groups (Lance 1957, Gifford and Tepper 1962, Corson and Gifford 1969). Inhibitors of mRNA and protein synthesis can block the transformation of the apex (Zeevart 1962, Galun, Grossel, and Keynan 1964).

From the physiological point of view, the work of C. Nitsch (1968) on *Plumbago indica* is highly significant. By inducing the entire flowering cycle *in vitro* from detached callus cultures, she has shown that the reproductive apex is an independent reaction system, which does not require specific influences from either roots or leaves. Furthermore, it differs markedly from the vegetative system with respect to the balance of hormones and growth substances. Concentrations of auxins and gibberellins reach a low level during the initiation of flowering, but rise later on. In contrast, the amounts of cytokinins, as well as of adenine, increase considerably, and flowering *in vitro* is promoted by adding these substances.

These physiological and biochemical changes undoubtedly reflect changes in the pattern of gene activity. The concept of differential gene action as the chief basis of morphogenetic differentiation is now generally recognized to be valid (Watson 1970, Heslop-Harrison 1967; see Chapter 6). On the basis of this concept, the transformation of the vegetative to the reproductive apex undoubtedly is accompanied by the suppression of the activity of some genes and the activation of others. The usual increase in the amount of protein formation during the transition suggests that more genes are active in the reproductive than in the vegetative meristem.

The bearing of differences in timing relations upon the structure of the inflorescence can best be recognized by comparing very different species. In *Michelia fuscata,* which has relatively large, solitary flowers, the histological transformation of the apex occurs after the differentiation of tepals has begun, before the differentiation of stamens (Tucker 1960). In various genera of Compositae, the transformation begins during the differentiation of involucral phyllaries, and the entire capitulum, including the florets, is differentiated under the influence of the reproductive condition (Popham and Chan 1952, Lance 1957, Rauh and Reznik 1953). An even earlier onset of the transformation relative to differentiation occurs in some grasses, such as *Triticum* and *Hordeum* (Weber 1939, Barnard 1955, and unpublished personal observations). In them, not only all of the spikelets but also the uppermost culm leaves are differentiated after the apex has acquired the reproductive condition. In general, changes can be detected in the apex at the time of initiation of the inflorescence, as in *Ananas* (Gifford 1969), *Beta* (Lance and Rondet 1958), *Carex* (Smith 1966), *Chenopodium* (Gifford and Tepper 1961), various Cruciferae (Bernier 1964, Lance-Nougarède 1961, Miksche and Brown 1965), *Datura* (Corson and Gifford 1969), Dipsacaceae (Philipson 1947*a*), *Musa* (Barker and Steward 1962), *Nicotiana* (Diomaiuto-Bonnand 1966), *Perilla* (Nougarède, Bernier, and Rondet 1964), *Plumbago* (Nitsch 1968), *Pharbitis* (Bhar and Radforth 1969), Rosaceae (Rauh and Reznik 1951), *Rubus* (Engard 1944), *Soja* (Borthwick and Parker 1938), *Teucrium* (Lance-Nougarède 1961), and *Valeriana* (Philipson 1947*b*). On the other hand, in *Garrya* (Reeve 1943), *Jussieua* (Michaux 1964), *Oenothera* (Bersillon 1957), and *Lochnera* (*"Vinca"*) *rosea* (Boke 1947), deviations from the histological appearance of the vegetative apex were not detected until some time between the initiation of the inflorescence and that of the individual flowers.

Species and genera can differ from one another with respect to the gradualness or suddenness of the transition. In *Aquilegia* and *Ranunculus* (Tepfer 1953), as well as in *Pharbitis* (Marushige 1965), the transition is very gradual, and appendages that are intermediate between leaves and bracts are differentiated during this stage. In Compositae and Gramineae, on the other hand, it is much more abrupt, and differentiation of appendages does not occur during the transitional period. Other genera are intermediate between these extremes. In *Nicotiana* (Diomaiuto-Bonnand 1966), the transition is gradual in *N. rustica,* but much more abrupt in *N. tabacum* and *N. nudicaulis.* In general, gradual transitions can be recognized in adult inflorescences by the presence of transitional forms of appendages.

Most important are differences with respect to the total amount of differentiation of the reproductive as compared with the vegetative apex. In our present state of knowledge, these are hard to estimate, because the necessary data, particularly with respect to cytological and biochemical characteristics, are not available. Nevertheless, even superficial examination reveals that the meristems that are responsible for the complex, highly derived inflorescences, such as the capitulum of the Compositae and the "panicle" of the Gramineae, are radically different from the vegetative apices of the species concerned, whereas in species that have simpler inflorescences, such as the racemes of *Oenothera,* the inflorescence apex differs little from the vegetative apex. One might speculate that in various lines of evolution the reproductive apex has become increasingly differentiated from the vegetative apex. Although this is true in general, there are undoubted exceptions. For instance, in inflorescences that have well-developed terminal flowers, the early reproductive apex loses all difference between a more active peripheral and a less active apical zone. On the other hand, indeterminate inflorescences retain an apical zone that may be nearly as well differentiated as that of the vegetative apex. Nevertheless, morphological evidence that has already been reviewed points strongly to the derived condition of the indeterminate inflorescence. With respect to the single character, differentiated vs. nondifferentiated apical zone, reversal of evolutionary direction is possible, as it is with most single characters.

When the inflorescence is strongly differentiated, its eventual shape is determined very early, during the initial stages of differentiation of the reproductive meristem. For instance, in Compositae like *Bellis* (Philipson 1946), the apex expands laterally to a great extent before differentiation of reproductive parts, except for the outermost phyllaries, has begun. This expansion is accompanied by numerous mitoses in which the spindles are oriented parallel to the surface of the future receptacle. In marked contrast is the initial floral meristem of grasses, particularly those species that form spikes, as in the tribe Hordeae. in them, the initial differentiation of the reproductive meristem is in a distal or vertical direction, and continues for some time before the beginning of spikelet differentiation. Less marked differences, as between the conical receptacle of the Compositae genus *Rudbeckia* and other genera having flat receptacles, are outlined by the direction of growth of the reproductive apex during its initial stages of expansion (Rauh and Reznik 1953). It seems likely that the number of flowers or (in grasses) of spikelets formed in these inflorescences is largely a function of the amount of expansion that takes place at this early stage, but quantitative data on this point are not yet available.

Further information on the comparative aspects of inflorescence development will be valuable chiefly as a means of estimating the probability with which morphological characters can reverse their direction of evolution. Differences that depend upon alterations of a single developmental process might be expected to be reversible, whereas irreversibility is most likely to characterize adult differences for which a combination of developmental differences is possible. In the former category are such characteristics as determinate vs. indeterminate inflorescences, as well as the number of flowers in a raceme or capitulum. On the other hand, differences that involve branching patterns, as between a cyme or thyrse and a raceme, are likely to be irreversible.

Adaptation, Selection, and the Evolution of Inflorescences

The evolutionary trends in the inflorescence illustrate very well the principles outlined in earlier chapters: increasing specialization in response to alternation of adaptive responses for survival under severe conditions followed by increased reproductive potential under favorable conditions, and the spread of broadly adaptive developmental patterns, once evolved, to habitats very different from those that brought them into existence.

The reduction from a leafy cyme or other branching type of inflorescence to a solitary flower is, in herbaceous groups, usually associated with habitats in which a more rapid flowering cycle has an adaptive advantage. The cycle of events associated with the origin of the scapose condition in *Agoseris* (Chapter 9) is representative. Similar cycles are probably responsible for the scapose, one-flowered condition present in various species of Ranunculaceae and Papaveraceae. In the genus *Paeonia*, subgenus *Paeon*, two independent origins of the one-flowered condition from the leafy cyme have different explanations. The species of this subgenus having a leafy cyme inhabit open, sunny habitats in mountainous regions where moisture is continuously available. The one-flowered species of the Mediterranean group (*P. daurica, corallina, Broteri,* and others, plus the derived polyploid *P. officinalis*) inhabit chiefly regions having a summer dry climate, in which there is an adaptive advantage for rapid blooming during the favorable spring conditions. The one-flowered Asiatic species, *P. obovata,* on the other hand, inhabits mesic forests, in which the adaptive value of early and quick flowering is associated with the presence of abundant light in spring, before the leafy canopy of the forest is fully developed, followed by an unfavorable period of deep shade.

The origin of the condition of solitary flowers in leaf axils, which is widespread in relatively primitive woody species, has a different explanation. The species concerned are for the most part undershrubs or small trees in broad-leaved-evergreen forests, and so are at all time subject to low light intensities. Consequently, there can never be a great accumulation of photosynthetic products. The most efficient use of available photosynthates for reproductive purposes is, therefore, to distribute the flowers evenly over the entire tree or shrub, so that each flower receives an equal supply of these products. One of the adaptive responses for reproduction in the understory of tropical and subtropical forests, therefore, is the reducton of inflorescences to a single flower and their distribution in leaf axils over much of the plant.

The origin of indeterminate from determinate inflorescences has been complex, and has involved at least two steps. No example exists in which the origin of a raceme can be traced directly from a simple cyme. Racemes have apparently originated from compound cymes in one of two ways: (1) via reduction in the number of flowers on the side branches of the inflorescence, followed by suppression of the terminal flowers; (2) via initial suppression of the terminal flower, to produce a thyrse, in which the main axis is indeterminate but the primary branches are determinate cymes, followed by the reduction of the flowers to one per primary branch. The racemes of *Prunus* have originated by the first pathway, since intermediate stages, exemplified by *Exochorda* and *Osmaronia,* have racemose inflorescences except for the presence of a terminal flower. The second pathway has most probably been followed in the Saxifragaceae, since the intermediate situation exists in *Heuchera,* which has a typical thyrse. It has also been followed in the Scrophulariaceae and their ancestors, since indeterminate thyrses are well developed in several genera of this family (*Scrophularia, Verbascum, Penstemon*). Intermediate pathways may have been taken in the Berberidaceae and Fumariaceae, in which some reduction of flower number took place in the side branches before the disappearance of the terminal flower.

Whatever may be the pathway of reduction that leads to the raceme, it is usually and perhaps always preceded by amplification leading from a leafy or simple bracteate cyme to a many-flowered compound cyme. The selective pressure favoring this amplification may be simply the advantage of higher reproductive potential under increasingly favorable conditions. In some instances, however, selection may favor an inflorescence having about the same reproductive capacity as the original one, but with more numerous flowers, each of which contains fewer seeds. As Burtt (1960) has pointed out with reference to Compositae, but as is equally valid in the present context, such an arrangement provides better protection against predatory insects that lay their eggs in a single flower.

The adaptive advantages of a raceme over a compound cyme under various conditions are as follows:

(1) As in the case of solitary axillary flowers, in woody plants that occupy the understory of forests, the production of many, few-flowered axillary racemes distributes the reproductive function more evenly over the plant as a whole than does the production of a few, many-flowered terminal inflorescences.

(2) In an indeterminate inflorescence, either a thyrse or a raceme but more definitely in the latter, the synthesis of reproductive material is extended over a relatively long period of time, so that the amount synthesized at any one time is less than that for a compound determinate inflorescence. If any one factor needed for synthetic activity, either light, moisture, or mineral nutrients, is present continuously but in limited quantity, this extension of the reproductive function has an adaptive advantage. In mesic cli-

mates, such conditions exist chiefly in herbs that grow in shady places. The examples listed in Table 11-2 illustrate a correlation that is probably widespread, though exceptions undoubtedly exist. In related genera having on the one hand compound determinate and on the other hand indeterminate inflorescences, the latter occur more frequently in genera or species that are shade loving.

(3) In species having flowers that are highly specialized for pollination by insects or other animal vectors, the raciform structure makes possible a sequence of blooming such that only one or two flowers of a plant are blooming at a particular time. In self-compatible species, this favors cross-pollination, whereas in self-incompatible species, it reduces the frequency of inefficient self-pollinations, in which the pollen that reaches the stigma is nonfunctional.

(4) Inherent in an inflorescence that is indeterminate either with respect to its main axis or because of repeated, sympodial branching is a flexibility that enables individual plants to form few or many flowers depending upon the length of the favorable season. Consequently, one might expect natural selection to

Table 11-2. Ecological distribution of related genera or species having compound determinate and axially indeterminate inflorescences.

Family	Compound determinate	Predominant habitat	Indeterminate	Predominant habitat
Ranunculaceae	*Clematis*	Mesic, open	*Delphinium*	Mesic, shade or open
			Aconitum	Mesic, shade
			Zanthoriza	Mesic, shade
			Actaea	Mesic, shade
Berberidaceae	*Nandina*	Mesic, open or shade		
	Berberis	Semixeric or mesic, open	*Berberis*	Mesic, shade
Fumariaceae	*Dicentra*	Semixeric or mesic, open or shade	*Corydalis*	Mesic, shade or open
Saxifragaceae	*Saxifraga*, sect. *boraphila*	Mesic, open or shade	*Heuchera*	Mesic, shade
			Mitella	Mesic, shade
			Tiarella	Mesic, shade
			Tellima, etc.	Mesic, shade
	Escallonia	Mesic, open or shade	*Ribes*	Mesic, shade
Rosaceae	*Spiraea*	Mesic, open	*Prunus*, sect. *Padus*	Mesic or semixeric, shade or open
Hippocastanaceae	*Aesculus californica* *A. Parryi*	Semixeric, open or shade	*A. parviflora.* *A. Hippocastanea*	Mesic, shade or open

favor this kind of inflorescence in a plant adapted to a semiarid climate having great variation in precipitation from one season to the next.

Two examples known to me illustrate such phenotypic flexibility in plants having inflorescences that are transitional from the determinate to the indeterminate condition. In *Pharbitis nil*, plants growing under natural conditions, which are exposed to short photoperiods only when they are fully grown, produce an indeterminate inflorescence consisting of flowers in the leaf axils of the main branch. If, however, plants are exposed artificially to short photoperiods at the cotyledon stage, an inflorescence of a similar nature is terminated by a single terminal flower (Marushige 1965, Bhar and Radforth 1969).

I have found a somewhat analogous situation in *Penstemon corymbosus* (Scrophulariaceae). Axially indeterminate inflorescences are almost universal in this family, including thyrses (*Scrophularia, Verbascum*), racemes (*Digitalis, Veronica*), and spikes (*Pedicularis, Euphrasia, Castilleja*). In the genus *Penstemon*, the inflorescence is either a thyrse or a bracteate raceme. A shrubby species of California, *P. corymbosus*, normally has a short thyrse. In a plant gathered in the wild (near Mount Saint Helena) and growing in my garden, the more vigorous shoots produced typical indeterminate thyrses, but the smaller ones produced determinate compound cymes having a definite terminal flower on the main axis (Fig. 11-6).

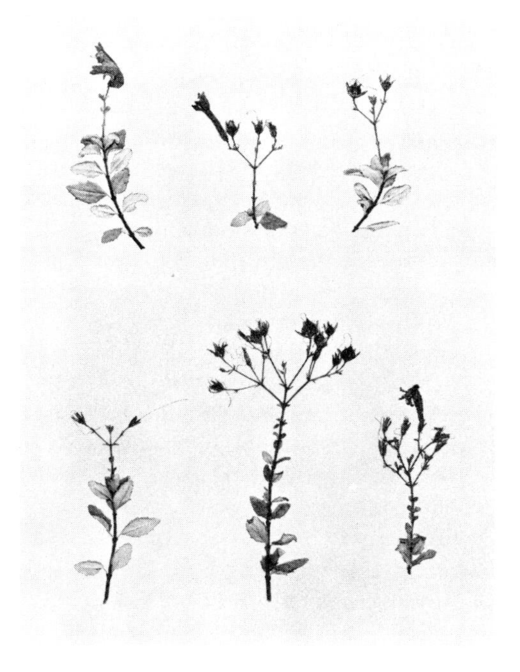

Fig. 11-6. Six different inflorescences from the same plant of *Penstemon corymbosus*, showing a series of transitions from a single terminal flower through corymbs having various degrees of development of the terminal flower, to indeterminate thyrses, in which no definite terminal flower is evident on the main axis. The last condition is the commonest one in this and most other species of *Penstemon*.

A noteworthy fact is that in both of these examples the presence of a terminal flower was associated with a less vigorous vegetative growth. In view of the discovery by Nitsch (1968) that flowering is associated with a low level of IAA and gibberellins, a possible explanation of these examples is that the weak vegetative growth did not permit a high enough concentration of these substances to suppress the differentiation of the terminal flower and allow the formation of the indeterminate inflorescence that is normal for the species. If this explanation is correct, then one might advance the hypothesis that the evolutionary trend from determinate to indeterminate inflorescences was brought about by genetic changes that altered the hormonal balance. Either greater amounts of hormones and gibberellins are produced in association with the beginning of flowering in species having indeterminate as compared with their relatives having determinate inflorescences, or the translocation of materials was altered in such a way that a larger proportion of these growth substances moves into the reproductive apex.

The conservation of organization and modification along the lines of least resistance are well illustrated by trends that take place in groups that have indeterminate inflorescences fixed and constant for the group. In them, the condition of many flowers at the ends of branches, if it is adaptive for a particular species or genus, is acquired most easily by means other than reversal of the direction of evolution and production again of a compound cyme or thyrse. The repeated branching of indeterminate racemes, or the suppression of axial growth of the raceme to produce a corymb, an umbel, or a succession of umbels, both require less modification of the developmental pattern than the return to a much-branched inflorescence bearing terminal flowers on the main axis and its branches. In the Leguminosae, amplification of the inflorescence is acquired either by additional branching of racemes to produce secondary and tertiary racemiform units, as in *Sophora japonica*, or by the arrangement of globose heads into racemes, as in many species of *Acacia* and other Mimosoideae. In the Umbellales, both suppression of the main axis and the appearance of extra branches have given rise to compound umbels.

Many trends from racemiform inflorescences to more reduced forms have taken place. Suppression of the pedicels of the individual flowers converts the raceme into a spike, whereas suppression of the main axis leads to the umbel. Suppression of both leads to the head or capitulum. In many instances of families in which both racemes and the more reduced forms are found in related genera, the latter tend to occur in genera that are best adapted to climates having a relatively short growing season. In the Scrophulariaceae, spicate inflorescences occur in the tribe Rhinanthoideae, the larger genera of which (*Pedicularis, Castilleja*) are either alpine and subalpine, semixeric, or both. Within the genus *Mimulus* in California there is a conspicuous difference in habitat between the species having pedicellate flowers, which occur in moist or mesic habitats, and those having sessile flowers, which are much more xeric in their ecological preferences. In the Leguminosae, the major centers of concentration for the genus *Trifolium* and related genera with capitate inflorescences are in regions having a Mediterranean type of climate, characterized by long, dry summers, so that plants that are basically mesophytic must produce flowers and seeds early in order to avoid the drought. In the Campanulaceae, the genera having capitate inflorescences are either alpine (*Phyteuma*) or Mediterranean (*Jasione*) in distribution. A similar difference in distribution separates the Dipsacaceae, which are primarily Mediterranean in distribution and have capitate inflorescences, from the Valerianaceae, which are more mesic in adaptation and have inflorescences with well-developed axes and branches.

A final form of inflorescence worthy of discussion is cauliflory. This highly specialized condition is widespread in trees that live in tropical rain forests. Inflorescences are produced from meristems formed on the main trunk or principal branches of the tree, under the leafy canopy, and in some instances, particularly in the family Lecythidiaceae, actually near the roots. The form of the inflorescence is, as a rule, highly irregular, and no definite architecture can be detected. In some form or other, it exists in genera belonging to many or most of the larger tropical families.

The adaptive significance of cauliflory appears to be associated chiefly with cross-pollination. The flowers, if not self-fertilized, are always pollinated by animal vectors, either insects or bats. In the tropics, the number of different tree species that are pollinated by such vectors is very large, and competition for pollinators is keen. Furthermore, the insect fauna of these forests is distributed in horizontal layers, so that different potential pollinators occur at various heights above the ground (Pittendrigh 1948, Richards 1952, Snow 1966). Since low shrubs and herbs are scarce in these tropical forests, adaptation to pollinators that live near the ground level is a niche that is most easily filled by cauliflorous trees.

If this reasoning is correct, the logical conclusion is that cauliflorous species have entered the rain forest secondarily, having been derived from ancestors that lived in more open habitats. Their adaptation to the specialized condition of cauliflory was in response to the presence of an unfilled niche, that of pollination by vectors living high in the canopy having been already filled. A test of this hypothesis would be provided by studying the immediate relationships of cauliflorous species and genera. Are their nearest relatives usually noncauliflorous species inhabiting rain forests and adapted to pollination by vectors in the canopy, or are these relatives found in more open regions that surround the rain forests?

Summary

The original trends of growth habit in angiosperms are believed to have been adaptive radiation from shrubs, in one direction to trees and in the other to herbs. Reversals of these trends have been frequent. Trends from herbs to shrubs and trees have occurred most often on oceanic islands, but occasionally in continents, particularly when semixeric herbs have given rise to more xeric shrubs. The deciduous condition of shrubs and trees has often evolved in plants adapted to stream banks in regions having a highly continental climate, as an adaptation to hardening the plant before the appearance of frosts. Reversals from the deciduous to the evergreen condition have been rare, but have probably occurred in some genera of angiosperms. Adaptive radiation of geophytes has been in response to specific kinds of climatic and edaphic changes, but once a particular kind of geophytic condition, such as the production of bulbs or corms, is acquired, the plants having it have often given rise to descendants that have retained the condition, even though they have altered radically their climatic adaptations. The origin of the spring-flowering bulb- or corm-forming geophytes found in mesic, temperate, deciduous woodlands is explained in this fashion. On the basis of both morphological and developmental information, trends of evolution in the inflorescence are followed. These include both reduction and amplification. The trend from simple, few-flowered inflorescences to larger, more complex ones is associated with an earlier onset of the histological and biochemical transformation of the vegetative-shoot apex into a reproductive apex, relative to the differentiation of the individual flowers. The structure of the inflorescence is also influenced by the suddenness or gradualness of the transition from the vegetative to the reproductive condition of the shoot apex. A review of various genera and families indicates that these transformations are initially associated with adaptations either to increasingly favorable or to deteriorating environments, but that many kinds of inflorescence structure are retained after the conditions that brought them into existence no longer prevail.

12 / Trends of Evolution in the Flower

Perianth and Stamens

The perianth of angiosperms has undergone principally four different kinds of change: *reduction,* particularly the suppression of the corolla and sometimes also of the calyx; *differentiation* of calyx and corolla; *intercalary concrescence,* producing "fusion" of sepals or petals and "adnation" of stamen filaments to the corolla tube; and *change of symmetry,* from radial to bilateral, producing zygomorphy or "irregular flowers." As has already been mentioned, each of these trends has occurred independently in many different evolutionary lines, and can be detected at the level of different species belonging to the same genus, as well as different genera in a family, and different families of an order.

The first stage of reduction in the perianth is from indefinite numbers of parts, spirally arranged, to definite numbers in distinct whorls. This stage occurred very early in the evolution of angiosperms, since oligomerous, cyclic perianths are found in several families that with respect to their total assemblage of characters are among the most primitive angiosperms (Austrobaileyaceae, Degeneriaceae, Himantandraceae, Annonaceae; see Table 10-2). It is probably a by-product of the reduction of the floral axis as a whole, which had the adaptive value both of placing the stamens and carpels near to each other, thus increasing the efficiency of insect pollination, and of speeding up floral development, in response to the demands of a seasonally unfavorable climate.

The second stage of reduction, suppression of the corolla or of the entire perianth, has taken place chiefly in association with the shift from insect to wind pollination (see Chapter 4). Its earliest occurrences, as in Cercidiphyllaceae, probably were in flowers that did not have a perianth differentiated into calyx and corolla. In many other groups, particularly those families formerly placed in the "order" Amentiferae (such as Salicales, Fagales, Juglandales), as well as other orders of the Hamamelidae (Hamamelidales, Urticales), the relationship between the apetalous families and their presumable perianth-bearing ancestors is so obscure that the actual course of reduction cannot be traced.

More instructive are examples of families in which related genera exist, some of which possess a corolla whereas others are apetalous, or of orders in which the difference exists between obviously related families. Two such examples are described in Chapter 4, *Thalictrum* in the Ranunculaceae and the tribe Poterieae in the Rosaceae. In them, as well as in the tribe Paronychieae of the Caryophyllaceae and the family Empetraceae of the Ericales, the loss of petals is associated with the shift from insect to wind pollination accompanying the entrance of the group into a region with a harsh climate, where insects are scarce and high winds prevail. In three other examples, *Stellaria* and *Sagina* in the Caryophyllaceae and *Glaux* in the Primulaceae, loss of petals may be associated with the shift from frequent outcrossing to predominant self-fertilization. These two kinds of adaptive shift are possibly responsible for the majority of examples of secondary loss of petals.

Differentiation of calyx and corolla

As stated in Chapter 9, the original angiosperm flower probably had a perianth consisting of spirally arranged, bractlike tepals, as in modern Eupomatiaceae, Illiciaceae, Calycanthaceae, and other relatively primitive families. In some families (Magnoliaceae, Winteraceae), the only change from this condition was the development of conspicuous color, making the flower more attractive to insects. The differentiation of the perianth into an outer, protective series (calyx) and an inner, attractive series (corolla) took place many times, and most probably in at least two different ways, as has been already discussed in Chapter 9.

A perianth that is differentiated into outer "sepals" and inner "petals" has probably originated by still another pathway in at least one family. Most genera of Portulacaceae have a perianth consisting of two "sepals" and five "petals." The family belongs to that group of Centrospermae or Caryophyllidae characterized by the presence of beta-anthocyanins (Mabry 1964). Since most other families of this group have a uniseriate perianth, that is, are "apetalous," this condition very probably existed in the ancestors of the Portulacaceae. In some genera, the two "sepals" resemble the uppermost leaves in both form and texture. Consequently, the "petals" of the Portulacaceae are probably transformed sepals, and the "sepals" are homologous to the uppermost stem leaves of the ancestral forms. A similar origin of "sepals" from uppermost stem leaves is possible in the Aizoaceae (Ficoideae).

Origin of the synsepalous calyx and the sympetalous corolla

The "fusion" of sepals to form a synsepalous calyx and that of petals to form a sympetalous corolla have both occurred repeatedly in many different evolutionary lines. Moreover, intercalary concrescence, at least in many instances, has taken place independently in the two perianth whorls. Although the majority of groups having united petals also have united sepals, many exceptions exist. Most of these have the sepals or their bases united into a cup-shaped, campanulate or cylindrical calyx, but with the petals separate. These include Papaveraceae (*Eschscholtzia*), Caryophyllaceae subf. Silenoideae, Bombacaceae, Malvaceae, Turneraceae, Malesherbiaceae, Frankeniaceae, Loasaceae, Pyrolaceae, Grossulariaceae, Leguminosae, Lythraceae, Thymeleaceae (Aquilarieae), Myrtaceae, Oliniaceae, Stackhousiaceae, Icacinaceae (some genera), and Erythroxylaceae. A smaller but still considerable number of groups combines the condition of separate sepals with united petals, for example, Theophrastaceae (some genera), Myrsinaceae (some genera), Fouquieriaceae, Loganiaceae (some genera), Gentianaceae (some genera), Convolvulaceae, Polemoniaceae (*Cobaea, Polemonium*), Plantaginaceae, Scrophulariaceae (*Verbascum, Penstemon*), and Acanthaceae (some genera).

On the basis of any hypothesis that postulates some internally controlled tendency toward union of parts, or some factors that cause different parts to become more specialized in association with one another so that the more specialized conditions are necessarily correlated, this diversity of combinations is very difficult to understand. On the other hand, all of the conditions that exist in the above-mentioned families might be explained on the basis of adaptive modification along the lines of least resistance. Selective pressures on the calyx would be expected to be very different from those acting on the corolla, because of their very different functions.

The function of the calyx is protection of the young floral organs against drought and cold as well as against the attacks of predatory insects. Increased efficiency in performing this function can be brought about in any one of several ways, such as increase in breadth of the growing sepals, so that they become imbricated or overlapping, thus providing several layers of covering; protection of entire inflorescences by specialized bracts, as in *Umbellularia* and other Lauraceae; and union of sepals, the extreme example of the latter being the formation of a calyptra, as in *Eucalyptus, Eschscholtzia,* and *Clarkia* (Onagraceae). Which trend will occur in any particular evolutionary line is likely to depend either upon the degree of preadaptation exhibited by the sepals or bracts at the time when increased selective pressure begins or upon the chance presence of certain genes in the initial population.

The function of the corolla is both attraction of animal pollen vectors and the efficient accomplishment of pollination, as discussed in Chapter 4. On this basis, the increase in tube length of an already sympetalous corolla is not difficult to understand. On the other hand, the initial formation of the intercalary meristem that brings about union of petal bases involves serious problems. Functionally, cup- or wheel-shaped flowers having the petals united only at their bases, as in *Ledum, Loiseleuria* (Ericaceae), *Sabatia, Frasera* (Gentianaceae), and *Lysimachia* (Primulaceae), are little if any different from flowers having similar shapes but with the petals completely separate. This condition may have arisen simply by the chance fixation of genes favoring an intercalary basal meristem in the corolla.

On the other hand, better knowledge of the physiological factors that favor the formation of intercalary meristems may lead to a different explanation. The relations of these meristems to the action of growth substances, discussed in Chapter 6, need to be studied on a comparative basis.

In this connection, it is interesting to note that in a number of instances a trend toward greater intercalary concrescence of petals is associated with an increase in corolla size, the initial stage being a rather small corolla and flower. Examples are Convolvulaceae (*Dichondra, Evolvulus, Convolvulus*), Primulaceae (*Lysimachia, Androsace, Primula*), Crassulaceae (*Crassula, Cotyledon*), and the *Brodiaea* complex of the Liliaceae (*Muilla maritima, Brodiaea hyacinthina, B. pulchella, B. coronaria, B. Ida-Maia*). One might imagine that, in a very small early floral meristem, localization of growth substances would be less likely to take place than in a larger one. The initial intercalary concrescence, therefore, might be simply the result of genes acting to promote more growth and greater adult size, and exerting their effects upon a floral meristem in which localization of growth substances had been largely suppressed because of its small size. This kind of pleiotropic action of genes, although it has not yet been demonstrated to exist, is mentioned because experiments could be devised to find out whether or not it is present.

The origin of zygomorphy

The shift from radial to bilateral symmetry of the flower has taken place independently in an even larger number of separate evolutionary lines than have other changes in perianth structure. Moreover, the great majority of these changes can be traced within the confines of a single family, or between two closely related families that many taxonomists merge into a single one. A review of the families of angiosperms has revealed only 10 separate origins at the level of modern families or orders, but 25 examples of differences among the genera of a single family, or even among the species of a single genus, as in *Saxifraga, Nicotiana,* and *Camassia.* This suggests that genetic differences favoring bilateral symmetry can arise more easily than those that favor union of parts.

The predominant selection pressure that favors the origin of zygomorphy has already been discussed in Chapter 4. In a plant that bears its flowers in terminal racemes or other elongate inflorescences that have most of the individual flowers placed laterally, adaptation to specialized pollinators usually involves placing the lower petals or corolla lobes in such a position that they form a "platform" upon which the pollinator can land, and often the upper ones in such a position that they are associated with the stamens and stigma, orienting these structures in a way that ensures efficient cross-pollination.

The high correlation of zygomorphic flowers with racemose and similar forms of inflorescences has already been recorded (Stebbins 1951). In this connection, the fact must be mentioned that terminal racemes or thyrses that bear laterally positioned but radially symmetrical flowers are rather frequent, and occur almost invariably in groups having closely related genera or species that differ from one another in floral symmetry, such as Cruciferae (*Thelypodium, Streptanthus*), Ericaceae (*Ledum, Rhododendron*), *Nicotiana,* and *Camassia.* This suggests that lateral positioning of the flowers is a necessary but by no means a sufficient preadaptation for the origin of zygomorphy. On the other hand, zygomorphic flowers that are terminally rather than laterally placed usually if not always occur in families having other genera that bear zygomorphic and laterally placed flowers. Examples are *Mimulus primuloides* and *Limosella aquatica* (Scrophulariaceae), *Pinguicula* (Lentibulariaceae), various genera of Gesneriaceae, *Thunbergia* (Acanthaceae), and *Cyprepidiumo acaule* (Orchidaceae). These species and genera have probably been derived from ancestors that had zygomorphic flowers positioned laterally, in racemes. In most instances, reduction of such racemes to a single flower would involve much less alteration of the developmental pattern than a harmonious reversion from bilateral to radial symmetry, so that the retention of the zygomorphic condition would be expected.

Trends in the stamens

The chief trends in the androecium of the flower have been: (1) the elongation of the stamen filaments through increased activity of individual intercalary meristems below each anther or microsporangial complex, combined with increased elongation of the filament cells; (2) changes in stamen number, both increases and decreases; (3) sterilization of stamens through suppression of the differentiation of archesporial tissue, and their conversion to staminodes of various sorts; (4) intercalary concrescence of filaments to produce staminal columns or elongate bundles; and (5) adnation of stamens to the corolla, by means of intercalary concrescence. All of these trends have occurred independently in several different evolutionary lines.

The elongation of stamen filaments in more specialized flowers can have either one of two different functions. In wind-pollinated flowers, it places the dehiscing anthers well above the level of the perianth, and so increases the efficiency of pollen dispersal. An extreme example, characteristic of cross-pollinating species and races of Gramineae, is familiar to most botanists. In flowers pollinated by nectar-seeking insects or other animal vectors, elongate filaments place the anthers in such a position that pollen is dusted onto particular parts of the animal's body during its visit.

Evidence for phylogenetic increases in stamen number is of two kinds. In some instances the genus having the largest number of stamens of any in its family is in other respects relatively specialized. An example is *Rosa,* most species of which have more than 100 stamens per flower, and some may have as many as 300. With respect to its cup-shaped receptacle and one-seeded carpels, *Rosa* is definitely more specialized than the genera of the tribe Spiraeoideae, most of which have from 15 to 50 stamens per flower. Evidence for secondary increase in stamen number is also afforded by developmental patterns, which are discussed in Chapter 6.

Most if not all flowers having very large numbers of stamens are relatively flat, and have little or no nectar. They are attractive to pollen-gathering vectors. Consequently, the adaptive value of an increased number of stamens is to provide an excess of pollen beyond that required for fertilization.

Trends toward reduction in stamen number are far more common than trends toward increase. This is an expected result of floral evolution based upon interactions between flowers and their pollinators. From the morphological point of view, four different kinds of trend can be recognized: (1) reduction in the number of stamen units per bundle or per phyllotactic unit; (2) reduction in the number of spirally arranged bundles; (3) suppression of individual stamens or of entire whorls from flowers that are already oligostemonous; (4) "fusion" of bundles through suppression of the differentiation of their primordial meristems.

The first trend is probably the commonest way in which flowers having few stamens of a definite number, usually the same as the number of sepals plus petals, have evolved from flowers having many stamens of an indefinite number. A good example is the Rosaceae, tribe *Potentilleae.* The most primitive species of the genus *Potentilla,* such as *P. fruticosa,* have 25–50 stamens arranged into 10 bundles or clusters (unpublished personal observation). An intermediate condition exists in *P.* (*Comarum*) *palustris,* in which five single stamens, opposite the petals, alternate with five clusters of three each, opposite the sepals (Murbeck 1914). Further reduction exists in *Horkelia,* the species of which have ten stamens, five antepetalous and five antesepalous. The final stage of reduction is reached in the genera *Ivesia, Purpusia,* and *Sibbaldis,* in which the antepetalous whorl of stamens is suppressed and only five stamens, opposite the sepals, persist.

Reduction in the number of spirally arranged bundles is probably rare, since in most flowers the stamens, or at least the early bundle primordia, are already in whorls. No clearly documented examples of this kind of reduction are known to me.

Elimination of stamen whorls is by far the commonest trend of reduction in flowers that are already cyclic and oligostemonous. Trends of this kind can sometimes be found among the species of a single genus (*Parvisedum, Ivesia*) and among closely related genera of a family (herbaceous Saxifragaceae). The difference between one and two stamen whorls (tetracyclic vs. pentacyclic flowers) is also a frequent diagnostic character of families and orders.

Suppression of individual stamens in a whorl is sometimes found in flowers that retain a radial symmetry, and in a few cases, for example, *Stellaria media* (Haskell 1949), may exist among the races of a single species. It is, however, most common in association with trends toward bilateral symmetry. In zygomorphic flowers such as those of Scrophulariaceae and Labiatae, the presence of four stamens rather than five makes possible a more efficient arrangement of the stamens with respect to the visits of pollinators.

Still another way in which stamen number can be reduced is through union, or suppression of differentiation of one or more stamen bundles. This condition has been described by Leins (1964*a*) in *Hypericum.* In *H. Hookerianum,* there are five sepals, five petals with five stamen bundles opposite them, and five carpels. In *H. aegyptiacum,* there are only three stamen bundles, and three carpels which alternate with them. Two of the stamen bundles, however, are larger than the third, and occupy positions relative to the sepals and petals which suggest that each of them is homologous to two of the five bundles found in *H. hookerianum.*

When they can be traced within a single genus or in closely related genera, each of these kinds of reduction is associated with reduction in overall size of the flower and its more rapid development. Probably, therefore, the selective pressure responsible for the trend was the onset of more severe environmental conditions, reducing the time available for flowering and fruiting, and placing a premium upon rapid development. The particular kind of response to such pressures was specific for each evolutionary line that underwent reduction in floral structure, depending partly upon modification along the lines of least resistance and partly upon chance.

If this interpretation of the origin of reductional trends in the androecium is correct, then flowers that have a small number of stamens but are, nevertheless, large in overall size must have evolved from smaller flowers through secondary increase. In a number of evolutionary lines, particularly in flowers having sympetalous corollas, suggestive evidence of this kind of increase exists. The species having larger flowers tend to have longer corolla tubes relative to the length of the lobes than those having smaller flowers. Examples are *Primula* vs. *Androsace* (Primulaceae), *Gentiana* vs. *Swertia* or *Pleurogyne* (Gentianaceae), *Convolvulus* or *Ipomoea* vs. *Cressa* or *Dichondra* (Convolvulaceae), and *Datura* or *Nicotiana* vs. *Lycium* (Solanaceae). The monocotyledons, which with the exception of the Alismatales have the stamen number reduced to six or three, have characteristically small flowers in those groups that are the most generalized with respect to both floral and vegetative structure, as in Liliaceae, tribes Narthecieae, Helonieae, and Veratreae, and in Scheuchzeriaceae, Petrosaviaceae, Rapateaceae, Xyridaceae, and Palmae. Monocotyledonous genera having large flowers are, for the most part, specialized either with respect to vegetative parts (Liliaceae, Tulipeae), corolla, gynoecium, or both (Amaryllidaceae, Iridaceae), or several of these (Musaceae, Cannaceae, Zingiberaceae, Orchidaceae). In monocotyledons, therefore, numerous trends toward increase in size of flowers have taken place, leaving the number of stamens unaltered.

The presence of staminodes that are homologous with stamens and apparently derived from them is characteristic of many flowers that are highly specialized for insect pollination. Some of these are described in Chapter 4. Since this kind of change has occurred rather sporadically in relatively specialized families, it would be more properly discussed in monographs of such groups than in a more general treatment like the present one.

The intercalary concrescence of stamen filaments to produce either visible, elongate bundles or a staminal column that surrounds the gynoecium has occurred rather infrequently, most often in tropical or subtropical families. Elongate or broad, fleshy stamen bundles are best known in the Guttiferae and in certain Australian genera of Myrtaceae, such as *Calothamnus.* Anatomically, they are quite different from the clusters or "bundles" that are described in Chapter 10, and are characteristic of many relatively primitive as well as moderately advanced families of angiosperms. In the primitive stamen clusters, the vascular trace to each stamen branches out from a common trunk bundle for the cluster (Figs. 10-2 and 10-3). In the bundles of the Guttiferae and Myrtaceae, on the other hand, the visible portion contains only traces leading to separate anthers, and the common trunk bundle is not more strongly developed than it is in related groups that do not possess visible bundles (Fig. 12-1). These bundles, therefore, appear to be produced by an intercalary meristem that begins its activity after the procambium for the traces to the separate anthers has already been differentiated. Such a complex pattern is almost certainly of secondary origin, as would be expected on the basis of the high degree of specialization with respect to other characteristics that is found in the groups concerned.

The origin of the stamen column of the Malvaceae and Sterculiaceae, the development of which has been well described by van Heel (1966), is probably an analogous trend of specialization. The same can

Fig. 12-1. (*A*) Staminate flower, and (*B*) individual stamen bundle, surface view, of *Garcinia multiflora* Champ (*C* and *D*) retouched photos of cleared stamens of *G. multiflora* (*C*) and *G. dulcis* (Roxb.) Kurz (*D*). (*A* and *B* from Hayata 1913; *C* and *D* original, from specimens in the herbarium of the University of California, Berkeley.)

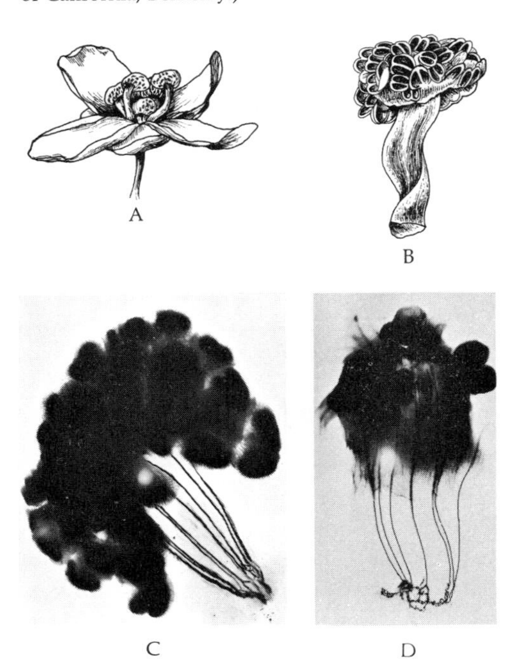

be said of stamen columns in families such as the Canellaceae, Cucurbitaceae, Meliaceae, and Myristicaceae, except that in these families the origin of the column was from an androecium having only five to ten stamens.

The adaptive significance of these floral specializations may become clearer when the pollination biology of the groups concerned is better known. In some instances (Malvaceae, Meliaceae), one might postulate that the lower part of the column served as a protecting envelope for the young gynoecium against damage from biting insects that were the original pollinators, as has been postulated by Grant (1950*b*) for the origin of epigyny. This explanation, however, cannot be applied to families such as the Myristicaceae and Cucurbitaceae, whose flowers are unisexual. The hypothesis that these rare, sporadically occurring examples of intercalary concrescence in the androecium are manifestations of some internally directed trend toward "fusion" is, in my opinion, even more difficult to maintain.

Adnation of stamens to the corolla tube is a common specialization that has occurred many times independently in various families. Its developmental basis is the suppression of differentiation between the intercalary meristem that is responsible for the corolla tube and that responsible for the stamen filaments. Such suppression could arise as a by-product of reduction that affected the early stages of development of perianth and androecium. The adaptive value of the condition, once it had evolved, was probably the precise positioning of the anthers with respect to the stigma and the insect pollinators.

Trends in the pollen grains

On account of their excellent fossil record, pollen grains have recently received much attention from botanists interested in angiosperm evolution. A thorough account of their comparative morphology has been presented by Erdtman (1966), while the succession of fossil pollen types has been discussed by Doyle (1969), Muller (1970), and others. Cronquist (1968) has presented a concise and clear account of the most probable evolutionary trends (see Fig. 12-2).

Fig. 12-2. Diagram showing the principal trends of specialization in the overall form of the pollen grain: (*upper left*) trends general for dicotyledons (from Takhtajan 1959); (*lower right*) trends found in the family Annonaceae (from Walker 1971); (*upper right*) monocotyledons (from Takhtajan 1959).

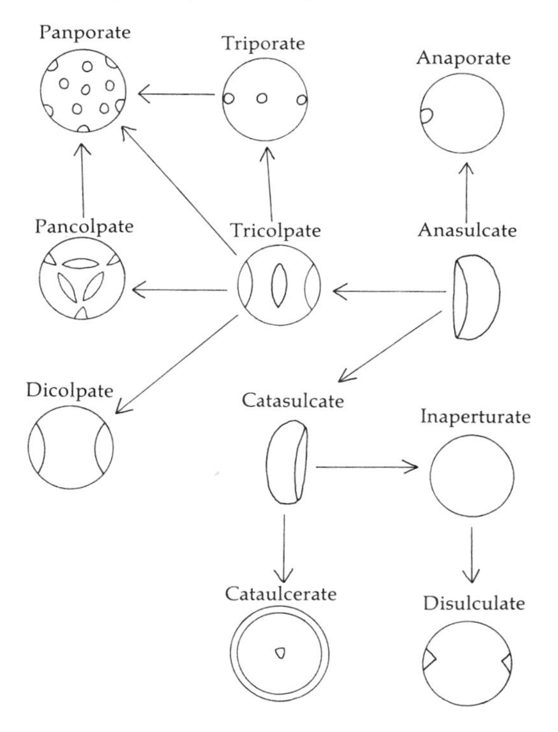

As stated in Chapter 10, the original angiosperms probably had pollen grains containing a single furrow, in the middle of which was a single pore, through which the pollen tube emerged. This kind of pollen grain persists in the majority of monocotyledons, but in the dicotyledons it is confined to the primitive Magnoliales and a few other primitive or archaic families. The shift from the one-furrowed (monocolpate) to the three-furrowed (tricolpate) condition cannot be recognized within the limits of any modern family or genus, although a suggestion of the tricolpate condition exists in the normally monocolpate Canellaceae (Wilson 1964). Apparently, it arose in some primitive, now extinct group of angiosperms having a general affinity to the class Magnoliidae, and ancestral to the Illiciales, Ranunculales, as well as the classes Caryophyllidae, Dillenidae, and Rosidae. Possibly the shift from the monocolpate to the tricolpate condition arose separately in different lines of evolution, but we have no indications of such parallelism.

In both monocolpate and tricolpate groups, trends toward reduction of the furrow and enlargement of the pore occurred repeatedly. Increase in numbers of pores has also been a frequently repeated trend, as in Amaranthaceae, Polemoniaceae, and Plantaginaceae (Erdtman 1966).

In addition to furrows and pores, the pollen grains of many angiosperms bear elaborate sculpturing on their outer coat or exine layer, consisting of various patterns of spines, ridges, and the like. These markings often have considerable taxonomic value, as in the Compositae, tribe Cichorieae (Stebbins 1953) and the genus *Saxifraga* (Ferguson and Webb 1970). The larger families of angiosperms differ greatly from one another with respect to the amount of variation found among their genera in the form and sculpturing of their pollen grains. Some, like the Lauraceae, Meliaceae, and all tribes of Compositae except the Cichorieae and Mutisieae, are monotonously constant with respect to this character, whereas others, such as the Euphorbiaceae, are enormously variable.

As Cronquist has stated, the adaptive significance of differences with respect to numbers of furrows and pores, as well as sculpturing, is very little understood. This lack of understanding, however, is due largely to the complexity of the problem and the paucity of information on a comparative basis. A large number of variables must be considered: (1) the amount of food reserves in the grain in relation to the distance that the pollen tube must grow to reach the ovule; (2) the rapidity and efficiency with which germination can begin when the grain reaches the stigma; (3) in species cross-pollinated by animal vectors, the ease with which the grain can adhere to the vector and be released when it reaches the stigma; (4) in wind- or water-pollinated species, the buoyancy of the grain and its reaction to different degrees of atmospheric humidity; (5) the relation between the size, shape, and sculpturing of the grain and its ability to adhere to the stigma; (6) the possible interactions between inner membranes or pores and enzymes excreted by the stigmatic cells; (7) factors controlling early development of the microspore wall (Heslop-Harrison 1971).

Some kinds of adaptive shifts in pollen-grain structure can be recognized fairly easily. For instance, in the majority of the family Compositae the strongly echinate exine aids in the clumping of the grains into masses that are transported with relative ease by the pollen vectors. In wind-pollinated species, on the other hand, such clumping is disadvantageous. Hence, the loss or reduction of spines in wind-pollinated genera of the tribe Ambrosieae and in the genus *Artemisia* (Erdtman 1966:123) is quite understandable.

The major shift from the monocolpate to the tricolpate condition is less easy to understand. Possibly, the presence of three germ pores enables the grain to germinate promptly no matter in what position it may lie when deposited on the stigma. If the stigma were large or deeply grooved, the single pore of a monocolpate grain would probably be placed in a position favorable for germination no matter how the grain was deposited; but in plants having small, convex stigmatic surfaces the assurance that at least one pore would be against the surface would have a much higher adaptive value.

In considering adaptive shifts in pollen-grain structure and physiology, one must recognize that selection is operating upon genetic differences at quite a different level than with respect to most other characters. As is pointed out in Chapter 1, individuals of cross-fertilizing species of plants are heterozygous at a large number of gene loci, so that their pollen grains are heterogeneous populations containing many different genotypes. Moreover, the pollen sample that is deposited on a stigma is usually derived from several individuals. Since the number of grains deposited on the stigma is always greatly in excess of the number of ovules that can be fertilized, strong competition exists between different pollen grains, so that selection will greatly favor those genotypes that permit the pollen tube to emerge promptly and to grow rapidly down the style and into the ovular micropyle. This selection will be independent of any benefit to the plant as a whole. As long as it does not reduce the overall adaptive value of the plant, any gene combination that permits pollen gametophytes to succeed consistently in competition with other grains that land on the same stigma will become incorporated into the genotype of the species.

Three trends in the cytology of pollen grains are worthy of mention. The first of these is from single grains to tetrads in which the separation of the individual microspores is suppressed, so that the entire tetrad is transported as a unit by the pollen vector. This condition is best known in the Winteraceae, most families of Ericales (Pyrolaceae, Ericaceae, Epacridaceae, Empetraceae), and Juncaceae. A further extension of this trend has taken place in the Cyperaceae and some genera of Epacridaceae, such as *Stypelia* (Smith-White 1955). In them, only one of the four nuclei produced by each meiotic division survives, while the other three degenerate soon after they are formed. This means that each pollen grain, in spite of having only one nucleus that enters the pollen tube, is nevertheless derived from an entire tetrad. As Smith-White has pointed out, this makes possible competition and selection between the haploid nuclei derived from a single diploid archesporial cell.

A second trend is from unicellular grains to multicellular grains or pollen masses. Multicellular pollen grains are found in many genera of Leguminosae subf. Mimosoideae (Davis 1966). In the more specialized genera of Orchidaceae, the pollen grains of tetrads adhere to each other to such an extent that the entire contents of an anther consist of 2, 4, 6, or 8 pollinia, which are transported as single units by the pollen vector. The adaptive value of this condition is quite understandable. Each seed capsule of an orchid flower contains tens or hundreds of thousands of ovules. If these are all to be fertilized, a corresponding number of pollen grains must be deposited on the stigma. Since orchid flowers are so specialized for cross-pollination that each flower is usually visited only once by a vector, a mechanism that facilitates the transport of a large number of pollen grains as a single unit would obviously have a great selective advantage in this family.

The third trend is from gametophytes that are binucleate when the pollen is shed to those that are trinucleate. In the latter case, the mitotic division that gives rise to the two fertilizing or male nuclei has become precocious, so that it takes place during the maturation of the pollen grain rather than during the growth of the pollen tube. Brewbaker (1967) has published an excellent review of this subject. Trinucleate grains exist in many orders and families, and are best known in the Caryophyllales and the family Gramineae. They are distinguished by their low degree of longevity when artificially stored and their sensitivity to various factors that promote germination. They are very difficult to germinate *in vitro,* and in cross-pollinations of an "illegitimate" kind, between plants having similar alleles of a self-compatibility system, trinucleate grains fail to germinate entirely, whereas in species having binucleate grains a small amount of pollen-tube growth takes place even on stigmas that are expected to exhibit an incompatibility reaction with the pollen grain. The trinucleate conditions, therefore, increase the speed and efficiency of cross-pollination at the cost of a diminished flexibility and ability to withstand unfavorable conditions. Maximal selection pressures for it would be expected in association with a shortening of the favorable period for flowering. Since the condition either is irreversible or is reversed only with a low probability (Cronquist 1968), it is likely to persist in an evolutionary line even after the selection pressure that brought it into being has ceased to operate.

Trends in the Gynoecium

The most significant trends of evolution in the angiosperm gynoecium have been union of carpels, shifts in the nature of the placentation of ovules, and the evolution of epigyny or the "inferior ovary" through adnation of the gynoecial walls to the calyx, the receptacle, or both of these structures.

In the great majority of angiosperms the gynoecium, through either its multilocular structure, the presence of two or more styles or stigmas, or its vascular anatomy, shows that it has been derived from an ancestral condition in which several separate carpels were present. This ancestral apocarpous condition survives in most of the Magnoliidae, the Dilleniales, several groups of Rosales, Fabales, and Alismatidae, and a few other isolated groups of monocotyledons. In some orders, such as the Santalales, Urticales, and Rafflesiales, the gynoecium has undergone such a high degree of reduction that no traces of the multicarpellate origin remain. Nevertheless, careful comparisons between members of these families and other less reduced families that resemble them with respect to a large number of diverse characteristics indicates that they, too, have followed the evolutionary pathway of carpellary union followed by reduction.

Union of carpels

Union of carpels has occurred independently in many different evolutionary lines. A review of the affinities of apocarpous groups to their nearest relatives having united carpels suggests at least 18 to 20 separate instances of carpellary union. In many instances, this trend can be detected among genera that on the basis of many other characteristics are recognized by all taxonomists as belonging to the same family. Families within which this trend has occurred are Magnoliaceae, Annonaceae, Ranunculaceae, Nymphaeaceae, Phytolaccaceae, Rosaceae, Saxifragaceae, Cunoniaceae, Anacardiaceae, Rutaceae, and Arecaceae (Palmae). In some instances, such as Phytolaccaceae, Rosaceae, Cunoniaceae, Anacardiaceae, and Rutaceae, the original syncarpous members of the family may have been the progenitors of other families and orders that are uniformly syncarpous. In other instances, such as Magnoliaceae, Annonaceae (*Annona*), Ranunculaceae (*Nigella*), and Nymphaeaceae, the syncarpous groups are end lines of evolution that have not led to anything else. Union of carpels, therefore, may offer opportunities for further evolutionary diversification, but it by no means guarantees that this will take place.

The estimate of 18 to 20 separate occurrences of carpellary union during angiosperm evolution is more likely to be an underestimate than an exaggeration. As a result of careful anatomical and developmental studies of the Rutaceae, Gut (1966) concluded that syncarpy originated in this family in a number of different ways. As stated in Chapter 2, the repetition of similar evolutionary trends in different lines, particularly when this repetition involves similar phenotypic trends that are based upon different modifications of gene action and developmental pattern, is difficult or impossible to explain except upon the assumption of a selective value for the trend. The most likely adaptive advantage of syncarpy may lie in the fact that a bilocular or multilocular syncarpous gynoecium elaborates a smaller amount of wall tissue compared with ovular tissue than do separate carpels that produce an equivalent seed mass. Hence in terms of the reproductive effort, as defined by Harper and Ogden (1970), a syncarpous gynoecium may be more efficient than an apocarpous one. This hypothesis can, of course, be tested by comparing the reproductive efforts of related species that differ from one another with respect to the degree of union of carpels.

A second functional significance of syncarpy has been suggested by Carr and Carr (1961). They point out that if the entire gynoecium has at its apex a single style base, even if the stigmas are separate, pollen that lands on any of several stigmas can send out pollen tubes capable of reaching and fertilizing any of the ovules. This advantage, however, is not realized in gynoecia of which the upper parts of the carpels are separate, as in species of *Saxifraga* and many genera of Rutaceae related to *Xanthoxylum.* The union of the basal portions of the carpels in such groups must be explained upon a different basis.

Trends in placentation

Evolutionary trends in the placentation of ovules have been well reviewed from the anatomical viewpoint by Puri (1952), and additional comments have been provided by Parkin (1955), Takhtajan (1959, 1969), and Cronquist (1968). These authors differ from one another with respect to the condition that is regarded as most primitive. Puri and Parkin hold to the "classical" opinion, that in apocarpous forms marginal placentation was the primitive condition, and that following the union of conduplicate carpels this led to the primitive sutural placentation in compound gynoecia, from which all other forms of placentation have been derived. Cronquist, following in part the lead of Bailey and Swamy (1951), believes that the primitive condition is laminar placentation, consisting of ovules scattered over the adaxial surface of the carpel, or in a few compound ovaries such as those of Gentianaceae (Lindsey 1940), over the inner surface of the ovary wall. Takhtajan, taking an intermediate position, believes laminar-lateral placentation to be primitive. From this condition both strictly laminar or laminar-diffuse and marginal or submarginal placentation are regarded as derived.

I agree in this respect with Takhtajan. Laminar-diffuse placentation is found only in a few groups, chiefly aquatic (Nymphaeaceae, Butomaceae, Hydrocharitaceae), which are specialized with respect to many morphological characteristics (Kaul 1969, 1970). A condition corresponding to it is found in a few genera having compound ovaries, such as certain Gentianaceae (Lindsey 1940), in which the ovules are scattered over the inner surface of the ovary wall. Considering these facts, a reasonable explanation of laminar-diffuse placentation is that it is one possible response to selection for increased ovule number in plants which because of their aquatic habitat or for other reasons do not rely for seed dispersal on precise dehiscence of the mature gynoecium.

According to the hypothesis accepted by Takhtajan and myself, marginal placentation evolved from the submarginal or lateral-laminar position in many different apocarpous groups, including all of those that later gave rise to syncarpous gynoecia. In a conduplicate carpel, the fused margins form the adaxial suture of the follicle, so that the placentae are best described as sutural (Fig. 12-3, *A*). In a syncarpous gynoecium derived from separate carpels having sutural placentation, the placentation is naturally axile (Fig. 12-3,

B). This is the principal reason for regarding axile placentation as the original condition in compound ovaries. Parietal placentation in which the placentae are longitudinal ridges along the inner surface of a compound ovary (Fig. 12-3, *C*) is regarded as derived. This opinion is fortified by the fact that in all of the families mentioned in Chapter 10 as having some apocarpous and some syncarpous genera, the latter genera have axial placentation. Possible exceptions cited by Cronquist are *Isoloma* and *Monodora* of the Annonaceae. These genera, however, differ so much from other genera of Annonaceae that they have been placed by Fries (1959) in a separate subfamily, the Monodoroideae. In a family that is relatively old, as

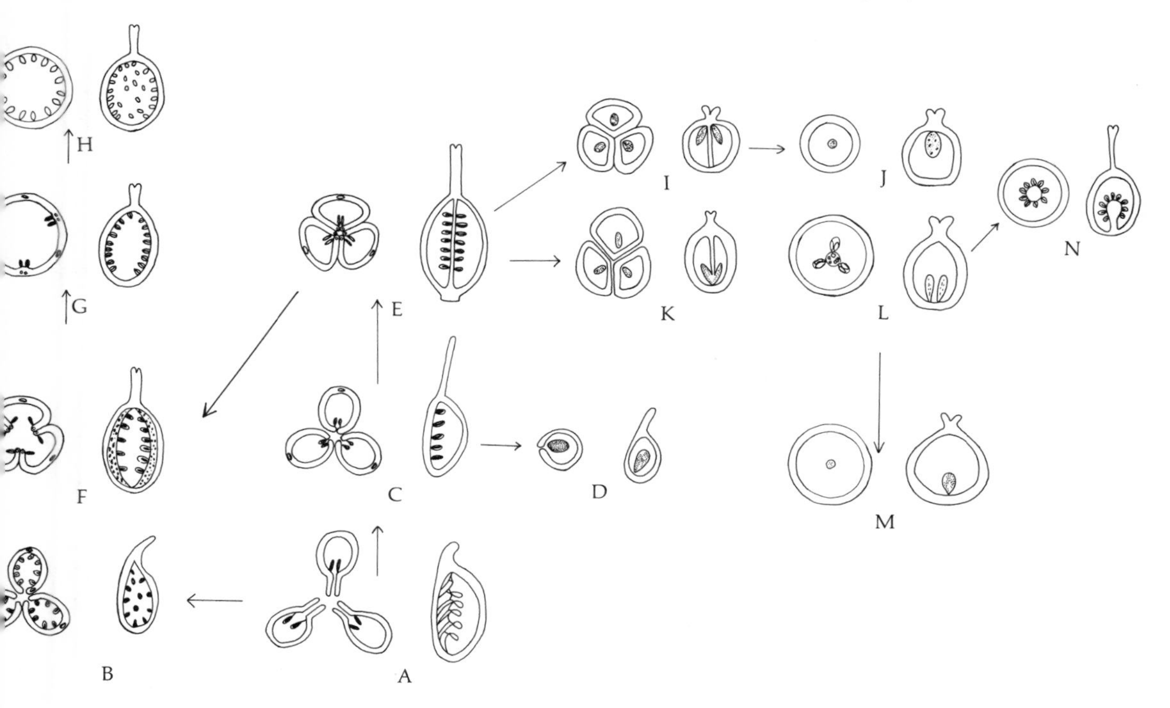

Fig. 12-3. Diagrammatic sections of gynoecia, showing the principal trends in placentation: (*A*) separate carpels, laminar-lateral placentation; (*B*) separate carpels, laminar placentation; (*C*) separate carpels, lateral or marginal placentation; (*D*) separate carpels (achenes), solitary basal ovule; (*E*) united carpels, trilocular ovary, axial placentation; (*F–H*) secondarily unilocular ovary, parietal placentation; (*I*) trilocular ovary, solitary pendulous ovule in each locule; (*J*) secondarily unilocular ovary, solitary pendulous ovule; (*K*) trilocular ovary, solitary basal ovule in each locule; (*L*) secondarily unilocular ovary, basal ovules; (*M*) secondary unilocular ovary, solitary basal ovule; (*N*) secondarily unilocular ovary, free central placentation. (Original.)

the Annonaceae must be, many connecting links between genera can be expected to have become extinct, so that the postulated existence of intermediate genera that had united carpels and axile placentation is not unreasonable. In this connection it is worth noting that the genera *Annona* and *Rollinia,* which also have the carpels united but are placed in the same subfamily with genera that have separate carpels, have axile or basal placentation.

Further support for the derived condition of parietal placentation comes from comparisons between genera or species belonging to families in which both conditions are present. Such families are Gutiferae, Elaeocarpaceae, Tiliaceae, Saxifragaceae (*sens. strict.*), and Grossulariaceae. In most of these families, the appearance of parietal placentation is associated with various degrees of reduction or adnation in other parts of the flower. A good example is the genus *Hypericum.* In most species of this genus the ovary is penta- or trilocular and the placentation axile. In the section *Brathys,* however, the ovary is unilocular with parietal placentation. Species of this section are mostly small plants having much-reduced leaves, inflorescences, and flowers.

In the Tiliaceae, Weibel (1945) has pointed out that axile and parietal placentation are by no means sharply distinct from each other. He lists seven genera—*Duboscia, Desplatzia, Tilia, Cantelea, Ancistrocarpus, Berrya,* and *Althoffia*—in which the lower part of the gynoecium is multilocular with axile placentation, whereas the upper part is unilocular with parietal placentation. He did not state specifically which condition he regarded as the more primitive, but a significant fact is that in the genus *Christiana,* in which the carpels are united in their lower parts but distinct in their upper parts, the placentation is strictly axile.

In the Gentianaceae, Lindsey (1940) postulated a series beginning with the subtribe Erythraeinae, which have parietal placentation, and leading to the bilocular ovary with axile placentation that exists in some genera of the subtribe Exacinae. He did not, however, give any clear reasons for reading the series in this direction. One reason for reading it in the opposite direction, and regarding the Exacinae as having the least-specialized gynoecia in the family, is the fact that in all of the other families of the order Gentianales, the Loganiaceae, the Apocynaceae, and the Asclepiadaceae, the ovary is bilocular and the placentation is axile. Since the Loganiaceae contain a large proportion of woody genera, whereas the Gentianaceae not only are almost entirely herbaceous but also contain many annual genera and a number of partially or entirely saprophytic plants, evidence at the family level would suggest that the Gentianaceae have been derived from ancestors similar to the Loganiaceae, and therefore having bilocular ovaries with axile placentation. This hypothesis is further supported by the fact that most genera of the Erythraeinae, which Lindsey regards as the least specialized, have their flowers reduced to the tetramerous condition, contrasted with the pentamerous condition that prevails in the great majority of the Gentianales. Furthermore, the condition of parietal placentae that are not at all intruded, which according to Lindsey's interpretation should be the least-specialized condition in the family, exists chiefly in the subtribe Gentianinae, which are in a number of respects fairly specialized. Although the family as a whole, like others of its order, is mainly tropical and subtropical, the Gentianinae are temperate or arctic-alpine. Their corollas often bear elaborate appendages, and in some genera specialized nectaries.

With respect to placentation, many species of *Gentiana, Frasera, Pleurogyne,* and *Crawfurdia* have the ovules scattered over the inner surface of the gynoecium, a condition similar to the laminal placentation found in a few apocarpous genera of Nymphaceaceae and Butomaceae, mentioned elsewhere in this chapter. The series in the Gentianaceae, therefore, can easily be read as beginning with a bilocular ovary having axile placentation, and proceeding through the unilocular ovary having intruded placentae to the unilocular condition with ovules scattered over the inner surface.

Cronquist (1968) has suggested that the family Canellaceae might be an exception to the rule and have a syncarpous ovary with initially parietal placentation. His conclusion is based upon his belief that this family does not "seem related to any group with axile placentation." Since, however, the Canellaceae do not show a close affinity to any other family, their origin is so obscure that they cannot serve as an example of a phylogenetic trend.

The hypothesis, based upon trends within families, that the origin of parietal placentation was associated with a general reduction of the flower as a whole apparently runs counter to the existence of parietal placentation in plants having relatively large flowers, such as *Bixa, Cochlospermum, Cistus, Mentzelia* spp., and *Passiflora.* This difficulty is removed if one assumes that characteristics of size and number of floral parts can evolve toward either increase or decrease with almost equal facility, depending upon the selection pressures that are present. The above-mentioned genera are, therefore, regarded as examples of secondary increase in flower size and ovule number in response to selection pressures for greater fecundity in relatively favorable habitats.

When species having parietal placentation are compared with their nearest relatives that have axile placentation, the parietal condition is found to be associated with the formation of a large number of ovules and many small seeds, sometimes in an ovary that is much reduced in size. This relation is evident in the Saxifragaceae (*Lithophragma, Mitella, Tiarella, Heuchera*), Gesneriaceae (Burtt 1970), Orobanchaceae, Lentibulariaceae, and the genus *Hypericum.* Hence parietal placentation may in most instances be a response to selection for high seed number in combination with small size of seeds and their rapid development. It is, however, an alternative rather than an inevitable response to such pressures, as is evident from the presence of many genera and families such as *Nicotiana, Mimulus,* and other Scrophulariaceae, and many genera of Orchidaceae, in which many small seeds are associated with a bilocular or trilocular ovary.

Basal and apical placentation (Fig. 12-3, *D, E*) are both the result of reductional trends, and are probably governed by similar selection pressures. As is general for reductional trends, the adaptive advantage of rapid development in association with a short season favorable for reproduction has probable significance here also. In addition, selection for increased seed size is likely to result in a correlated reduction in ovule number (Stebbins 1950:132, Harper, Lovell, and Moore 1970). The way in which the reduction takes place, leading to either the apical or the basal position of the remaining ovule, may be largely a result of the chance establishment of certain initial mutations.

The condition of free central placentation (Fig. 12-3, *F*) is very rare, and deserves special mention. Eames (1961) and Cronquist (1968) have suggested that it is derived from axile placentation, and cite the Caryophyllaceae as a probable example. In this family, however, the placentation is not strictly free central, since septa between locules are found in the lower part of the ovary in most genera (Thomson 1942, Eames 1961). Since the Caryophyllaceae are not closely related to the Primulales, in which free central placentation is well developed, the former family is not well chosen as a model to show how the condition in the latter order arose.

The most characteristic and widely cited example of free central placentation is in the genus *Primula*. Consequently, the situation in both the family Primulaceae and the order Primulales deserves special attention. With respect to number of ovules per gynoecium, it is as follows. The woody and tropical families Theophrastaceae and Myrsinaceae are both generally regarded as less specialized than the herbaceous and temperate Primulaceae. In the Theophrastaceae, the presence of five staminodes that alternate with five stamens, that is, the incomplete reduction of the androecium to a single staminal whorl, is another indication of a low degree of specialization. In both of these families the gynoecium contains only a few ovules, and in most of their genera all but one of these abort, so that the fruit is one-seeded. An exception is the genus *Maesa*, in which the ovary is half inferior, an unusual specialization for this order.

Within the Primulaceae, one can first consider such genera as *Lysimachia*, *Anagallis*, and *Glaux*, which occupy a great variety of habitats, and are relatively generalized with respect to growth habit. In these genera the number of seeds per capsule is relatively low, ranging from 3 (*Glaux*) to 20–30 (*Lysimachia* spp.).* Next comes the genus *Androsace*, which is specialized with respect to its rosette habit of growth and in having its flowers aggregated into an involucrate umbel, but in which the corolla tube is relatively short and the lobes are long, so that it is less specialized than *Primula* itself. Perennial species of Androsace have 15–20 ovules per ovary, numbers not significantly different from those in *Lysimachia*. The genus *Primula* is characterized not only by a specialized growth habit and inflorescence similar to that of *Androsace*, but in addition its species have a well-developed floral dimorphism in the form of heterostyly. Numbers of ovules per capsule for representative species are: *P. veris*, 30–45; *P. farinosa*, 50–60; *P. Parryi*,

*Data on ovule and seed numbers reported in this section are based upon personal observations of material from specimens in the University of California Herbarium, and from garden plants identified by me.

140–150. Finally, the genus *Cyclamen* is perhaps the most specialized in the family on the basis of its highly developed root-crown tuber, its flowers on scapes that become coiled after anthesis, and its complex plicate corolla with reflexed lobes. In *Cyclamen europaeum* the number of ovules per capsule varies from about 100 to about 150.

These figures suggest strongly that in the Primulales there has been a general trend toward increase in number of ovules per gynoecium, which may have progressed independently in different phylogenetic lines. On the basis of anatomical evidence, Dickson (1936) concluded that the vascular supply to the ovules in this family was derived originally from the ventral bundles of a conventional multilocular ovary, a conclusion with which I agree. Since these ventral bundles must have become much reduced in the ancestral Primulales that lost the partitions between ovary loculi, the number of ovules supplied by them would be expected to have been small. A response to selection pressure for increased fecundity in herbaceous groups having small seeds was apparently the reactivation of meristematic activity in the differentiating ovary, in a form that conferred on the resulting placenta an anatomical structure suggesting a partly axial origin (Dickson 1936). This meristem developed progressively into the foundation from which are differentiated the free central "knob" of placental tissue and the numerous ovules attached to it that are characteristic of advanced genera such as *Primula, Dodecatheon,* and *Cyclamen.*

In its typical form, therefore, free central placentation is apparently the result of two reversals in the action of natural selection. Starting with a form having a multilocular gynoecium and axile placentation, the first trend may have been toward an adaptive syndrome of rapid development combined with numerous small seeds, a condition that is fulfilled by a gynoecium having parietal placentation. This trend probably took place prior to the differentiation of the order Primulales. It is paralleled by trends within the family Styracaceae, which belongs to the order Ebenales, regarded by both Takhtajan and Cronquist as ancestral to or closely related to the ancestors of the Primulales.

The second trend, toward basal placentation, probably took place during the differentiation of the Primulales from their Ebenaleslike ancestor. This was probably a response to selection for reduction in seed number in association with an increase in seed size and seedling vigor. The third trend, from basal to free central placentation, probably took place within the order Primulales and chiefly the family Primulaceae, and has already been described. Such reversals in the trend of natural selection are to be expected on the basis of the hypothesis that angiosperm phylogeny is based upon successions of adaptive radiations and adaptive shifts. Furthermore, conditions that require for their realization two successive shifts would be expected to be rare, as is free central placentation.

Trends toward perigyny and epigyny

In the most primitive, original flower, and in many modern families, the sepals, petals, stamens, and gynoecium are inserted separately on the receptacle and the flower is said to be *hypogynous.* Some trends of specialization involve the activation of an intercalary meristem that produces the union of sepals into a calyx tube, combined with adnation of petal bases and stamen filaments to this tube. This common floral "cup" or "tube," the *hypanthium,* renders the flowers *perigynous.* It is best known in the Rosales (Rosaceae, Saxifragaceae, Leguminosae), and the perigyny found in different families of this order may have had a single origin.

Increased activity of the same kind of

Fig. 12-4. Diagrams showing degrees of epigyny in species of the family Rosaceae, in which there is a well-developed hypanthium: (*A*) *Physocarpus opulifolia* (L.) Maxim., ovary superior; (*B*) *Sorbaronia sorbifolia* (Poir.) C. K. Schneid (*Sorbus americana* M. March × *Aronia melanocarpa* Ell.) and (*C*) *Spiraea Vanhouttei* (Briot) (Zbl.) (*S. cantoniensis* Lour × *S. trilobata* L.), ovary semiinferior; (*D*) *Malus pumila* Mill., ovary fully inferior (From Eames and MacDaniels 1947. Used with permission of McGraw-Hill Book Company.)

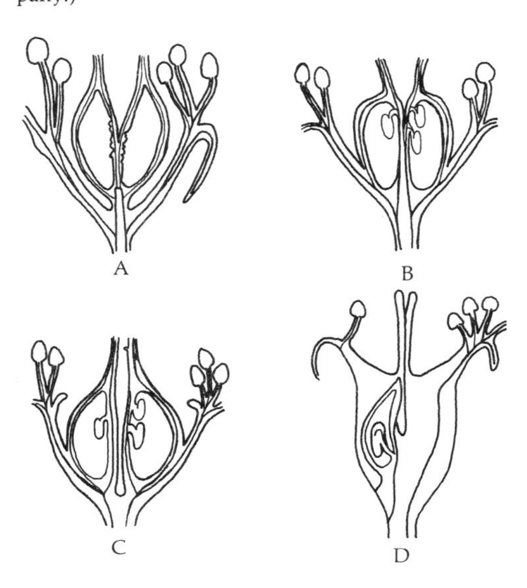

intercalary meristem brings about adnation of the hypanthium with the gynoecium, so that the ovary becomes semi-inferior (Fig. 12-4, *B*) or inferior (Fig. 12-4, *C*). This latter condition is known as epigyny. Separate trends from perigyny to epigyny can be traced in different families of Rosales (Rosaceae, Saxifragaceae, Grossulariaceae, Hydrangeaceae). Among the species of the genus *Saxifraga,* one can find every possible intermediate condition from perigyny with superior ovaries, through half-inferior ovaries, to complete epigyny, indicating that the trend to epigyny has taken place within the confines of this single genus (Morf 1950).

The Myrtales, which are probably related to the Rosales, contain both perigynous and epigynous families, as well as some families (Rhizophoraceae, Melastomataceae) that contain both perigynous and epigynous genera. Their perigyny may have been derived from a common ancestry with the Rosales, but in both orders the trend from perigyny to epigyny appears to have taken place many times.

The trend from hypogyny to epigyny does not always include perigyny as an intermediate stage. In the Ericaceae, Campanulaceae, and the closely related pair of families, Loganiaceae-Rubiaceae, some genera have hypogynous flowers in which the calyx tube and corolla tube are inserted separately on the flat receptacle and the stamens are adnate to the corolla. In the more specialized genera of these families, including the subfamily Vaccinioideae of Ericaceae and nearly all genera of Campanulaceae and Rubiaceae, the flowers are epigynous, with the calyx tube and corolla tube both adnate to the gynoecium, so that the ovary is inferior (Fig. 12-4, *C*).

In the monocotyledons, the trend to the inferior ovary takes place somewhat differently from either of the two abovementioned examples. In most hypogynous flowers of monocotyledons, the perianth is not at all or only weakly differentiated into calyx and corolla, so that intercalary concrescence between perianth members produces a floral tube having a single series of free perianth lobes, as in *Hyacinthus, Maianthemum,* and *Kniphofia.* In some of these genera, the stamen filaments are adnate to the perianth tube, so that the structure is somewhat but not completely analogous to the hypanthium of the Rosales. The epigynous flowers found in most of the Liliaceae (*sensu* Cronquist, including Alstroemeriaceae and Amaryllidaceae) are probably derived from such "perigynous" flowers, but the direct origin of epigynous from truly hypogynous flowers, at least in the monocotyledons, is certainly a possibility.

Finally, there is the highly controversial phenomenon of the origin of epigyny partly or entirely through adnation between floral parts and a concave or cup-shaped receptacle, so that part or all of the wall of the inferior ovary is regarded as "axial" in origin. This question has been carefully discussed by Puri (1952), Douglas (1957), Takhtajan (1959), Kaplan (1967), and Cronquist (1968), so that a further discussion of it here would seem to be superfluous. As is implied by the remarks that have already been made, I agree with the conclusions of those just mentioned that most epigynous flowers are entirely appendicular with respect to the homologies of their floral parts. Nevertheless, the comparative anatomy of the flowers of Cactaceae (Boke 1964), Aizoaceae (Takhtajan 1959:109), Santalaceae, and the genus *Rosa* suggests that in all of these groups the intercalary meristem that is responsible for the epigyny has produced at least some degree of adnation between ovary, floral tube, and receptacle. On the other hand, as Kaplan (1967) has pointed out, any attempts to distinguish between the regions in these flowers that are homologous to the ancestral receptacle, and so are to be regarded as "axial" in nature, and those that are homologous to the floral tube, and therefore "appendicular," are futile and meaningless. Intercalary meristems that bring about adnations also suppress differentiation to such an extent that parts which are separate in the ancestral form become so united in the derived form that they entirely lose their individuality. The epigynous flower is an integrated, unitary structure that has its own developmental pattern. Attempts to regard such a structure as a mosaic of different parts, some having "axillary" and others "appendicular" homologies, are a holdover from the kind of idealistic morphology that confuses rather than clarifies our understanding of evolution on the basis of genetic alterations of developmental patterns.

From the review that has just been presented, the conclusion can be reached that trends toward epigyny have taken place many times in different groups of angiosperms, and have followed a number of different pathways. The repeated occurrence of these trends, plus the fact that each separate trend involves a regular sequence of gene-controlled alterations of the developmental pattern, essentially precludes the possibility that epigyny has originated through chance fixation of random mutations. On the other hand, the hypothesis of internally directed trends toward epigyny is incompatible with the evidence indicating that this condition has been acquired in a number of different ways. If, however, epigyny can be regarded as a modification of floral structure that provides superior adaptation to certain environmental conditions, its repeated origin and the diverse pathways by which it has been evolved can be explained. The fact that predominantly or entirely epigynous families, such as Myrtaceae, Onagraceae, Umbelliferae, Rubiaceae, and Compositae, now occupy a great variety of ecological niches would then be explained on the basis of conservation of organization and modification along the lines of least resistance (see Chapter 2). The developmental pattern of an epigynous flower is so complex that reversion to the perigynous or hypogynous condition is difficult to achieve. If successful epigynous groups invade new ecological niches, adaptive modification along the lines of least resistance is for them any one of a number of modifications of the epigynous pattern, rather than its abandonment. As is discussed in Chapter 5 epigynous flowers can give rise to dehiscent capsules, many-seeded berries, stone fruits, or fruits that are adapted to wind dispersal or to external dispersal by animals. These structures are functionally analogous but not at all homologous to various kinds of fruits produced by hypogynous or perigynous flowers.

Nevertheless, the actual adaptive significance of the trend from hypogyny to perigyny and epigyny is still problematical. One suggestion, offered by Grant (1950*b*), is that epigyny provides better protection from insect or bird pollinators having biting mouth parts. He noted that pollination by birds or beetles is present in about 93 percent of the angiospermous families with inferior ovaries. The data available, however, do not permit one to decide whether a shift to this kind of pollination actually accompanied the evolution of epigyny. Furthermore, small-flowered plants such as species of *Saxifraga* and other Saxifragaceae are difficult to explain on this basis. Their flowers are pollinated by a variety of small insects, chiefly Diptera and Hymenoptera, that have sucking mouth parts (Knuth 1895–1905).

Two suggestions can be offered for the adaptive advantage of epigyny in these smaller, more delicate flowers. One is that the inferior ovary is surrounded by a firmer, thicker wall than is the superior ovary of related species, and so the tender developing ovules within it are better protected against heat, cold shocks, or drought to which the flower might be exposed between anthesis and seed maturity. In support of this hypothesis with respect to the genus *Saxifraga* is the fact that many of its species that have superior ovaries (*S. arguta, S. rotundifolia, S. stellaris, S. hederacea, S. cymbalaria*) grow in moist, shady sites, whereas those having inferior ovaries (*S. tridactylites, S. adscendens, S. aizoon*, et aff) more often inhabit sunny, drier situations (Engler 1928). Another hypothesis is that adnation of the ovary with the hypanthium produces a structure in which the proportion of seed mass is higher relative to the rest of the floral tissue, so that an equivalent amount of seed material can be elaborated with a smaller total production of plant material, thus making possible a more rapid development of the flower and maturation of the seed capsule. Both of these hypotheses could be tested by suitable comparisons of related species.

None of the hypotheses to explain the origin of epigyny are mutually exclusive, nor can any of them be regarded as the only way by which the proposed adaptive advantage could be achieved. Different trends to epigyny may well have been guided by entirely different selective pressures. Protection of developing flowers against biting insects, as well as from environmental shocks, has almost certainly been achieved in a variety of different ways, as is also true of increased efficiency and rapidity of development. With respect to epigyny as well as other evolutionary trends of the flower, we must guard against either the acceptance or the categorical rejection of unitary, generalized explanations, based upon one particular kind of selection pressure.

Trends in the ovules and seeds

As is stated in Chapter 10, the ovules of the original angiosperms were probably anatropous and bitegmentary, having a well-developed outer integument that was markedly differentiated from the inner integument, so that the openings of the two integuments did not coincide with each other, and the condition known as "zigzag micropyle" existed. Ovules of this kind persist in many families of Magnoliales and Dilleniales, as well as in some families belonging to more specialized orders, such as Capparales, Malvales, and Rosales. From a review of ovule structure in nearly all families of angiosperms, based upon the compilations of Davis (1966), Takhtajan (1966), and Cronquist (1968), a number of different trends away from this condition can be recognized. These consist entirely of reductions. As with all of the trends recorded in this and preceding chapters, each of the reductional trends in the ovule has taken place independently many times.

The trend from anatropous to hemitropous or orthotropous ovules, consisting of the reduction or elimination of the funiculus, is often associated with the reduction of the ovules to a single one per gynoecium. Orthotropous ovules exist in the Piperales, Platanaceae of the Hamamelidales, Juglandales, Myricales, Polygonales, some genera of Proteaceae and Santalaceae, a few families of monocotyledons (Potamogetonaceae, Zosteraceae, Xyridaceae, Restionaceae, Rusceae of Liliaceae), and isolated genera of various families. It is, however, a rather uncommon trend, and the reduction of the ovules to one per gynoecium has taken place far more often with the retention of the anatropous orientation than with the shift to the orthotropous orientation. In a gynoecium having a single ovule that fills the entire space within the ovary wall, the elimination of the funiculus would seem to provide a more economical growth pattern and a more efficient distribution of tissues in the mature gynoecium. Apparently, however, the mutations or other genetic changes that could bring about this morphological change have appeared rather infrequently.

Much more frequent have been reductions or intercalary concrescence involving the ovular integuments. The reduction of the outer integument so that the micropyle is formed only by the inner integument has taken place in a majority of those families such as the Winteraceae, Degeneriaceae, and Annonaceae. Reduction or "fusion" (intercalary concrescence) leading to a single ovular integument (the unitegmic condition) is somewhat less common, but this condition nevertheless exists in all families of the orders Ericales, Umbellales, and the subclass Asteridae, as well as in most families of the orders Cornales and Santalales, and in scattered families belonging to other orders (Piperaceae, Juglandaceae, Myricaceae, Actinidiaceae, Loasaceae, Sapotaceae, Symplocaceae, various Rosales, Limnanthaceae, and others).

Another kind of reduction has been the trend to the tenuinucellate condition. This is defined by Davis (1966) as the development of a single archesporial cell directly into the megaspore mother cell, without formation of an archesporial tissue or cutting off of a primary parietal cell. This ultimate stage of the reduction of archesporial development is associated with the unitegmic condition in the orders Ericales and Santalales, the superorder Asteridae, the families Sarraceniaceae, Actinidiaceae, Loasaceae, Sapotaceae, Symplocaceae, Pittosporaceae, Hydrangeaceae, Limnanthaceae, and Umbelliferae, and some genera or subfamilies belonging to other families.

The association between the unitegmic and the tenuinucellar conditions occurs far more frequently than would be expected on the basis of chance alone. Of the 354 families recognized by Cronquist, 89 (25 percent) have unitegmentary ovules and 102 (29 percent) have either the tenuinucellar condition or the condition characterized by Davis (1966:17) as pseudocrassinucellar, which is probably a secondary derivative of the tenuinucellar condition, since it also has a single archesporial cell that develops directly into the megaspore cell. If the unitegmentary and tenui- plus pseudocrassinucellar conditions were randomly associated with each other, we would expect them to appear simultaneously in 26 families ($0.25 \times 0.29 = 0.0725$; $0.0725 \times 354 = 26$). Actually, the combination of these two conditions appears in 65 families, including some of the largest in the flowering plants (Compositae, Rubiaceae, Scrophulariaceae, Labiatae, Umbelliferae). It must be regarded, therefore, as a highly successful combination of characters.

With respect to the development of the megaspore and female gametophyte, the great majority of angiosperms possess the normal sequence in which meiosis forms a tetrad of megaspores, which may be either linear or T-shaped, and the chalazal megaspore develops into an octonucleate embryo sac, termed by Maheshwari (1950), Davis (1966), and others the *Polygonum* type. Deviations from this sequence occur in many different families, usually in isolated genera or groups of genera, and few if any families possess uniformly any one of the numerous recorded deviations. Perhaps the most consistent deviation in a family is the possession in all members of the Onagraceae that have been investigated of a tetra-nucleate, "*Oenothera*-type" embryo sac. More frequent, however, are deviations that affect the formation of megaspores, particularly the formation of a dyad rather than a tetrad, so that the second division of meiosis becomes the first nuclear division to form the embryo sac. Less common, but still appearing in isolated genera belonging to a number of families, is the complete suppression of megaspore formation, so that the four nuclei that are the products of meiosis each divide only once more to produce the octonucleate embryo sac. Still less common, but well known because they occur in species that are often used in botany classes, are various nuclear fusions that give rise to the polyploid condition, particularly in the antipodal nuclei of the embryo sac ("*Fritillaria*" type of development).

These variations of megaspore and gametophyte development can hardly be regarded as evolutionary trends. This is evident both from their sporadic occurrence and from the fact that several different kinds of sequence can occur in the same species or even in the same individual plant (Hjelmquist 1964).

With respect to the development of the embryo itself, Johansen (1950) has recognized six types, three of which (Asterad, Onagrad, and Solanad) are almost equally common (Davis 1966:25). Admittedly, however, many intermediate conditions between these "types" of development exist, and in some species or genera two different types may occur simultaneously. Since, moreover, no obvious correlation can be seen between the distribution of these types among angiosperms and any other characteristics, and since between a third and a half of angiosperm families have not been investigated with respect to embryo development, early stages of embryogeny, at least in the present state of our knowledge, are of little or no value for determining evolutionary trends. In the monocotyledons, later stages of embryogeny are of some significance, and are discussed in Chapter 13.

Of considerably greater importance are trends involving the relation between embryo and endosperm. As is stated in Chapter 10, one of the most distinctive features of angiosperms as compared with other seed plants is the fertilization of a diploid maternal nucleus by a male gamete, giving rise to a triploid, heterozygous nucleus, and the subsequent rapid division of this nucleus to form a considerable amount of endosperm even before zygotic division and embryogenesis begin. Although this condition is almost universal among angiosperms, the later development of the endosperm relative to the embryo differs greatly between families or orders, and not infrequently between different genera of the same family. In the majority of Magnoliales, Ranunculales, and Dilleniales, as well as in many of the more advanced groups, the mature seed contains abundant endosperm combined with a very small, sometimes rather poorly differentiated embryo. In other groups, however, the endosperm ceases to develop at a relatively early stage in embryogeny, and the mature seed contains a large embryo, the cells of which are often filled with stored food, and with little or no endosperm. This condition is widespread in the subclass Hamamelidae, particularly in the orders Juglandales, Myricales, Fagales, and Casuarinales; in the orders Myrtales, Proteales, and several families of Sapindales, Geraniales, Polygalales, and Rosales in the Rosidae; in most families of the order Theales in the Dillenidae; in the subclass Alismatidae among monocotyledons; and in various scattered families belonging to other orders and subclasses. It has obviously arisen many times during angiosperm evolution, most probably in response to different selective pressures in different groups. Most families and genera having this condition consist of woody plants that possess vigorous seedlings, suggesting that the substitution of food material in the cotyledons for that in the endosperm somehow increases the speed and efficiency of germination and the vigor of the seedling. In the monocotyledons, however, the absence of endosperm is associated not with seedling vigor in terrestrial species but with the aquatic environment. Finally, there are some families, such as Begoniaceae, Datiscaceae, Podostemaceae, Lentibulariaceae, and Orchidaceae, in which the absence of endosperm in the mature seed is associated with very small seed size and high seed numbers per flower.

Correlations Between Trends

The trends of specialization with respect to the different characters described in this and preceding chapters do not occur independently of one another. Correlations between trends that result in associations between the advanced or specialized states of different characters, as well as, not infrequently, between the advanced states of some characters and the primitive states of others, were pointed out by me many years ago (Stebbins 1951). In that study, primary emphasis was placed upon the eight characters that taxonomists use most frequently in their diagnostic keys for recognizing angiosperm families. Table 12-1 shows the significant associations that were found. A noteworthy fact brought out by this chart is that, in addition to thirteen highly significant associations between the advanced states of two different characters, there are six examples of highly significant associations between the advanced state with respect to one character and the primitive state with respect to another.

One possible explanation of the latter associations, which is either expressly stated or implied by Sporne (1954) in his review of my work, is that I and other angiosperm taxonomists are mistaken in our judgment as to which character states are primitive and which are advanced, and that, in fact, associations are always between two or more advanced states. This assumption is, however, contradicted by the chart itself. For instance, there exists a highly significant positive association between the woody habit and a reduced number of ovules per gynoecium, but an equally significant negative association between the woody habit and a reduced number of stamens per flower. One might argue on this basis that the presence of these associations indicates that the condition of having many ovules, because of a negative association with the woody habit, is, in fact, an advanced rather than a primitive state as compared with the condition of having few ovules. This assumption is, however, incompatible with the existence of a positive association between reduced ovule number and reduced stamen number. Clearly, the existing associations between character states cannot be explained on the assumption that they all reflect the tendency of different characters to become more specialized in association with each other. The only possible assumption that can explain all of the recorded associations is that, occasionally, the presence of an advanced or specialized state with respect to one character renders *less* likely the existence in the same group of advanced states with respect to other characters.

The existing associations can, however, be explained on the assumption that the trends toward specialized states in various characters take place as components of adaptive syndromes and in association with adaptive shifts. None of these adaptive syndromes consists entirely of advanced or specialized states with respect to all characteristics. For instance, with respect to the eight reproductive characters that I scored (Stebbins 1951), no family possesses the specialized state with respect to all of them, and only 9 of the 290 families scored are specialized with respect to seven of the eight characters. The great majority of families, 227 in all, are specialized with respect to two, three,

Table 12-1. Degree of association with respect to the advanced states of eight characters that are widely used in diagnostic keys to angiosperm families. (From Stebbins 1951.)

	Woody habit	Epigyny	Parietal basal placent.	Red. no. ovules	Syncarpy	Red. no. stamens	Zygomorphy	Sympetaly
Apetaly	0	0	++	++	0	+	−−	0
Sympetaly	−−	++	0	0	++	++	++	
Zygomorphy	−−	0	0	−	0	++		
Reduced no. of stamens	−−	0	+	++	++			
Syncarpy	−	++	++	++				
Reduced no. of ovules	++	0	+					
Parietal or basal placentation	−−	0						
Epigyny	−−							

EXPLANATION OF SYMBOLS:

0, no significant association

+, significant (p = 0.01 to 0.05) positive association

++, highly significant ($p < 0.01$) positive association

−, significant negative association

−−, highly significant negative association

four, or five of the eight characters. This is exactly what would be expected on the assmption of adaptive syndromes and the maintenance by natural selection of primitive or generalized states with respect to those characters of which the specialized state does not form part of the adaptive syndrome.

Several well-marked adaptive syndromes or "peaks" can be recognized. One of them, which is most characteristic of woody tropical or subtropical families, is the combination of actinomorphic flowers, containing separate sepals, petals, and stamens in regular alternating whorls, with the stamen number equal to or less than the number of sepals and petals, and a gynoecium consisting of united carpels that contain few ovules placed in an axile or basal position in separate loculi and seeds with little or no endosperm. A similar "peak" is one in which the flowers differ from those just described only with respect to the gynoecium, which has numerous ovules on parietal placentae in a single ovary loculus, and seeds with abundant endosperm plus a relatively small embryo. Although this peak is represented by a few tropical or subtropical woody families (Flacourtiaceae, Pittosporaceae), it is most characteristic of temperate, herbaceous families. Even more characteristic of herbaceous families are flowers having the characteristics mentioned above plus sympetaly, zygomorphy, epigyny, or a combination of two or all three of these characters. This association of the herbaceous habit with large numbers of specialized floral characters, particularly those of the perianth and the androecium, need not be explained by assuming that the trend from the woody to the herbaceous habit automatically brings about or is accompanied by a change in floral structure. An explanation that is much more compatible with developmental genetics as well as with present knowledge about the basis of major evolutionary changes in other groups of organisms is that in short-lived herbs, which are strongly dependent upon seed reproduction for their survival and have short generation times, selection pressures for more specialized and efficient flower structures are likely to be stronger than they are in long-lived trees.

The diversity of adaptive peaks is illustrated further by the existence of a syndrome that is in many ways the opposite of those characterized by sympetaly, zygomorphy, and numerous ovules. This is the combination of flowers actinomorphic, perianth monochlamydeous or wanting, stamens few, ovules few (usually one) per gynoecium, flowers hypogynous. This syndrome is most characteristic of woody plants of temperate regions and of wind-pollinated herbs. Entomophilous tropical or subtropical families having this syndrome, such as the Moraceae and Araceae, probably acquired their entomophily secondarily.

The fact must be emphasized that these adaptive syndromes or "peaks" are far fewer than the adaptive "valleys," that is, combinations of specialized floral characteristics that are not possessed by any group of plants. Even with respect to combinations of only two characteristics, the valleys (17) outnumber the peaks (11). The "peaks" that combine six or more specialized characters either are relatively "low," containing only a few small families, or if "high" are dominated by a single large, highly specialized family, such as the Compositae, Cucurbitaceae, Gesneriaceae, or Loranthaceae.

Furthermore, the highest adaptive "peaks" are occupied by several different families or groups of families which, on the basis of the sum total of their vegetative and floral characteristics, cannot be regarded as closely related to or derived from one another. These "peaks" have apparently been "climbed" several times by different evolutionary lines.

More evidence on the existence and nature of adaptive peaks with respect to floral characteristics is badly needed. Nevertheless, the distribution of specialized conditions and combinations of characters in angiosperms is more easily understood on the basis of adaptive syndromes acquired by the action of natural selection than on the basis of any other hypothesis.

Summary

The principal trends in the corolla are reduction, union of parts, and the shift from actinomorphy to zygomorphy. The terminal stages of these trends are obviously adaptive in relation to different methods of pollination, but the origin of their earliest stages is difficult to understand. The principal trends in the stamens are changes in number, elongation of filaments, and in some families secondary aggregation of stamens into bundles or unions of filaments to form staminal sheaths. Flowers that are large but have few stamens are believed to be the products of reduction in early stages of their evolutionary lines, followed by secondary increase. The most common trends in pollen grains are increases in numbers of furrows or pores, or both, but other kinds of modification are associated with the shift from insect to wind pollination.

The principal trends in the gynoecium have been union of carpels, the shift from axile to parietal placentation, and that from hypogyny to perigyny and epigyny. The rare condition of free central placentation is believed to have evolved via an indirect route involving (1) a shift from axile to parietal placentation, (2) reduction in number of ovules to one or a few per gynoecium and (3) secondary increase in ovule number with development of the free central placenta. The trend from hypogyny to perigyny and epigyny usually involves intercalary concrescence of perianth, androecium, and gynoecium, but in some instances there has been concrescence between the gynoecium and a concave floral receptacle. All trends in the gynoecium have occurred many times in different lines of evolution.

Trends in the ovule have been the shift from the anatropous to the campylotropous and orthotropous condition, from two integuments to a single integument, and from the crassinucellate to the tenuinucellate condition. The principal trend in the seed has been from copious to scanty endosperm, with transfer of stored material to the cotyledons. Trends of the embryo sac are of minor importance.

Correlations between different morphological trends support the hypothesis of adaptive "peaks" of characters and the inadaptive nature of the great majority of possible combinations between various specialized characteristics.

13 / Evolution of the Monocotyledons

For more than a century botanists have recognized two major subgroups of angiosperms, the monocotyledons and the dicotyledons.* The differentiation between them is a prime example of the kind of differentiation that exists between any two major groups of higher plants. In the first place, no single character is found exclusively in one or the other of these groups. Secondly, each group is more advanced than the other with respect to the modal condition of some of its characters, and less so with respect to others; that is, each group is a mosaic of relatively specialized and relatively unspecialized character states or attributes.

The distinctive characteristics of monocotyledons as compared with dicotyledons are usually listed as follows (cf. Cronquist 1968:128): cotyledon of embryo one or undifferentiated; leaves mostly parallel veined; vascular bundles without cambium or secondary vascular tissue; vascular bundles scattered; floral parts usually in sets of three; pollen grains uniaperturate or uniaperturate-derived; mature root system wholly adventitious.

Of these characters, those of leaf venation, activity of cambium, and symmetry of flowers are not uniform among monocotyledons; the others are found occasionally in various groups of dicotyledons. The characterization of the leaf venation of monocotyledons as "parallel" in contradistinction to "net-veined," al-

* Although the systematic treatment in this book has in general followed the lead of Takhtajan and Cronquist, there seems to be no good reason here for taking up their unfamiliar names Magnoliatae and Liliatae in place of the well-established and widely used names dicotyledons and monocotyledons.

though useful for identification, is not strictly accurate. Although the principal veins in the leaves of monocotyledons run parallel to one another from one end of the leaf to the other, they are not completely independent of one another. Even in such strongly parallel-veined leaves as those of grasses, the longitudinal veins are connected to one another by short, secondary, transverse vascular bundles, so that the leaf is actually net-veined. An equally distinctive characteristic of the leaves of monocotyledons is the tendency of their main veins to converge at the tip of the leaf. This is produced in development by the early maturation of the tissues of the leaf apex and the formation of the main body of the leaf by means of the activity of a strongly developed intercalary meristem. This condition is found also in dicotyledons—occasionally in leaves, as in species of the genus *Plantago,* and much more often in bud scales or cataphylls (Arber 1925:99, 113, 115). The significance of this condition with reference to the homology of the leaves of monocotyledons is discussed below. Although the presence of free endings of veinlets in dicotyledons and their absence in monocotyledons is usually a good distinction, free veinlet endings are absent in some dicotyledons (*Quiina,* Foster 1952) and are found in some monocotyledons (*Arisaema,* Ertl 1932).

The absence of cambium within the vascular bundles, and consequently of secondary xylem and phloem, is characteristic of nearly all monocotyledons. Nevertheless, as Arber (1925:41, 82) points out, a small amount of cambial activity exists in the bundles of some forms, both in the axes and in the leaves. No monocotyledon, however, has a ring of vascular bundles connected to one another by interfascicular cambium, which is the usual condition in the stems of herbaceous dicotyledons. In this respect, however, the stems of a few dicotyledons, such as the Nymphaeaceae, resemble those of monocotyledons.

With respect to floral characteristics, the two groups are by no means distinct from each other. Although the flowers of monocotyledons are usually trimerous, species having tetramerous flowers are scattered through various orders. Moreover, trimerous flowers occur regularly in some families of dicotyledons, such as the Annonaceae and the Aristolochiaceae. The usually monocolpate pollen grains of monocotyledons are distinctive, but are matched by those of the Magnoliales, the Nymphaeales, and some Piperales, that is, of relatively unspecialized Magnoliidae. The presence of exclusively adventitious root systems in the adult plant is characteristic of many herbaceous and rhizomatous dicotyledons.

Even though they have no completely distinctive characteristics, the monocotyledons apparently had a single origin that was separate and distinct from that of any modern groups of dicotyledons. The three best reasons for believing this are the nature of the embryo, of the pollen grains, and of the inflorescence. The embryo is distinctive not only because of its single cotyledon, but also because of its growth pattern, which begins in early embryogeny (Suessenguth 1920, Boyd 1932, Johansen 1945, Haccius 1952, Souèges 1954). It is bilaterally symmetrical in only one plane, whereas the embryo of dicotyledons is symmetrical in two planes. Moreover, although a number of different types of embryo and embryogeny can be recognized, they are all connected with one another by intermediate conditions (Boyd 1932).

The monocolpate-derived form of the pollen grains is not approached by any herbaceous dicotyledons, except for the Nymphaeaceae. This fact makes highly unlikely the hypothesis previously advanced by many botanists, that the monocotyledons were derived from herbaceous dicotyledons such as the Ranunculales. This hypothesis is made even more unlikely by the nature of the xylem vessels of monocotyledons, which is discussed below.

Another unifying feature of the majority of monocotyledons, which has been largely overlooked, is the nature of their inflorescence. With the exception of the Alismatales and some of the palms (Tomlinson 1970, Tomlinson and Moore 1968), the flowers of monocotyledons are grouped into inflorescences that are basically racemose or spicate, or, if they are solitary or umbellate, a derivation from the racemose condition can be recognized. Except for the palms, the compound inflorescences of monocotyledons are aggregates of racemes or spikes. Compound inflorescences that consist of aggregates of determinate, dichasial units (see Chapter 11), which are the commonest kind in the less specialized orders of dicotyledons, exist only in palms such as *Nannorhops* (Tomlinson and Moore 1968). This suggests that the shift from the determinate to the indeterminate condition of the inflorescence, which in dicotyledons took place in various evolutionary lines and different stages of advancement, in the monocotyledons took place generally and at an earlier stage.

On the basis of the characteristics that have just been mentioned, the monocotyledons must be regarded as containing a mosaic of relatively specialized and more generalized attributes. The structure and development of the embryo, the reduction in intrafascicular cambial activity, the scattered vascular bundles of the axis, the either reduced or indeterminate nature of the inflorescence, and the relatively small number of parts of the individual flower (except in Alismatales), all indicate a higher degree of specialization than in the more generalized orders of dicotyledons. On the other hand, the monocolpate pollen grains and the relatively unspecialized xylem vessels (Cheadle 1942) indicate a stage of advancement with respect to these characters that is less than that found in herbaceous dicotyledons. This condition is best explained by assuming that the ancestral stock of monocotyledons branched off from the line or from one of the lines leading to dicotyledons at a relatively early stage of advancement, and then became rapidly modified with respect to certain characteristics.

Isolated Positions of the Less-Specialized Groups of Monocotyledons

The fact has been emphasized elsewhere (Chapter 10) that the least-specialized orders of woody dicotyledons are very different from one another in a number of respects and can be grouped together only because they possess in common an assemblage of characters that indicate a low degree of specialization. The same situation exists to an even more marked degree with respect to the least-specialized families of monocotyledons. The presence of separate carpels is certainly a less-specialized condition than their union to form a compound ovary. In the dicotyledons, the apocarpous families are a reasonably homogeneous group, in spite of marked differences between individual families. Through the Lardizabalaceae and Menispermaceae, the woody and herbaceous orders of the subclass Magnoliidae are connected with one another to a certain degree, while the Lardizabalaceae at least partly bridge the gap between Magnoliidae and Dillenidae.

On the other hand, the gaps between the different orders of apocarpous monocotyledons are wide and unbridgeable by any existing groups. Takhtajan (1969:108) lists as monocotyledons with apocarpous gynoecia the Alismatales (Butomaceae, Limnocharitaceae, Alismataceae), Scheuchzeriaceae, Potamogetonaceae, Triuridaceae, and some palms. To this list might be added the genus *Petrosavia,* which Cronquist places in a separate family belonging to the order Triuridales, a position reaffirmed by the careful investigations of Stant (1970), and the tribe Tofieldeae of the Liliaceae, which, though having a syncarpous ovary, are the nearest approach to apocarpy that exists in this large and complex order.

With respect to other characteristics than apocarpy, these groups form as heterogeneous an assemblage as one could imagine. The Alismatales are aquatic or subaquatic plants, which are less specialized than other monocotyledons with respect to their determinate inflorescences, but are more specialized than the other apocarpous groups with respect to their perianth, which is differentiated into calyx and corolla, their trinucleate pollen grains, and their mature seeds, which lack endosperm. *Scheuchzeria palustris,* which is the only species of the monogeneric family Scheuchzeriaceae, is a boreal marsh plant having a simple, racemose inflorescence, an undifferentiated perianth, and, like the Alismatales, trinucleate pollen and seeds without endosperm. The Potamogetonaceae are fully and strictly aquatic. In addition to trinucleate pollen and seeds without endosperm, they have a much reduced gynoecium containing a single ovule, as well as highly specialized vegetative parts. With the exception of the Butomaceae in the Alismatales, all of these groups have pollen grains that are specialized in being either without pores or with two to several pores.

The Triuridales (including *Petrosavia*) are tropical saprophytes, of which the roots depend upon symbiotic mycorhizal fungi. Their leaves are reduced to scales. Their flowers are racemose and much reduced in size. Although they agree with the Alismatales in having trinucleate pollen grains that are either monocolpate or acolpate, their seeds are less specialized in possessing abundant endosperm. The palms are (probably secondarily) arboreal or, in the case of a number of the apocarpous genera, plants that form clusters of large rosettes from a branched caudex, and at maturity they have highly specialized palmately or pinnately cleft leaves. Their inflorescences are highly indeterminate aggregates of elongate thyrses or spikes, with the individual flowers either in clusters of three or solitary, and sessile or very shortly pedicelled. They have chiefly monocolpate, always binucleate pollen grains and seeds with abundant endosperm. Finally, the least specialized of the Liliales, the tribe Tofieldieae, are boreal or temperate marsh plants that form short rhizomes. Their flowers are in simple racemes; their pollen grains are monocolpate and binucleate; and their seeds possess abundant endosperm.

One searches in vain among this heterogeneous assemblage of diverse plants for any group that could by the widest stretch of the imagination be regarded as a common denominator, which might serve as their common ancestor. If one classifies them with respect to other characteristics that might be supposed to serve as guidelines, the ambiguity becomes even greater. All of them have six separate perianth parts, which in the Alismatales are differentiated into green sepals and showy petals but in the other groups are undifferentiated. Other than the pollen grains, there are no distinctive features of the androecium.

Among vegetative characteristics, the nature of the vascular tissue has received considerable attention. According to Cheadle (1942, 1943, 1953), the least specialized of the above-mentioned groups with respect to xylem are *Scheuchzeria*, the Potamogetonaceae, and the Liliaceae Tofieldieae, which have vessels only in the roots and these have scalariform perforation plates. Palms with apocarpous gynoecia (*Phoenix*) Cheadle regards as more advanced, since they have vessels in both roots and shoots, but all of these have scalariform perforations. More advanced in a different direction are the Alismatales, which have vessels only in the roots, but these have simple as well as scalariform end perforations. The scanty xylem found in the saprophytic Triuridales is relatively unspecialized (Stant 1970).

Two other characteristics, the life form and the histology of the stomatal apparatus, deserve consideration, since these characters are highly correlated with each other. As mentioned in Chapter 10, the least-specialized life form in angiosperms is that of phanerophytes, in which the vegetative organs are active, green, and photosynthetic throughout the year. Stebbins and Khush (1961), after reviewing the condition of the stomatal apparatus in the leaves of 192 species of monocotyledons, concluded that a high degree of correlation exists between stomata having many subsidiary cells surrounding the guard cells, the phanerophytic growth habit, and tropical or subtropical distribution. This combination is most prevalent in the subclass Arecidae, but it exists also in various families belonging to the Commelinidae (Commelinaceae, Bromeliaceae, most families of Zingiberales). It probably represents the least-specialized condition with respect to these characters. Among the apocarpous monocotyledons, only certain genera of palms have it. In addition, some palms have the unspecialized conditions of undifferentiated perianth, monocolpate and binucleate pollen grains, and seeds with abundant endosperm. Consequently, the apocarpous palms have a larger number of unspecialized characteristics than any other single group of monocotyledons. Nevertheless, the least-specialized phanerophytic members of the Commelinaceae and Bromeliaceae are equally or even more generalized than these palms with respect to most of their vegetative characteristics, whereas the Alismatales are less specialized with respect to their inflorescences and gynoecia.

The facts mentioned above lead inevitably to one of two hypotheses: either the monocotyledons are highly polyphyletic, or the modern groups are end lines, descendants from a large complex of ancestral forms that are now completely extinct. The hypothesis of polyphylesis has already been rejected as improbable. The postulate of a large complex of extinct forms would explain both the discontinuous and heterogeneous nature of modern apocarpous monocotyledons and the lack of any connecting links between modern monocotyledons and dicotyledons. From the ecological point of view, this hypothesis has much to recommend it. Those woody dicotyledons that form connecting links between modern families, orders, and subclasses, exist predominantly in tropical rain or cloud forests, where the permissive and constant nature of the environment has enabled them to persist as relics for very long periods of time (Chapter 8). The herbaceous ancestral monocotyledons, however, lived most probably in open, seasonally dry habitats, as postulated above. Such habitats support a much smaller number of perennial species, and can often be dominated, even in the tropics, by one or two vigorously growing species. Under such conditions, highly evolved species, which are vigorous in vegetative growth and have highly efficient reproduction from seeds, could easily crowd out and eliminate their less efficient forerunners. At present, the herbaceous flora of seasonally dry tropical savannas is dominated by grasses and sedges, These families are among the most highly evolved and recent of the monocotyledons. Their predecessors, which occupied comparable habitats during the Cretaceous Period, may well have been the primitive monocotyledons that formed the connecting links between modern orders and are now extinct.

The Origin of the Monocotyledons

Contemporary botanists who are familiar with angiosperms as a whole, such as Hutchinson, Takhtajan, Cronquist, and Thorne, agree that the monocotyledons have been derived from primitive dicotyledons. As mentioned in Chapter 10, numerous morphological and cytological characters attest to the basic unity of flowering plants, rendering highly improbable the origin of monocotyledons independent of dicotyledons from some other group of nonangiospermous seed plants. Nevertheless, all attempts to derive the monocotyledons from any living order of dicotyledons have failed. The opinion once held by many botanists, that monocotyledons are connected to dicotyledons via a relationship between the Alismatales and the Ranunculales, can now be rejected on the basis of several lines of evidence, derived from both vegetative and reproductive characteristics (Meyer 1932,

Cheadle 1942, 1953, Stant 1964, Takhtajan 1969).

Both Takhtajan and Cronquist, while stating specifically that monocotyledons could not have been derived from any group of modern Nymphaeales, nevertheless suggest that the affinity of the Nymphaeales "with the Alismatales cannot be doubted" (Takhtajan 1969:112) or that "the premonocotyledonous dicots were probably something like the modern Nymphaeales" (Cronquist 1968:315).

In my opinion, both of these statements can certainly be doubted. There is little more reason for suggesting an affinity between primitive monocotyledons and Nymphaeales than for the now discredited hypothesis of an affinity with herbaceous Ranunculales. Granted that the Nymphaeales are the only herbaceous dicotyledons that have monocolpate pollen as do the monocotyledons; the significance of this fact is diminished if one considers, as both of these authors do, that monocolpate pollen was present in all primitive dicotyledons. All of the other resemblances between Nymphaeales and Alismatales could be explained on the basis of convergence due to their independent acquisition of the aquatic habitat, and I believe that this is the most probable explanation. The reasons for this are discussed below.

My opinion, therefore, is that no living order of dicotyledons is clearly related to the ancestor of monocotyledons. This hypothesis, unsatisfactory as it is, becomes the most plausible one if one accepts two other hypotheses that are maintained in this book: that the common ancestors of all modern orders of dicotyledons are completely extinct, and that, as discussed earlier in this chapter, modern orders of monocotyledons are separately descended from different members of a large series of now extinct monocotyledons, which included the earliest members of the subclass. As is mentioned in Chapter 7, the lesson to be learned from all groups of animals and plants that possess a reasonably good fossil record is that the nature of the ancestors of ancient and widespread modern classes can never be deduced by direct comparisons between modern forms.

Origin of monocotyledons as a major adaptive shift

Botanists who have speculated about the origin of the monocotyledons have adopted one of two rather divergent viewpoints. One group, headed by Agnes Arber (1925), believed that the drastic alterations of external morphology that characterized this origin took place through internally directed changes, without benefit of natural selection, and that the adaptive features of the group are secondary and of minor importance. In her own words: "But now that biologists have gradually and painfully learned that the Natural Selection hypothesis is not the master key to the mysteries of the organic world, the centre of significance of morphological study has shifted, and we look to our results as representing the raw material whence the laws of evolution may eventually be deduced" (Arber 1925:10).

During the almost half century that has elapsed since these words were written, the kind of comparative morphology that Arber championed, based as it was upon typological concepts of adult structures (on p. 3 of her book she explicitly states her allegiance to the type concepts of Goethe and de Candolle), has failed to uncover any laws or processes of evolution. On the other hand, extensive experimental research on plants, animals, microorganisms, and even viruses has shown that, as Darwin supposed, natural selection is at least one of the master keys to understanding evolution. Consequently, although the factual observations of Arber, as well as those of such botanists as Bews (1929), Holttum (1955), and Gatin (1906), are of great value to the modern evolutionist, their interpretations must be reevaluated in the light of modern knowledge. The same must be said of the introductory pages to the otherwise admirable review by Tomlinson (1970), since he completely discards all attempts to explain the evolutionary origin of a group. He states specifically (p. 211) that problems of origin can be explained completely by a description of the developmental pattern that gives rise to a particular structure in an individual plant. This is possible only if one rejects evolution altogether, and, like the classical morphologists of the early 19th century, believes in special creation. I obviously cannot agree with such a point of view.

Other botanists have regarded the origin of the monocotyledons as an adaptive shift of major proportions. These include Sargant (1903), Henslow (1911), Takhtajan (1959, 1969), Tzveliov (1969), and myself. Their opinions differ, however, as to what was the nature of this shift. Henslow looked upon the distinctive features of monocotyledons as the result of a primary adaptation to an aquatic habitat. His concept of evolutionary change, however, was based upon a Lamarckian belief in the inheritance of acquired modifications, and his rambling, anecdotal discussion is unconvincing to a modern botanist. The other botanists mentioned, including myself, recognize that adaptation to wet conditions must have played a role in the major adaptive shift that led to the monocotyledons, but reject the hypothesis of a strictly aquatic origin. A comparison and discussion of their hypotheses can best be made after the most significant of the distinctive characteristics of monocotyledons have been described in greater detail.

General features of monocotyledons and their analogues among dicotyledons

Since no living group exists that might serve as a reference point, we cannot hope to do more than seek a balance of probabilities, as was done for the origin of the angiosperms as a whole. The best starting point is a consideration of those characteristics that are most generally found throughout the class, are most closely connected with their basic pattern of growth, and are most distinctively different from the modal growth patterns found in dicotyledons. These are the unidimensional symmetry of the young embryo and the strong development of intercalary meristems, which may occur in the cotyledon, shoot axes, leaves, and inflorescence axes, or in all of these organs. Following the principle of genetical uniformitarianism (see Chapter 2), the adaptive significance of these characteristics can best be understood by using carefully selected analogies with dicotyledons in which similar characteristics have arisen independently and more recently.

The development of the embryo and seedling are of first importance, since, as stated in Chapter 5, seed germination and seedling establishment are subjected to stronger selection pressures than any other stage of the life cycle. The unidimensional symmetry of early embryo development is probably adaptive as a prelude to later events, since it makes possible the formation of a cotyledonary tube with an opening on one side, through which the first leaf and, later, the shoot can emerge. The adaptiveness of the extensive intercalary meristem that is formed in the cotyledon of seedlings belonging to types I and II as recognized by Boyd (1932; see Fig. 13-1) can be appreciated by considering the method of germination and early seedling growth found in those plants in which it is well developed. These include *Erythronium* (Arber 1925:152), various palms (*ibid.*:157), and *Colchicum* (Galil 1968), as well as dicotyledons in which a cotyledonary tube has arisen independently, such as *Marah* (Schlising 1969). In these plants, the elongation of the cotyledon or cotyledonary tube by means of an intercalary meristem, combined with a positive geotropism of the hypocotyl, serves to "plant" the hypocotyl and plumule far into the ground. This is most easily accomplished when the ground is saturated with water and is relatively soft, but its adaptive advantage is greatest in regions characterized by long dry periods, since the "planted" hypocotyl can become deeply rooted quickly and easily, while the subterranean plumule, surrounded as it is by the firm mem-

Fig. 13-1. A series of seedlings of monocotyledons: (I) a simple epigeal cotyledon, which may have either one or two vascular bundles; (II) a hypogeal cotyledon having two vascular bundles, and elongation of the hypocotyl; types I, II, and intermediate forms are predominant in the Liliales, Alismatales, and Arecales; type II is believed to represent the original one among monocotyledons. Types III (Commelinaceae), IV, and V (both in Zingiberaceae) represent shortening in the course of vascular strands in the cotyledon, associated with adaptation to mesic tropical habitats. In type VI (Gramineae), the cotyledon has become a haustorial organ and no longer emerges from the seed. (From Boyd 1932.)

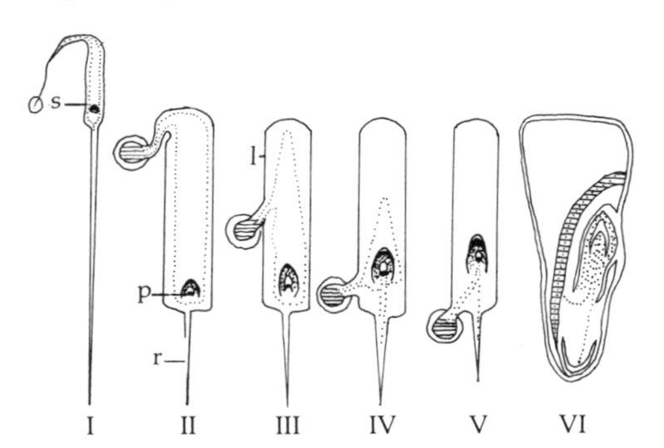

brane of the cotyledonary tube, has acquired maximum protection from the effects of seasonal drought. Consequently, the maximum selective pressure for the origin of the monocotyledonous type of embryo and seedling development would be expected in a climate having a strong alternation between favorable and unfavorable seasons. This point was strongly emphasized by Sargant (1903), and I fully accept it.

Sargant reviewed examples of 31 species of dicotyledonous seedlings having a cotyledonary tube, which represent 20 different genera in 10 families of dicotyledons, and an additional list of 12 species of dicotyledons in 6 families that have pseudomonocotyledonous seedlings. She noted that nearly all of these are geophytes, in which the seedling enters a prolonged period of dormancy soon after or even before the first leaf has matured. These examples were nearly all taken from temperate, boreal, or subalpine floras, so that the unfavorable season to which their dormancy is adapted is cold rather than drought. Under such conditions, the geophytic life form is the only possible one. On the other hand, in habitats characterized by an unfavorable season of drought rather than cold, the seedling may be able to survive this season without becoming completely dormant, if the leaves are sufficiently drought resistant and the tender meristems are sufficiently well protected. Such is the case in many monocotyledons such as *Yucca*, *Agave*, and those palms that inhabit seasonally dry regions. These plants, although they possess a well-developed cotyledonary tube, are nevertheless rosette-forming phanerophytes rather than geophytes. The primitive monocotyledons, therefore, could have been either geophytes or rosette-forming phanerophytes. The latter condition would be compatible with the hypothesis advanced by Sargant, that the elongate, parallel-veined leaf of monocotyledons has an advantage under these conditions, since its well-protected basal intercalary meristem enables it to regenerate quickly if its tip either becomes desiccated or is nipped off by an animal. This form of regeneration can easily be seen in the spontaneous seedlings that surround trees of *Phoenix canariensis*, wherever this species is cultivated.

If the principle of genetic uniformitarianism and the use of analogies to modern dicotyledons is applied to deduce a hypothesis concerning the origin of the distinctive stem anatomy and leaf venation found in monocotyledons, the conclusion is reached that the selection pressures that brought about these changes may have been somewhat different from those that brought about the shift in seedling structure. Dicotyledons having "parallel-veined" leaves similar to those of monocotyledons are scattered through various families and orders. Some of them, such as the Epacridaceae, are shrubs, but the majority are rosette-forming herbs. Examples are various species of *Ranunculus* (*R. lingua*, *R. alismaefolius*, *R. flammula*), the genera *Eryngium* (Umbelliferae) and *Plantago*, and various Gentianaceae and Compositae (*Scorzonera*, *Tragopogon*). So far as I am aware, none of the species involved have cotyledonary tubes or pseudomonocotyledonous seedlings. Most of these rosette-forming herbs occupy habitats characterized by a soil that is either seasonally or continuously saturated with water.

Dicotyledons that resemble monocotyledons in having, in young plants, a stem structure characterized by separate vascular bundles and a reduction or absence of interfascicular cambium are most widespread in the order Caryophyllales, particularly the families Phytolaccaceae, Nyctaginaceae, Chenopodiaceae, and Amaranthaceae. With the exception of the highly specialized Chenopodiaceae, these families are characteristically tropical or subtropical, and are abundant in arid or seasonally dry climates.

The hypothesis that the earliest monocotyledons were aquatic is not supported by analogies with aquatic dicotyledons, based upon the principle of genetic uniformitarianism. Species that are strictly or facultatively aquatic are scattered throughout the dicotyledons, and in some instances represent extreme adaptive radiations within genera that are characterized by a variety of ecological preferences. Examples are *Ranunculus* subg. *Batrachium, Polygonum amphibium, Lobelia Dortmanna, Veronica Anagallis-aquatica* and *V. americana,* and the genera *Subularia, Nasturtium, Elatine, Callitriche, Myriophyllum, Jussiaea, Ludwigia, Hydrocotyle, Oenanthe,* and *Utricularia.* Although most of these genera have a reduced amount of vascular tissue, they do not characteristically have many separate vascular bundles as in the stems of monocotyledons. Their leaf blades are sometimes reduced, but when the venation is well developed, it is not of the parallel, "closed" type that is characteristic of monocotyledons. Finally, their embryo and seedling development, where known, is typically dicotyledonous, and shows no sign of asymmetry, development of a cotyledonary tube, or reduction of the cotyledon blades. The same is true of the embryos and seedlings of the Nymphaeaceae.

The suggestions of affinity between primitive monocotyledons and Nymphaeaceae are weakened by the fact that much of the apparent similarity is between monocotyledons such as the Alismataceae, which at least with respect to vascular anatomy and seeds without endosperm are relatively specialized, and genera of Nymphaeaceae such as *Nuphar, Nymphaea,* and *Victoria,* which at least with respect to the overall architecture of the plant, including the arrangement of the flowers, are equally specialized. As has been clearly shown by Dormer and Cutter (1959), the thick rhizomes of these genera produced alternately leaves that have no flowers or branches in their axils and flower stalks that are not situated in the axils of either foliage leaves or bracts. This condition is unique in the flowering plants, and might be regarded as a highly specialized consequence of the adaptation of these plants to living in deep water. The condition in the genera *Brasenia* and *Cabomba,* which have solitary long pedunculate flowers in the axils of foliage leaves, is widespread in dicotyledons, and is found in other aquatic genera, such as *Elatine, Callitriche, Jussiaea, Ludwigia,* and *Ranunculus* subg. *Batrachium.* Consequently, solitary and axillary flowers are probably the least-specialized condition in the Nymphaeales. Significantly, this condition is shared by two of the three families that are ordinarily placed in the order, the Nymphaeaceae (*sens. lat.*) and the Ceratophyllaceae. This arrangement of the flowers is consistently associated with elongate stems and the absence of the rosette habit, which, therefore, is most probably the original condition in the Nymphaeales, but is not found in the Alismatales.

On this basis, I regard the origin of the growth habit in the Nymphaeales as analogous to and governed by selective pressures similar to those that govern the origin of *Ranunculus* subg. *Batrachium, Elatine, Callitriche, Jussiaea,* and *Ludwigia.* This kind of adaptive shift has little or nothing in common with the adaptive shift that gave rise to the distinctive vegetative characteristics of the monocotyledons.

The adaptive shift that led to the monocotyledons

Based upon the comparisons that have just been made, an admittedly speculative reconstruction of the adaptive shift that led to the monocotyledons is as follows. The dicotyledons that were ancestral to the monocotyledons are postulated as belonging to the now extinct complex discussed in Chapter 10. They were probably shrubs or subshrubs similar to the smaller species of *Drimys* subg. *Tasmannia,* having numerous closely spaced leaves separated by short internodes, few branches, and wood that either was vesselless or had very primitive vessels. Their flowers were in cymose inflorescences and had an undifferentiated perianth that may already have consisted of two alternating trimerous whorls. Their stamens were six, nine, or perhaps more in number, and already had well-differentiated filaments and anthers as in most modern angiosperms. The pollen grains were monocolpate and binucleate. The gynoecium consisted of a single whorl of three carpels, each containing relatively few ovules. The seeds had a small embryo and abundant endosperm. They had epigeal germination, the blades of the two cotyledons being raised above the ground.

From this generalized dicotyledonous ancestor, the adaptive shift probably took place in two stages. The first was the origin of rosette-forming herbs, having elliptic, entire leaves with the beginnings of parallel venation, a trend analogous to the one that can be seen in the Gentianaceae with reference to the less-specialized family Loganiaceae. By analogy, we can suppose that this first stage was associated with occupancy of the moist shores of streams or lakes. The evolutionary line that underwent this shift may well have diverged further toward the aquatic habitat, and have given rise to the Nymphaeales.

The second stage, accompanied by the occupation of pools or shores that were flooded during the wet season and desiccated during the dry season, included the shift to hypogeal germination, the evolution of a cotyledonary tube accompanied by the elimination through reduction of the blades of the original cotyledons, and the enlargement of the leaf rosette and the central crown, giving rise to the condition of many separate vascular bundles in both caudex and leaf bases.

The nature of the earliest monocotyledons

The earliest monocotyledons, which evolved as a result of this adaptive shift, might be reconstructed as follows. They were rosette-forming phanerophytes with a branching caudex, adapted to tropical or subtropical climates with seasonal rainfall. They occupied sites that were saturated or flooded during the wet season and desiccated during the dry season. Their wood either lacked vessels or had vessels the elements of which had primitive, scalariform end perforations. Their leaves were elliptic and entire, and the venation was of the closed type, with numerous parallel main veins converging at the tip, as well as abundant cross veins, forming a network, but without free endings of veinlets, as in most leaves of dicotyledons. Leaves of Cyclanthaceae, Liliales such as *Aspidistra* and *Curculigo,* Aroids such as *Spathiphyllum,* and seedling leaves of many palms could serve as models. The stems had scattered vascular bundles, which probably had a vestigial intrafascicular cambium, but no other kind of cambium existed.

Their inflorescence was probably intermediate between the dichasial or cymose and the indeterminate, racemose organization, as is true of some palms. Their flowers were trimerous, containing an undifferentiated perianth of six tepals in two whorls, six or nine stamens, or possibly more in some species of the ancestral complex. The pollen grains were monocolpate and binucleate. The gynoecium consisted of three separate carpels, each of which contained relatively few ovules. The seeds had abundant endosperm. Germination was hypogeal, the distal end of the cotyledons remaining in the seed, and its proximal end forming a cotyledonary tube that buried the hypocotyl and radicle into the ground (type II germination of Boyd; see Fig. 13-1).

The evolution of vessels

Three points of this hypothetical reconstruction require further discussion. The first is the question whether vessels originated separately in monocotyledons and dicotyledons, and whether, as Cheadle (1953) has postulated, they appeared in the monocotyledons first in the roots and only later in the shoots. In my opinion, this question is unanswerable. As Cheadle himself recognizes, the only monocotyledons that are completely devoid of vessels are extreme aquatic forms in which the vascular tissue is very much reduced. These forms are almost certainly derived, and, as Cronquist (1968) has correctly pointed out, they very likely would have lost their vessels as a consequence of this reduction, even if their terrestrial ancestors had them. Cheadle's second group, monocotyledons that have vessels only in the roots and not in the shoot axes or leaves, are a heterogeneous group, some of which have relatively primitive and others much more specialized vessel elements. They include both terrestrial and aquatic forms. Finally, some of the plants, such as palms, that have vessels in both roots and axes, nevertheless have vessel elements that are as primitive as those found in species having vessels only in the roots.

These facts permit the formulation of two hypotheses, which can be regarded as equally probable. Possibly, as Cheadle suggests, the dicotyledonous ancestor of the monocotyledons was vesselless, and vessels appeared in the roots during the adaptive shift postulated above. Their adaptive value could have been as a means of securing prompt and rapid translocation of water and minerals throughout the root system and into the bases of the leaf rosettes, at the time when the plants were resuming growth after the dry period. Later, some monocotyledons evolved the ability to produce vessels in shoots as well as roots, whereas others never did. Judging from Cheadle's lists of families, no obvious correlation exists between the presence vs. absence of vessels in the shoots and either growth habit, climatic distribution, or specialization with respect to the flowers.

An alternative hypothesis is that the dicotyledons, both woody and herbaceous, that were ancestral to monocotyledons, as well as the earliest monocotyledons themselves, resembled the more primitive modern palms in possessing primitive vessel elements throughout the plant. In those later evolutionary lines that retained the terrestrial, phanerophytic growth habit, as well as in many of the more advanced hemicryptophytes and cryptophytes, vessels were retained and their elements became more specialized. In many of the more advanced lines that were characterized by reduction in size or persistence of the shoots, and particularly in those that became aquatic, vessels were lost from either the shoots or the entire plant as a result of reduction in the amount of vascular tissue that was elaborated.

Since the forms in which these primary changes occurred are believed to be completely extinct, the only possible way of deciding between these two hypotheses would be through the discovery of ancient fossil monocotyledons in which the vascular tissue was preserved.

Syncotyly vs. heterocotyly

Since the 19th century, botanists have been divided as to the way in which the single cotyledon of the monocotyledons evolved from the paired cotyledons that are characteristic not only of nearly all dicotyledons but also of most gymnosperms, except for certain conifers, and that probably existed in the ancestors of the angiosperms as well as in the earliest representatives of the phylum. One group, led in more recent years by Arber (1925), Metcalfe (1936), Eames (1961), and Takhtajan (1969), maintains that the single cotyledon of monocotyledons is homologous to one of the two cotyledons found in dicotyledons, and that the second cotyledon gradually became reduced and aborted. This is the viewpoint held by most modern botanists who have considered the question. The alternative hypothesis, elaborated by Sargant (1903) and maintained by Johansen (1945), Souèges (1954), and, in a modified form, by Cronquist (1968), is that the single cotyledon of monocotyledons represents the union, through intercalary concrescence, of the petioles of the two ancestral cotyledons, the blades having become reduced and suppressed.

I favor the second hypothesis, that of syncotyly, for the following reasons. In the first place, intercalary meristems are a characteristic feature of monocotyledons, and probably played an important role in the evolution of the distinctive form of seedling development that exists in the class. Hence the modification of the cotyledons to form a single cotyledonary tube via intercalary concrescence of their margins would have been according to adaptive modification along the lines of least resistance. A much more complicated sequence would be the abortion of one cotyledon, followed by intercalary concrescence of the margins of the single refaining cotyledon, to form the cotyledonary tube.

The simplicity of this change from the developmental point of view is well illustrated by the experiment of Haccius and Trompeter (1960) on *Eranthis hiemalis*, a dicotyledon that has a pronounced cotyledonary tube. Through the application of 2,4-dichlorphenoxyacetic acid (2,4-D), they were able to convert the cotyledons phenotypically into a single cotyledonary tube, with suppression of the blades. Although the authors deny that this conversion is at all related to the evolution of the condition in monocotyledons, my own interpretation is that it is highly significant. Anatomically, the cotyledonary tube produced by 2,4-D action in *Eranthis* is very much like the normal cotyledon of monocotyledons such as *Anemarrhena*.

Second, the anatomical evidence of Sargant is to me much more convincing than other authors, particularly Arber, have been willing to admit, and Arber's objections can be refuted. Sargant's evidence for syncotyly is by no means confined to the fact that in many monocotyledons the single cotyledon contains two vascular bundles throughout its entire length. More important is the fact that in seedlings such as that of *Anemarrhena* the two bundles in the basal part of the cotyledonary sheath lie opposite each other, and join with the bundles of the tetrarch root in a regular, alternating fashion. This is exactly similar to the way in which the two bundles of the united cotyledonary petioles join with the root bundles in dicotyledons such as *Eranthis*, in which a cotyledonary sheath has been formed by their intercalary concrescence (Fig. 13-2). The fact that many cotyledons of dicotyledonous plants have two vascular bundles in their petioles is well known, but is irrelevant to Sargant's argument. As Metcalfe (1936) has shown in the case of *Ranunculus ficaria*, which has probably acquired monocotyly by means of abortion of one cotyledon, the two bundles of the single cotyledon do not assume the opposite position that exists in the sheath that is formed by the fused petioles in other Ranunculaceae. Moreover, in forms like certain genera of palms, such as *Thrinax*, *Chamaerops*, and *Phoenix*, the more numerous bundles are of an even number, 4, 8, or 10.

Arber was fully aware of these facts, but she rejected this evidence by an argument that was based entirely upon idealistic morphology, and is incompatible with any hypothesis that is based upon the concept of evolutionary modification of adult structure through genetic alterations of developmental patterns. In her words (1925:175): "The significance of the vascular duality in the cotyledon, and also of the behavior of the strands in the hypocotyl, has, I think, been misunderstood, simply because Sargant and certain later workers . . . have traced the bundles from above downwards—as in the description cited above—and have hence been led into the fundamental error of treating the root-stele as *made up of* cotyledonary and plumular strands. De Candolle, in 1827, laid down the principle that each organ

Fig. 13-2. Abnormal syncotylous and monocotylous seedlings obtained by treating developing embryos of *Eranthis hiemalis* with growth substances: (*A–C*) normal embryo; (*B*) cross section at base of hypocotyl; (*C*) three-dimensional reconstruction of anatomy of the upper part of the hypocotyl and the base of the cotyledons; (*D–F* and *G–I*) corresponding parts of seedlings derived from seeds that were treated at the stage of proembryos 0.1 mm long with 2,4-dichlorphenoxylacetic acid (2,4-D); (*J*) cross section of the lower portion of a seedling of *Anemarrhena asphodeloides* Bunge, a monocotyledon, showing strong resemblance to (*F*); (*K*) similar cross section of the lower part of the cotyledon of a seedling of *Zygadenus elegans* Pursh, another monocotyledon, showing strong resemblance to (*I*); (*L–P*) monocotylous seedling and embryos obtained by treating seeds of *Eranthis hiemalis* at the same stage with phenyl boric acid, which induces heterocotyly. In (*L*), note absence of marginal fusion that is evident in (*D*) and (*G*), and groove (*g*) at apex of cotyledonary petiole; also note differences between (*N, P*), and (*J, K*). Symbols: *m*, midvein of cotyledon, l_1, l_2, lateral veins; *pl*, plumule; co_1, first cotyledon; co_2, second cotyledon; lf_1, first leaf. (*A–I* from Haccius and Trompeter 1960; *J, K* from Sargant 1903; *L–P* from Haccius 1960.)

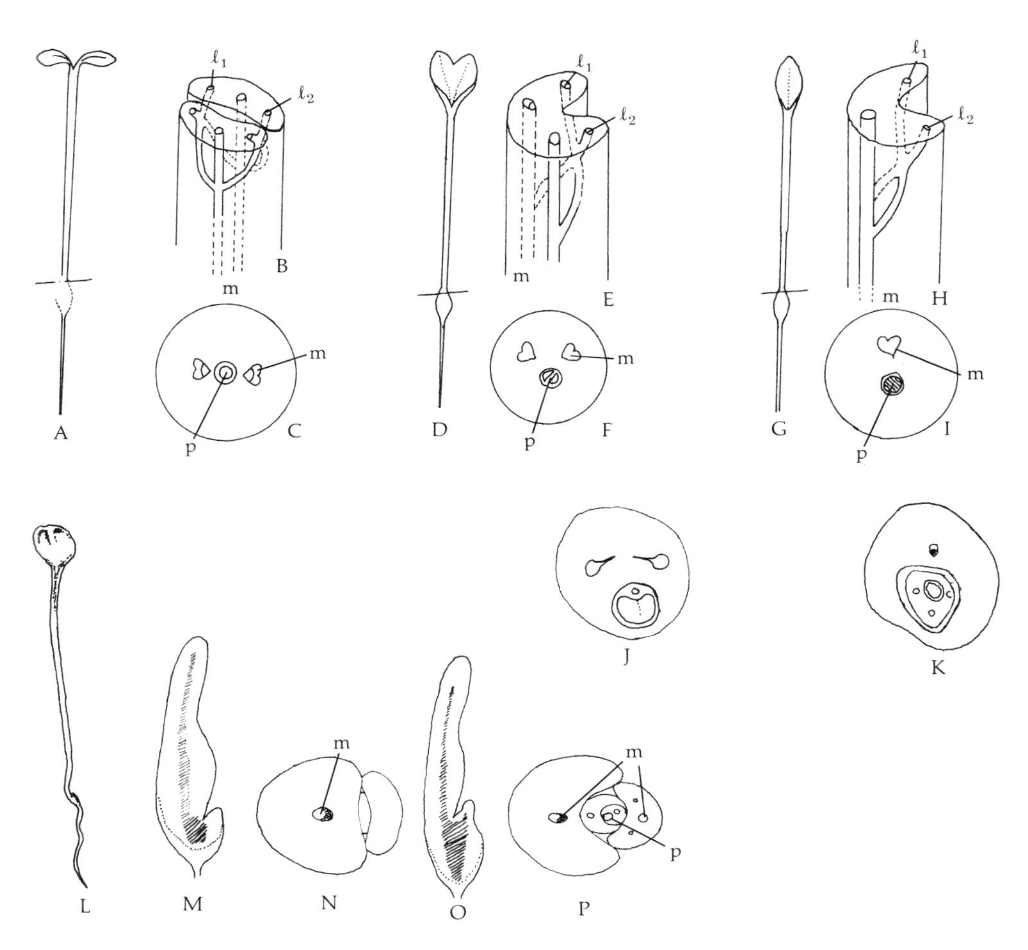

should be considered 'comme se développant ou sortant de celui qui sert de support immédiat.'"

This is, I believe, a misinterpretation both of Sargant's deductions and of the true significance of her observations. Whether one views a structure from the bottom upward or from the apex downward, its adult structural pattern must be regarded as the outcome of a complex pattern of development. If, therefore, two adult patterns are so similar that a very similar and complex developmental sequence can be observed or inferred, the probability is high that they represent genetic modifications of a single ancestral pattern. I believe that this is true of the remarkable similarity between the hypocotyl and the basal portion of the cotyledonary tube or sheath in a great number of monocotyledons, including *Anemarrhena,* and in many dicotyledons in which a cotyledonary sheath has obviously been formed by intercalary concrescence of the petiole margins of two otherwise separate and well-developed cotyledons.

Further evidence in favor of the homology of the cotyledon of monocotyledons to the petioles of the two cotyledons of dicotyledons has been presented by Suessenguth (1920) and Souèges (1954) and is discussed by Johansen (1954). This is the fact that in many young embryos of monocotyledons, the terminal zone from which the single cotyledon is differentiated corresponds to the zone or group of cells from which both cotyledons are differentiated in the young embryo of dicotyledons. Although Haccius (1952) and Baude (1956) have discarded this evidence on the ground that in various embryos of monocotyledons this zone may appear to be terminal, lateral, or in some intermediate position, the basic fact of comparison and similarity between the zones of initiation in monocotyledonous and dicotyledonous embryos still remains.

Arber's final conclusion (1925:179) is that we need not assume evolutionary homology between cotyledons of dicotyledons and monocotyledons. This is clearly the interpretation of an idealistic morphologist who, basically, did not believe in either natural selection or evolutionary continuity. If, as I and many of my contemporary botanical colleagues believe, evolution was a continuous process and the two classes of angiosperms had a common ancestry, then the evolutionary homology of the first appendage or appendages produced by the embryo is a necessary consequence of this assumed fact. Furthermore, embryos having a pattern of development similar to those of monocotyledons do not exist in any other seed plants, whereas the dicotyledonous pattern is similar to that of many gymnosperms belonging to various classes. Consequently, the evolution in one way or another of the cotyledon of monocotyledons from the two cotyledons of ancestral seed plants, as postulated in this chapter, is by far the most reasonable and probable explanation of its origin.

The homologies of various two-bundled organs in monocotyledons

If the single cotyledon of monocotyledons has resulted from intercalary concrescence of two cotyledonary petioles, then what are the homologies of other foliar organs in the monocotyledons that contain two nearly equal principal vascular bundles, rather than a single midrib? The structures included are some of the prophylls that subtend branches of side shoots in many different species belonging to various families, though by no means all of them, as pointed out by Blaser (1944). In addition, they include the coleoptile that is a characteristic feature of the embryo and young seedling of grasses, and the palea of the grass flower. The first argument advanced for the dual nature of the cotyledon, that it is the natural consequence of the highly developed intercalary meristem in monocotyledons and of adaptive modification along the lines of least resistance, applies equally well to all of these organs.

The prophyll is functionally analogous to the cataphylls of dicotyledons, since it serves as a protection for the young bud, probably against the attacks of insects. Its two bundles may either converge at the apex (*Tradescantia,* most grasses) or, as in certain grasses (*Zea, Festuca*), as well as in *Potamogeton natans, Calla* spp., *Tofieldia,* and *Asparagus Sprengeri* (Rüter 1918), the two principal bundles of veins may end separately in a bifid apex (Fig. 13-3). In either case, the bundles lie almost or quite opposite each other near the base of the prophyll and enter the vascular system of the principal shoot from opposite directions. Their orientation is, therefore, comparable to that of the two cotyledonary bundles.

Arber (1925:131–135) has argued in favor of the unitary nature of the prophyll on the following grounds. First, she considers that the two-keeled form is the result of pressures during development that inhibit its growth. This argument can be shown to be invalid by even a cursory examination of the early development of the prophyll. The procambium for its two bundles is laid down when the primordium is at most a few hundred microns in length, and before it has come into contact with any other structures. Second, she refers to anatomical studies by Gravis of the prophyll of *Tradescantia* in which additional vascular bundles and two others, one of them nearly as prominent as the so-called "median" bundle and the other much less prominent, are regarded as laterals. Nevertheless, even in these prophylls the "median" and the nearly equivalent "lateral" bundles occupy positions nearly or quite opposite to each other, and so correspond to the two keel-forming bundles that exist in other prophylls. Unfortunately, the relation of these bundles in the *Tradescantia* prophyll to those of the main shoot has not been described.

A third argument advanced by Arber is based upon the fact that one of the two main bundles of the prophyll in *Tradescan-*

Fig. 13-3. Bifid apex of prophyll in (*A*) *Zea mays* L. and (*B*) *Coix lachryma-jovi* L., showing termination of the two principal veins in separate points. (From Rüter 1918.)

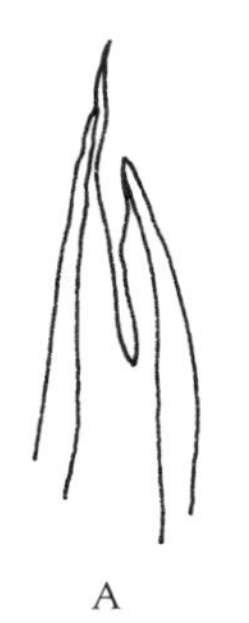

tia—as well as in *Avena* and other grasses (unpublished personal observations)—subtends an axillary bud, whereas the other does not. She argues that, if each of the two main bundles is homologous to the median bundle of an ancestral foliar bract, both of them should subtend axillary buds. This, however, is by no means necessary. In most if not all monocotyledons, and certainly in *Tradescantia* and various grasses, many leaves do not subtend axillary buds at all.

Consequently, although the homology of the prophyll of monocotyledons to two concrescent bracts has certainly not yet been clearly demonstrated, no convincing evidence against this interpretation has been produced. The coleoptile of grasses, which is a specialized form of prophyll, is subject to the same arguments.

With respect to the palea of the grass floret, the evidence for its dual origin via intercalary concrescence is more convincing (J. Schuster 1910, Janchen 1938, Butzin 1965). The principal evidence for it is the fact that in *Streptochaeta*, a relictual monotypic genus of grasses that combines several specialized with many other obviously primitive characteristics, the "palea" consists of two separate organs (Page 1951, Butzin 1965). In addition, the genus *Ampelodesmos*, which is primitive with respect to its many-flowered spikelets, undifferentiated lemmas, and three lodicules, has a palea that is more strongly bifid than in most grasses, and in the very young floret appears as two nearly separate primordia, each of which already contains a procambial strand (unpublished personal observations). Finally, in the palea of nearly all species of grasses, the two veins end separately in the bifid apex, and do not converge as is characteristic of the veins of normal foliage leaves. For all of these reasons, the interpretation of the palea as the result of intercalary concrescence between two sepals of the ancestral flower, as postulated by Schuster, Janchen, and Butzin, as well as by Takhtajan (1970), appears to me the most plausible one. If this interpretation is accepted, then the grass flower must be regarded as primitively trimerous with a two-whorled perianth, the outer whorl being represented by the palea and the inner by the lodicules. In most grasses the third member of each whorl, that nearest to the axis of the spikelet, has been suppressed, but in bamboos and some other grasses having relatively unspecialized flowers all three members of the inner whorl, but only two of the outer whorl, are present.

The phyllode theory of the origin of the leaf

Starting with A. P. de Candolle in the early 19th century, a number of botanists have suggested that the leaf of monocotyledons is not truly homologous to that of dicotyledons, but to its sheathing base or petiole. This "phyllode theory" was elaborated by Arber (1925) and has been accepted by Cronquist (1968), in connection with his belief that the earliest monocotyledons were aquatic. In my opinion, it is unnecessary. Arber's elaboration of it is a typical example of reasoning based upon typological morphology that disregards evolutionary continuity. The examples of dicotyledonous leaves that she chose for her comparisons—*Petasites, Oxalis, Clema-*

tis, Acacia, various Umbelliferae—all have a high degree of differentiation between blade and petiole, much more than exists in *Drimys* subg. *Tasmannia, Hibbertia* sect. *Cyclandra,* and other dicotyledons that are believed to approach more nearly to the primitive ancestors of the class. Moreover, the structure and development of the typical dicotyledonous leaf is more elaborate and specialized than that of most other leaves of vascular plants. The condition of tertiary and quaternary veins and veinlets diverging from one another at right angles, as well as the innumerable free endings of veinlets, are specializations that exist elsewhere only in *Gnetum* and a few ferns that have highly specialized leaves, such as the genus *Tectaria* (Polypodiaceae). They are, moreover, absent from the leaves of a few dicotyledons, such as *Quiina pteridophylla* (Foster 1952), and from most monocotyledons (W. Schuster 1910, Ertl 1932). Networks of tertiary and quaternary veinlets, including many free endings, are found in a few monocotyledons, such as *Arisaema* (Araceae), *Smilax* (Liliaceae), and some species of *Dioscorea,* but these are almost certainly specialized end lines, and with respect to other anatomical characters, including those of the petioles and other parts of the leaves, they diverge widely from all dicotyledons.

Kaplan (1970) has recently revived the phyllode theory as a result of his developmental studies of the leaf of *Acorus calamus* (Araceae). He bases his interpretation largely upon a striking and undeniable resemblance between the developmental patterns of this unifacial foliage leaf and the phyllode of *Acacia.* The latter organ is unquestionably derived by modification of the petiole and rachis of a compound leaf, through alterations of its developmental pattern. However, a direct evolutionary homology between foliar organs of this specialized legume and the equally specialized leaves of any monocotyledon would be very difficult to maintain. The two structures have been derived independently by similar modifications of initial foliar organs belonging to very distantly related plants. Consequently, the fact that the phyllode of *Acacia* was derived from a petiole plus rachis is not *ipso facto* to be regarded as indicating necessarily that the less differentiated forerunner of the unifacial leaf of *Acorus* was also of this nature.

In view of these facts, the following hypothesis appears to me to be more probable than the phyllode theory. The leaves of the original angiosperms are believed to have been elliptical, obovate, or spatulate in outline, and tapered at the base to an indistinct petiole. They had a netted venation which lacked free endings of veinlets and in which the ultimate veinlets were differentiated almost simultaneously, as in *Quiina.* From this generalized condition, evolution in the earliest dicotyledons proceeded rapidly toward a stronger differentiation of blade and petiole, a greater differentiation between primary, secondary, tertiary, and quaternary veins associated with the generalized distribution and long persistence of plate meristem (Pray 1955*a*), and the general occurrence of numerous free endings of veinlets. In the monocotyledons, on the other hand, the predominant trend was toward localization of the intercalary meristem near the base of the leaf and its persistence only in this basal position, and consequently less development of veinlets (Pray 1955*b*). As a result, the extensive branching of veins and veinlets, as well as the hierarchy of veins and veinlets, was developed in monocotyledons to a much lesser degree than in dicotyledons, and the leaf venation became dominated by the numerous unbranched primary veins. The resemblance between the monocotyledonous leaf and the petioles, rachises, sheaths, and cataphylls of some dicotyledons is explained on the basis of this hypothesis as evolutionary convergence, which took place in some of the more advanced members of the class, such as Leguminosae, Umbelliferae, and Compositae.

Trends of Evolution Within the Monocotyledons

The most distinctive features of monocotyledons, in addition to the development of their embryos and seedlings, are associated with the anatomical structure and method of growth of the adult plant. Consequently, evolutionary trends within the class are associated with vegetative differentiation to a far greater degree than in dicotyledons. As has been clearly shown by Holttum (1955) and Madison (1970), the directions of these trends have been determined by the distinctive characteristics that must have arisen when the members of the class first became differentiated from their generalized angiospermous ancestors. They are, therefore, prime examples of evolutionary canalization based upon the conservation of organization and adaptive modification along the lines of least resistance.

Trends in the vegetative structure

The absence or poor development of cambial meristem has three consequences. In the first place, it greatly restricts the ability of the stem to increase in thickness. Such increase can come about only through the expansion of leaf bases at a relatively early stage in ontogeny, as in palms (Tomlinson 1970), or through the origin of a generalized cambial tissue that permits the differentiation of new vascular bundles, as in *Dracaena.* The palm method apparently suppresses or greatly restricts the ability of the stem to form branches, whereas that found in *Dracaena* is associated with much slower growth than that of arboreal angiosperms. Secondly, the absence of a solid vascular cylinder makes the young stems so weak that they can grow upward successfully only if they have previously acquired great thickness, as in the larger palms, or if they are supported by massive surrounding leaf sheaths, as in Musaceae and the larger grasses, including bamboos.

The third consequence of the absence of cambium is that the shoots of monocotyledons can elongate more rapidly than those of most dicotyledons. In the tropics, this favors evolution in the direction of herbaceous climbers, as in Araceae, Commelinaceae, the genus *Freycinetia,* the rattan palms (*Calamus*), and the climbing bamboos (*Chusquea*). In both tropical and temperate regions, the evolution of slender aquatics is favored. It is no accident that the only seed plants that compete with marine algae are monocotyledons such as *Zostera, Phyllospadix,* and *Posidonia,* which have greatly elongated internodes and leaves. Finally, the capacity for rapid shoot growth preadapts monocotyledons for extensive vegetative multiplication by means of rhizomes or stolons. Species of Gramineae, Cyperaceae, and Juncaceae are the principal binders of sand dunes, as well as the pioneers that enable the land to encroach upon lakes, ponds, and estuaries. Rhizomes of species belonging to these families also form the firm foundation of marsh and bog vegetation, while those of grasses, as long as they are not disturbed by man and his domestic animals, keep dry plains and savannas from becoming "dust bowls."

As Holttum has pointed out, all of these more specialized growth forms of monocotyledons have most probably been derived by adaptive radiation from ancestral phanerophytes of tropical distribution, of which the relatively thick underground stems branched in a sympodial fashion. The predominant trend in monocotyledons has been from thicker and shorter to thinner and more elongate stems, with respect to both the aerial and the subterranean parts of the plant. The largest of the climbing phanerophytes, such as *Calamus, Freycinetia, Philodendron, Monstera,* and *Chusquea,* are relatively specialized with respect to their reproductive structures. The same is true of the deep-water aquatics, such as the majority of Najadales, and of the most extensively rhizomatous monocotyledons, such as *Typha, Juncus balticus,* the *Scirpus validus* complex, *Carex* sect. *Vesicariae,* and grasses such as *Phragmites, Ammophila, Cynodon, Stenotaphrum,* and *Pennisetum clandestinum.*

This trend toward slenderness, however, is by no means unidirectional and irreversible. Though palms are certainly not derived from more slender Liliales, they may have had less massive ancestors among the primitive extinct monocotyledons. On the other hand, the other "rosette trees" that exist among monocotyledons, such as *Yucca brevifolia, Nolina* spp., and the genus *Dracaena,* have so many characteristics in common with Liliales with respect to both leaves and flowers that they are almost certainly derived from smaller, more slender representatives of this order. Also, as Arber (1925) has pointed out, the larger bamboos are a unique growth form that shows many specialized characteristics. Their massive rhizomes and thick stems are relatively late specializations. The ancestral members of the Gramineae may have been relatively small plants, slender with loosely branching sympodial bases, that occupied semiarid, perhaps montane habitats (Stebbins 1956*a*, Tzveliov 1969, Serebryakova 1971).

Another series of vegetative specializations in monocotyledons are a consequence of the presence of numerous vascular bundles, particularly in the leaf bases, and the development of leaves with broad, sheathing bases. This condition can serve as an initial modification that promotes evolution of bulb and corm geophytes, in response to highly seasonal climates that have a long unfavorable period of cold or drought. The overwhelming majority of angiosperms that possess bulbs or corms are monocotyledons, most of them belonging to the single order Liliales. In this order, the trend toward geophytes is associated with the appearance of stomatal complexes in which the guard cells are not surrounded by subsidiary cells (Stebbins and Khush 1961). This reduction of subsidiary cells, in turn, is associated with the rapid elongation of the epidermal cells that surround the stomatal complexes, which is a by-product of selection for rapid growth of young leaves as growth is resumed after the dormant period. In leaves that produce subsidiary cells in their stomatal complexes, these cells are differentiated through interaction between the guard-cell mother cell and the neighboring cells of the epidermis (Stebbins and Shah 1960). Most probably, the rapid elongation of the epidermal cells in most Liliales causes them to be in a nonreceptive condition at the time when they would normally be recieving the stimulus for division from the guard-cell mother cell. Genera of Liliales that do not show this rapid elongation of epidermal cells, but still lack subsidiaries, are believed to have been descended from ancestors that did have this elongation, and in which the genetic capacity for interaction was lost by mutation, since it could not be expressed.

An interesting specialization in certain bulb geophytes of the Liliaceae, particularly the genera *Tulipa, Erythronium,* and *Gagea,* is the formation of "droppers" (Arber 1925). In these plants, a young axillary bud becomes adnate to the base of a leaf or prophyll, which then elongates with positive geotropism and forms a long tube that buries the bud farther in the ground. This bud then forms new bulb scales and adventitious roots, and so de-

Fig. 13-4. "Droppers" in *Tulipa:* (*A*) young plant in August, showing beginning of dropper formation; (*B*) young plant in March, showing well-developed "dropper." (From Arber 1925.)

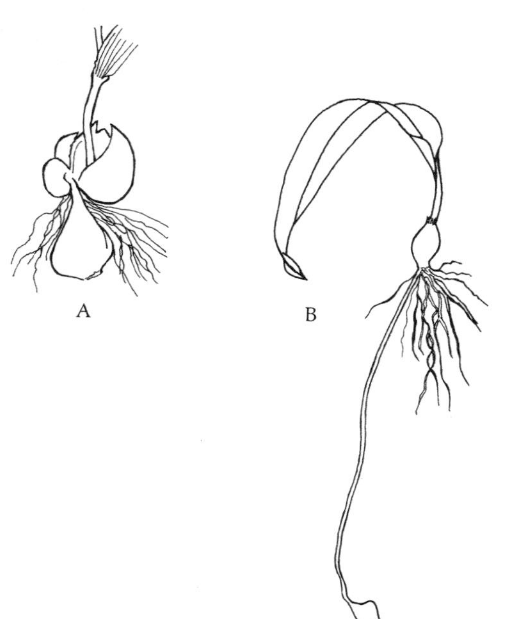

velops into a new bulb (Fig. 13-4). The adaptive value of this complex mechanism is obvious, since it affords additional protection for the young bulb from both desiccation and the attacks of animal predators.

The developmental sequence involved in the formation of droppers forms a most interesting parallel to the sequence by which the plumule is buried by the geotropic action of the cotyledonary tube, as described above. One is tempted to speculate that the genetic mechanism for dropper development is actually homologous to that for early seedling growth and involves the same sequence of activation and inhibition of the growth substances.

Trends with respect to germination and seedling development

The various modifications of early seedling development are of particular significance, since they affect the developmental stage at which selective pressures would be expected to be maximal (see Chapter 5). They have been well described in a great variety of monocotyledons by Boyd (1932). Starting with the tubular, geotropically elongating cotyledon, which buries the plumule in the ground without emerging, so that germination is hypogeal, five divergent trends can be recognized. The first of these is retention of the geotropic cotyledonary tube but increase in elongation of the cotyledon, so that its upper portion emerges from the ground, carrying with it the seed coat, and germination is epigeal. This kind of germination can easily be seen in cultivated species of *Allium,* and is found in most bulb- and corm- or rhizome-forming temperate genera of Liliales. The anatomical modifications from the more generalized hypogeal type are minimal.

The second trend likewise involves the shift from hypogeal to epigeal germination, but also includes great reduction and elimination of the cotyledonary tube as well as of the geotropic elongation of the cotyledon. This condition is found in aquatic and marsh plants. It exists in all of the Alismatidae, as well as the Phylidraceae, Pontederiaceae, Eriocaulaceae, and Juncaceae. In addition, it has evolved independently in the epiphytic genera of Bromeliaceae. This trend involves a marked reduction in the width of the cotyledon, and the reduction of the cotyledonary bundles from two to one.

A somewhat similar trend is the suppression of the cotyledonary tube and the development of a broad, ligulate cotyledon in the genus *Dioscorea* (Lawton and Lawton 1967). In view of the specialized nature of this genus, and the absence of similar conditions in other groups of monocotyledons, I cannot accept the hypothesis that the first seedling leaf of *Dioscorea* is homologous to a modified second cotyledon.

The fourth trend includes the retention of hypogeal germination, but the elimination of geotropic elongation, so that the plumule remains near the surface of the ground. This condition is found chiefly in tropical, mesic groups. It is characteristic of the Commelinaceae, Haemodoraceae, Cannaceae, and Zingiberaceae, and exists also in some tropical and subtropical Liliales, such as *Kniphofia, Chlorophytum,* and *Gloriosa.* Anatomical modifications associated with this shift are absent or minimal.

Finally, there exist highly specialized families of monocotyledons in which germination is hypogeal and the cotyledon remains in the seed, where it apparently serves the function of aiding in the digestion of the stored food in the endosperm and supplying the growing seedling with both nutritives and growth substances. This condition, which is developed primarily in the Cyperaceae and Gramineae, also involves a reduction in the anatomy of the cotyledon, which usually contains only a single bundle.

Trends in the inflorescence and flowers

With respect to the inflorescences and flowers, three principal kinds of trends can be recognized, associated, respectively, with the three different methods of pollination—animal vectors, water, and wind. These trends have each been repeated separately in a number of different evolutionary lines. Furthermore, they are not sharply distinct from one another, and many intermediate conditions can be recognized. In addition, they have been subject to frequent reversals. As mentioned in Chapter 4, insect pollination has arisen frequently in basically wind-pollinated groups, whereas individual species and genera belonging to basically insect-pollinated groups have from time to time shifted to wind pollination. Nevertheless, these trends are sufficiently different from one another, and associated with different syndromes of morphological characteristics, that they are worth mentioning.

Adaptation for pollination by insects and other animal vectors, associated with conspicuous perianths, often with increase in size of anthers, and usually, though by no means always, with gynoecia having large numbers of ovules, has been most conspicuously developed in three lines: one leading from primitive Liliaceae to more advanced members of this same family (in the broadest sense), and from thence to the Iridaceae and Orchidales; a second leading from unknown ancestors to the Musaceae, Zingiberaceae, Marantaceae, and related families; and a third represented by the Commelinaceae. Each of these trends has been accompanied by extensive ramification into various side lines. More restricted trends in the same direction are represented by the Bromeliaceae, Alismatales, and a few other families. The situation in the Araceae is somewhat anomalous, since most if not all of them are insect pollinated, but nevertheless their individual flowers have either very inconspicuous perianths or none at all, and the gynoecium consistently possesses a single ovule. In many species of aroids, the inflorescence is conspicuous because it is subtended by a colored bract or spathe, and the flowers are so crowded on their axis or spadix that with respect to pollinators they could function like a single flower. This situation could be explained on the assumption that the ancestors of the Araceae were wind pollinated, and that their insect pollination is a secondary reversion. In other families of the subclass Arecidae, such as the Palmae, Pandanaceae, and Cyclanthaceae, those species that are pollinated by insects are probably also descended from wind-pollinated ancestors.

Two conspicuous differences can be detected between the trends toward specialization for insect pollination in monocotyledons and in dicotyledons. In the first place, the differentiation of the perianth into a protective green calyx and a showy, attractive corolla is much less frequent in mono- than in dicotyledons. Although present in the overwhelming majority of animal-pollinated dicotyledons, it is found in monocotyledons only in the Alismataceae, Limnocharitaceae, Hydrocharitaceae, Bromeliaceae, Commelinaceae, Xyridaceae, and some genera of Liliales, such as *Calochortus* and *Trillium.* This difference can be explained on the assumption that in both classes the original perianth was not differentiated into calyx and corolla, and that the selective pressure favoring such differentiation was stronger in most groups of dicotyledons than in monocotyledons. This is understandable when one realizes that in monocotyledons the weak young stems and inflorescences are often surrounded by protective leaves and sheathing bracts, so that the adaptive value of separate protective coverings for flower buds is less.

A second difference, which is harder to understand, lies in the relation between zygomorphy and epigyny. In both classes the most specialized of the insect- or bird-pollinated groups consist partly or entirely of species that are both zygomorphic and epigynous. On the other hand, the dicotyledons include several large families in which the majority of species combine a zygomorphic flower, either polypetalous or sympetalous, with a completely superior ovary. The principal groups are Caesalpinoideae and Papilionoideae of the Leguminosae, Violaceae, Sapindaceae, Vochysiaceae, Polygalaceae, Tropaeolaceae, Balsaminaceae, Labiatae, Scrophulariaceae, Acanthaceae, Orobanchaceae, Gesneriaceae, and Lentibulariaceae. These families belong to diverse orders and subclasses, and in some instances the zygomorphy appears to have arisen within the family. Consequently, in dicotyledons, groups having superior ovaries have repeatedly given rise, without accompanying changes in the gynoecium, to lines having zygomorphic flowers. In monocotyledons, on the other hand, the only flowers that are both zygomorphic and hypogynous are found in a few members of the Commelinaceae and the Pontederiaceae, the latter being only weakly zygomorphic. In the Liliales, where the trend from actinomorphy to zygomorphy has taken place repeatedly, and can be followed with relative ease, the nearest relatives of the zygomorphic forms are actinomorphic and epigynous. No explanation can be suggested for this puzzling difference.

The trend toward water pollination has taken place chiefly in the Najadales. It is associated with reduction in all parts of the flower and the frequent presence of unisexual flowers. The inflorescences are usually simple racemes or spikes, and may be reduced to small cymes or solitary flowers. It is also associated with trinucleate pollen grains and with seeds that lack endosperm.

The third series of trends, those toward wind pollination, are found chiefly in two very different groups of orders, one of them relatively primitive and the other including some of the most advanced and successful members of the class. As expected, wind pollination is associated with reduction of the flowers, particularly the perianth but also the gynoecium, which is commonly reduced to a single ovule and seed. In some groups the inflorescence is a simple raceme or spike, as in Typhales and Pandanales, but more commonly the wind-pollinated monocotyledons have evolved extensive compound inflorescences, which are umbellate or paniculate.

The first group of primarily wind-pollinated orders includes the palms and Pandanales, which, although they are rather distantly related to each other, nevertheless may well have been descended from a wind-pollinated ancestor. In fact, the presence of wind pollination in these primitive monocotyledons raises the question whether wind pollination may not have been the original method in the subclass as a whole. In this connection, the small size and inconspicuous nature of the individual flowers in the most primitive Liliales may be significant, as also the strong indication that a perianth which is differentiated into calyx and corolla is a secondary acquisition that has evolved independently in relatively few members of the subclass.

The remaining two orders of Arecidae, the Cyclanthales and Arales, are entomophilous so far as is known (see Harling 1958), but the structure of their inflorescences and flowers renders highly plausible the suggestion that they are descended from wind-pollinated ancestors. As has already been mentioned, the shift from wind to insect pollination has probably taken place several times during the evolution of the palms.

The second and larger group of wind-pollinated monocotyledons are specialized orders of the subclass Commelinidae, the Restionales, Juncales, Cyperales, and Poales. They have probably been derived from more primitive insect-pollinated members of the subclass, perhaps in two different lines, one represented by the Restionales and Poales, and the other by the Juncales and Cyperales (Takhtajan 1969). In contrast to the wind-pollinated orders of Arecidae, which generally have binucleate pollen, the wind-pollinated Commelinidae, except for some of the Restionaceae and perhaps smaller families that have not yet been investigated, have trinucleate pollen. The Juncaceae form an exception to the general rule of one ovule and seed per gynoecium, since their capsules have many ovules and seeds. The suggestion offered by Walker (1971) with respect to the Annonaceae, that the retention of the mature pollen in tetrads that are shed as units is an adaptation to a secondary increase in ovule number, since it increases the probability that a small amount of pollen will provide enough male nuclei to fertilize many ovules, applies well to the Juncaceae. The combination of wind pollination, pollen in tetrads, and many ovules per gynoecium is a syndrome of characters that among the monocotyledons is unique to this family.

The most specialized of these families, the Cyperaceae and Gramineae, are among the most successful and actively evolving of all flowering plants. The grasses, in particular, illustrate especially well the cardinal principles of evolutionary success in angiosperms (Arber 1934). They combine great flexibility and adaptability in their vegetative parts with a structural ground plan that permits adaptive radiation to the evolution of treelike forms, intermediate tufts that can compete with secondary undergrowth in forest or savanna, horizontally spreading perennials that can form either dense sods or loose ground cover in swamps or in the forest floor, and annuals that can form aggressive weeds in disturbed and cultivated areas. With respect to their reproduction, some species have evolved elaborate rhizomes and stolons that can perpetuate individual genotypes for an indefinite period of time, and can spread them over vast areas. Many other grasses rely principally upon seed reproduction, and have evolved a variety of mechanisms for seed dispersal by means of any one of the available agents. Finally, their seeds, which possess abundant highly nutrititive endosperms and large embryos, equipped with a haustorial and metabolically highly active cotyledon, as well as embryonic shoots and roots that can develop very rapidly upon germination, are ideally adapted for rapid and certain establishment of new seedlings in a variety of habitats. In my opinion, the climax of flowering-plant evolution is represented by the grasses, which, in addition, are the most useful to man of all families.

Summary

Although the monocotyledons do not possess any single characteristic that is absent from all groups of dicotyledons, their distinctive combination of characteristics renders highly probable the hypothesis that they originated as a single evolutionary line, distinct from any modern group of dicotyledons. The common ancestor of monocotyledons and dicotyledons was probably some shrubby primitive angiosperm having vesselless xylem and primitive floral structure. This group is now completely extinct, and cannot be reconstructed on the basis of any modern family or order. Moreover, the wide differences that exist between modern monocotyledons that have apocarpous gynoecia indicate that none of them are similar to the extinct ancestral monocotyledons. The ecological conditions that favored the evolution of monocotyledons were probably temporary swamps or pools in a climate having a marked dry season. This hypothesis explains the distinctive method of seed germination which is believed to have been original in the group. Aquatic monocotyledons, as well as xeric forms and those having an arboreal growth habit, are regarded as secondary derivatives. The single cotyledon is believed to have originated by intercalary concrescence of the petioles of two original cotyledons, and the loss of their blades. Most prophylls, as well as the palea of the grass flower, are also regarded as products of intercalary concrescence. The phyllode theory of the origin of the leaf of monocotyledons is rejected.

The principal trends in the vegetative structures of monocotyledons, which have been repeated in several parallel evolutionary lines, are: (1) development of "rosette trees," such as palms, *Dracaena*, and so forth; (2) extension of the rhizome, to form more slender plants having many adventitious roots; and (3) reduction of underground stems to form corms or bulbs, usually surrounded by sheathing leaf bases. Distinctive trends have also occurred with respect to early development of seedlings. Trends in reproductive parts have been similar to those in dicotyledons.

14 / A Glimpse into the Future

Biologists who are not primarily evolutionists, as well as laymen, are more interested in major evolutionary trends than in processes at the level of populations and species. By many people, research of a microevolutionary nature is regarded as a means to the end of understanding transspecific evolution. Consequently, in spite of the hazards and uncertainties of extrapolation from the population to the transspecific level, some research of this nature must be carried out if biologists are to obtain a full and satisfying understanding of the whole of evolution. As a guide to such extrapolation, the concept of genetic uniformitarianism, with all of its corollaries and implications, is indispensable.

Ideally, the extrapolation should be carried out on organisms having the following advantages: (1) populations that can be analyzed in a quantitative fashion, with respect to gene content, chromosomal variability, and breeding system; (2) subspecies, species, and at least some groups that qualify morphologically as genera that can be intercrossed and progeny obtained for genetic analysis; and (3) a fossil record by means of which significant evolutionary changes can be followed.

Unfortunately, no such group of organisms exists. Among modern organisms, significant fossil records exist only for mammals, molluscs, conifers, marine echinoderms and coelenterates, and some groups of unicellular organisms having hard parts, such as foraminifera and diatoms. Unfortunately, none of these organisms fully satisfy requirements (1) and (2) above. Small mammals, particularly rodents, perhaps come nearest, but many species of rodents are difficult to rear in captivity in sufficiently large numbers for genetic analysis. Insects and higher plants which, along with rodents, have been the chief materials for research on the genetics and breeding systems of populations and species, both have notoriously poor fossil records.

When these facts are considered, angiosperms qualify as well as any other organisms for making the necessary extrapolations from populational to transspecific levels. For genetic analyses of populations, they are as favorable as any multicellular organisms except for some groups of insects, such as *Drosophila.* For comparing subspecies, species, and occasionally genera, they are superior to any group of animals, since hybridization is easier, and barriers of reproductive isolation between morphologically and ecologically diverse populations are, in general, less strongly developed. The poverty of their fossil record is to some extent compensated by the presence of intercrossable populations that differ with respect to at least some characters similar to those that separate major groups. Consequently, research on the borderline between the population and the transspecific approach needs to be much more strongly developed than at present. It can become both a fascinating and a rewarding branch of evolutionary science.

Genetic Analyses of Analogues to Differences Between Higher Categories

The point illustrated by Table 2-1, that individual morphological characters ordinarily regarded as being significant at the transspecific level sometimes vary at the level of species and perhaps even of populations within species, needs to be explored further. A systematic search for such examples might uncover favorable groups of plants that have not yet been recognized. If a complete list of them were available, clues might be found leading toward an answer to the question: Why should some groups of plants vary significantly with respect to these transspecific characters, whereas others consistently display a constant pattern of reproductive structures, subject only to quantitative variations in emphasis?

At the experimental level, research of the kind begun by Huether (1966, 1968, 1969) needs to be developed further. A relatively small amount of experimentation might provide an answer to the question: Can populations of *Linanthus* or any other genus of Polemoniaceae be synthesized that consist of vigorous, adaptively successful plants that are consistently or at least modally tetramerous, hexamerous, or heptamerous?

In a somewhat different direction, but also in connection with points brought up in Chapter 2, the hypothesis of the conservation of organization might be tested experimentally, using the stomatal complex of Gramineae. The questions could be asked: Are subsidiary cells a necessary component of stomatal complexes in grass leaves? Is it possible to obtain grasses having stomata that lack subsidiary cells? If so, are they vigorous and adaptive? Since complexes without subsidiaries can be obtained by developmental shocks, one might apply the genetic-assimilation technique of Waddington

(1962) to a genetically variable population of a grass species, in order to obtain if possible a race having in its leaves, as many species have in their coleoptiles, stomatal complexes in which the guard cells are not flanked by subsidiary cells.

Also in connection with the hypothesis of conservation of organization, developmental studies need to be made of natural deviations from the normal architecture of the plant. In Chapter 2, the suggestion was made that the hooded gene in barley has effects analogous to those exerted by the genes responsible for inflorescences that are borne on the adaxial surface of leaves. This hypothesis needs to be tested by careful developmental studies of such genera as *Phyllonema*, *Chailletia*, and *Chisocheton*, along the lines of those carried out by Yagil and Stebbins (1969) on hooded barley.

Ecological and Physiological Studies of Adaptation

Research that aims toward a synthesis of facts derived from morphology, autecology, physiology, biochemistry, and genetics holds particularly attractive opportunities for botanists interested in transspecific evolution. What are the distinctive features of their biochemical pathways and cellular organization that restrict most Ericaceae to acid soils and many Caryophyllaceae and Cruciferae to neutral or alkaline soils with a high nitrate content? Where exceptions to these rules occur, are they recent specializations, or possible clues to the origins of the families in question? Along similar lines, how did the Labiatae evolve relative constancy with respect to the synthesis of menthol and related aromatic compounds, while in Oxalidaceae and Polygonaceae oxalic acid is predominant and in Ranunculaceae the probable defense mechanisms against insects consist largely of toxic and bitter alkaloids? In order to answer these questions, the kind of research on secondary substances which is now current, and which is aimed mainly at identification and comparison of particular substances, will have to be supplemented by detailed comparative studies of biosynthetic pathways.

Although angiosperm leaves are so variable in form and structure that they are very poor guides to evolutionary trends, nevertheless a full understanding of the adaptive value, or possible neutrality, of such differences should provide a valuable guide to understanding evolutionary trends. Consequently, an expansion of the kind of physiological and ecological analysis of leaves that is reviewed in Chapter 3 would be highly desirable.

Ecological Studies of Reproductive Cycles

As discussed in Chapter 4, research on pollination biology and breeding systems of angiosperms is now highly developed, and is progressing rapidly. Guidelines for such research have been admirably set forth by Cruden (Baker and Hurd 1968). This research has already shown that many differences in floral structure have arisen through natural selection for different pollen vectors and for different methods of pollination, and botanists will surely understand these selection pressures much better in the near future.

On the other hand, systematic studies of seed and seedling biology are badly needed. The research of Harper and his associates, discussed in Chapter 5, serves as a model of what might be done, but is not oriented toward understanding evolutionary trends. In no instance have these workers attempted to compare an entire group of closely related species, or to analyze differences between races or ecotypes of the same species with respect to seed and seedling biology. The following are some of the critical questions to which answers are unavailable or incomplete:

1. What are the adaptive advantages of increasing seed size, and under what conditions are they most strongly evident?
2. How strong is the correlation between increased seed size and increased time required between flowering and seed maturity?
3. What effect, if any, does the shift from hypogyny to epigyny have upon the degree to which developing ovules are protected from extremes of temperature, insects or other predators, or both of these hazards?
4. What are the physiological differences with respect to germination between endosperm-containing and exalbuminous seeds? Does the shift of food storage from endosperm to cotyledons always increase germination efficiency? If not, under what conditions is the adaptive value of this shift relatively high, and when is it inconsequential?

Answers to these questions require research of two different kinds. First, comparative explorations of a descriptive nature must be made in order to identify groups in which these problems can be effectively studied. Second, particular situations must be analyzed by combined physiologic and genetic experiments. Research of this kind must be carried out on many groups before botanists can hope to answer intelligently the broader question: To what extent are the differences in the gynoecium that so often differentiate families and other major categories the result of the direct action of natural selection, and what kinds of differences must be ascribed to indirect action or to chance?

Research on seed and seedling biology may provide one of the best ways of testing the validity of one of the major hypotheses elaborated in this book—Ganong's hypothesis of adaptive modification along the lines of least resistance. One might, for instance, select for study a genus such as *Plantago* or *Phacelia,* in which there exist species that differ markedly from one another with respect to the number of seeds per gynoecium. If the hypothesis is correct, the effect on the entire reproductive system of artificial selection for the same characteristic, such as increased seed size or seed number, should be quite different in species having few seeds per gynoecium and in those having many. Furthermore, the differences between related species with respect to their reactions should be paralleled in different, unrelated genera.

An aspect of seed and seedling biology that has been relatively well studied is seed dispersal. Nevertheless, many questions about this process are still unanswered. What kinds of genetic changes must be established to bring about the shift from wind to animal dispersal, or vice versa? To what extent and under what conditions are shifts from one kind of dispersal to another reversible? Can shifts in methods of seed dispersal be usually associated with phylogenetic trends, as suggested in Chapter 5 for *Geum* and other genera? Comparative studies of various kinds on groups of related species are needed to provide answers to such questions.

Developmental Basis of Evolutionary Trends

Because of the complexity of gene action in development, analyses of developmental patterns are an essential prelude to a full understanding of the genetic basis of evolutionary trends. For these analyses, emphasis must be shifted away from such generalized concepts as recapitulation, embryonic similarity, and neoteny, and toward comparisons of the mechanisms of development at the cellular and tissue level. As is suggested in Chapter 7, here lies the clue to valid estimates of the reversibility or irreversibility of individual evolutionary trends. The guidelines for such research must consist in the application to developmental processes of the principle established by Simpson for irreversibility with respect to combinations of adult characteristics. The greater is the number of individual gene-controlled changes required to bring about a trend in one direction, the lower is the probability that this trend can be reversed. At the level of developmental patterns, the changes that must be considered are shifts in timing or intensity of hormone synthesis, distribution of hormones or their inhibitors, cell polarity and differentiation, membrane permeability, stimulation or inhibition of mitosis, and the like.

Specifically, experiments like those of Torrey (1957) on alteration of protoxylem points in the pea root need to be performed on a variety of objects, particularly on genetically controlled material. In another direction, the physiological basis of intercalary meristems needs to be compared with that of apical or other kinds of meristems. Do intercalary meristems require a more elaborate system of controls for their activation and suppression than other kinds of meristems? If so, a rational basis would exist for the assumption that seems reasonable on the basis of comparative studies: intercalary meristems are relatively specialized, and in any particular group their strong development usually indicates a higher degree of specialization than their absence.

The Evolutionary Significance of Vascular Anatomy

In my opinion, correct evaluation of vascular anatomy as a guide to evolutionary trends depends upon solutions of developmental problems similar to those mentioned in the previous section. As I said in Chapters 6 and 7, I believe, following Carlquist, that the concept of vestigial bundles has been greatly overemphasized in the past, and should be either drastically modified or abandoned altogether. On the other hand, I cannot agree with Carlquist that vascular patterns reflect nothing more than the physiological demands of the organs concerned, and are no more valuable for indicating evolutionary trends than are comparisons of gross morphological features.

The intermediate position that I advocate is one based upon development, taking as its cue the demonstration of Wetmore and Rier (1963) that in organ cultures the differentiation of vascular elements bears a direct relation to the condition of the developing organ at the time when procambial tissue becomes differentiated. The experimental basis for this position is still meager, and more experiments with organ cultures should be performed, particularly on a comparative basis. Nevertheless, the probability that Wetmore and Rier's observations have general significance is strong enough that comparative studies of procambial differentiation in relation to size and vasculation of adult structures are well justified. One could ask such questions as: Is the number of vascular bundles in an organ directly related to the number of meristematic cells present at the stage when procambium becomes differentiated? If this relation exists, then observed differences between related species with respect to the ratio of size of organ to number of vascular bundles could be ascribed largely to the action of genes late in development, after the procambium has been laid down. The significance of such a conclusion would depend, in turn, upon the validity of the hypothesis stated in Chapter 6, that genes acting late in development have a greater chance of becoming established than those acting earlier. If this is so, deviations from the usual pattern that are due to late-acting genes are more likely to be specializations than vestiges of primitive conditions.

Another hypothesis that could be tested by comparative studies of anatomy and development is that the existence of several vascular bundles running parallel to one another through an organ is a result of the activity of an intercalary meristem during its later development, after the differentiation of procambium. If this is so, then the primitiveness or specialized state of many vascular patterns would depend upon conclusions with respect to the phylogenetic significance of intercalary meristems, which is discussed earlier in this chapter.

Ecological Comparisons as Aids to Determining Evolutionary Trends

The hypotheses presented in Chapters 8 and 9 need to be tested by a whole series of investigations in which facts from systematics, phytogeography, and ecology are synthesized and directed toward phylogenetic problems. The relative phylogenetic position of related species and genera existing around the borders of tropical rain and cloud forests needs to be carefully studied, in order to provide more definite evidence of migration either into or away from them. I believe that the ecophyletic chart, of which examples are presented in Chapter 9, is a useful device for this purpose. In well-known families, particularly those that show evidence of relatively rapid evolution in recent times, evidence should be sought for increased reproductive efficiency in groups that have had an evolutionary history of migration into more severe habitats followed by secondary migration to more favorable environments.

Investigation of Particular Phylogenetic Trends and of Relations Between Major Categories

As the reader should now fully realize, the interpretation of individual phylogenies in the absence of a fossil record involves so many uncertainties that satisfactory results probably will never be achieved. All of the methods that are now being used for this purpose involve making certain assumptions, which are questionable at best. Some taxonomists have suggested that if a few major assumptions are made at the beginning, and the rest of the deductions are performed by a computer according to a set program, the results will be more reliable. This method has one advantage over the older, completely subjective methods. In order to program the computer, the investigator must state clearly what his assumptions are. Other investigators can, therefore, evaluate the work on the basis of the probability that these assumptions are justified. As is stated elsewhere in this book, I have not been able to find an example of computerized phylogeny for which the initial assumptions are justified on the basis of known facts. For this reason, I believe that a closer approach to understanding phylogeny will be made by concentrating first on reasearch that aims to gather facts upon which more valid assumptions can be made.

These facts are of three kinds. First, attempts should be made to improve the known fossil record, both by discovering new fossils of a critical nature and by learning how to interpret better the available fossils, particularly the abundant microfossils. Second, methods for determining biochemical affinity, particularly with respect to proteins and nucleic acids, need to be refined, and applied to critical examples. Finally, the whole field of comparative morphogenesis, which on a systematic level hardly even exists, needs to be fully developed, and to be connected much more with developmental physiology as well as, where possible, with genetics.

Epilogue

The prospects for arriving at a full understanding of evolutionary trends in flowering plants are by no means hopeless. Research in this field is for botanists who are somewhat impatient with studying in ever-increasing detail further aspects of known phenomena. Imagination, boldness, and resourcefulness in ferreting out new avenues of approach are the chief characteristics required for success. Above all, anyone entering this field must be able to combine a feeling of adventure with a respect for tradition and for the important discoveries made by his colleagues and by previous workers. A good, complete synthesis is the all-important achievement. Its only reward will be the satisfaction of having put together a harmonious, self-satisfying synthesis of previously unrelated facts. For botanists of this disposition, the field is wide open, and a world of adventure lies ahead.

Appendix / The Orders and Families of Angiosperms

For this list, I have followed chiefly the system of Cronquist (1968), with modifications according to the systems of Takhtajan (1966, 1970), Thorne (1968), and Hutchinson (1959) where these seem to be more in line with my own observations. A question mark (?) before the name of a group indicates that authorities disagree with respect to its relationships.

After the name of each family, the number of genera and the approximate number of species that it contains are given; these are followed by abbreviations that indicate its geographic distribution, as given by Takhtajan (1966). The abbreviations are as follows:

As	Asia
Aus	Australia
E/As	Eastern Asia
E/N/Am	Eastern North America
Euras	Eurasia
Mad	Madagascar
Mal	Malaysian region
Med	Mediterranean region
n	New World
N/Am	North America
N/Cal	New Caledonia
N/Guin	New Guinea
N/Z	New Zealand
Pac Ids	islands of the Pacific Ocean
S/Afr	South Africa
S/te	south temperate, both hemispheres
SE/As	Southeast Asia
SW/As	Southwest Asia
te/aa	temperate and arctic alpine
te/tr	chiefly temperate, some in tropics
te/Euras	temperate Eurasia
te/S/Am	temperate South America
tr	tropical, both hemispheres
tr/Afr	tropical Africa
tr/As	tropical Asia
tr/n	tropics, New World
tr/o	tropics, Old World
tr/te	chiefly tropical, some in temperate regions
W/N/Am	Western North America
ww	world wide

Since actual phyletic relationships of genera can never be accurately determined, so that orders and families must of necessity be to some extent arbitrary groupings made for the sake of convenience, the number recognized in this list has been kept to the minimum that is compatible with the opinions of contemporary taxonomists.

A. SUBCLASS DICOTYLEDONES

I. Superorder Magnoliidae

1. Order Magnoliales

1. Magnoliaceae 12 210 tr/te/E/As Am
2. Winteraceae 7 70 tr/te Mal Aus N/Z n
3. Degeneriaceae 1 1 Fiji
4. Himantandraceae 1 3 N/Guin Aus
5. Annonaceae 122 2000 tr te
6. Myristicaceae 16 380 tr
7. Canellaceae 5 10 tr/Afr/n

2. Order Illiciales

1. Illiciaceae 1 42 tr/te/As, Mal, tr/te/N/Am
2. Schisandraceae 2 47 te/tr/As Mal N/Am

3. Order Laurales

1. Austrobaileyaceae 1 2 tr/Aus
2. Lactoridaceae 1 1 Juan Fernandez

3. Eupomatiaceae 1 2 tr/Aus N/Guin
4. Amborellaceae 1 1 N/Cal
5. Trimeniaceae 2 7 Aus N/Guin N/Cal Fiji
6. Monimiaceae 31 450 tr/te/As Aus N/Z N/Cal S/Am
7. Gomortegaceae 1 1 Chile
8. Calycanthaceae 2 5 te E/As N/Am
9. Lauraceae 50 2000 tr te
10. Hernandiaceae 4 70 tr
?11. Chloranthaceae 5 75 tr/As Mal Pac Ids N/Z Mad n

4. Order Piperales
1. Saururaceae 4 6 te/As N/Am
2. Piperaceae 10 2000 tr

5. Order Aristolochiales
1. Aristolochiaceae 10 600 tr te
?2. Nepenthaceae 1 75 tr/As Mal Aus N/Cal Mad

6. Order Nymphaeales
1. Ceratophyllaceae 1 6 te tr
2. Nymphaeaceae 7 70 ww
?3. Nelumbonaceae 1 2 te/tr/As N/Am

7. Order Ranunculales
1. Ranunculaceae 45 2000 ww
2. Circaeasteraceae 2 2 te/As
3. Berberidaceae 14 660 te/Euras N/Am S/Am
4. Lardizabalaceae 9 30 te/tr/As S/Am
5. Menispermaceae 70 450 tr/te

8. Order Papaverales
1. Papaveraceae 27 470 te/tr/Euras N/Am
2. Fumariaceae 16 400 te

?9. Order Sarraceniales
1. Sarraceniaceae 3 16 te/tr/Am

II. Superorder Hamamelidae

1. Order Trochodendrales
1. Tetracentraceae 1 1 te/E/As
2. Trochodendraceae 1 1 te/E/As

2. Order Hamamelidales
1. Cercidiphyllaceae 1 2 te/E/As
2. Eupteleaceae 1 1 te/E/As
3. Platanaceae 1 6 te/tr/Euras N/Am
4. Didymelaceae 1 2 Mad
5. Hamamelidaceae 26 112 Euras N/Am
6. Myrothamnaceae 1 2 S/Afr Mad

3. Order Eucommiales
1. Eucommiaceae 1 1 E/As

4. Order Leitneriales
1. Leitneriaceae 1 1 E/N/Am

5. Order Myricales
1. Myricaceae 3 60 te tr

6. Order Fagales
1. Balanopaceae 1 10 tr/Aus N/Cal Fiji
2. Fagaceae 8 900 te tr
3. Betulaceae 6 130 te/aa/Euras N/Am S/Am

7. Order Casuarinales
1. Casuarinaceae 1 60 Aus N/Guin N/Cal SE/As Mal

III. Superorder Caryophyllidae

1. Order Caryophyllales
1. Phytolaccaceae 24 143 tr te
2. Nyctaginaceae 30 300 tr te
3. Didieraceae 4 11 Mad
4. Cactaceae 150 2000 tr/te/Am tr/Afr
5. Aizoaceae 11 2300 tr te
6. Molluginaceae 14 10 tr te
7. Caryophyllaceae 80 2100 tr aa
8. Portulacaceae 20 500 te tr
9. Basellaceae 5 22 tr te
10. Chenopodiaceae 100 1500 te tr
11. Amaranthaceae 65 900 tr te

2. Order Batales
1. Bataceae 1 2 tr/n N/Guin

3. Order Polygonales
1. Polygonaceae 40 900 ww

4. Order Plumbaginales
1. Plumbaginaceae 15 500 ww

IV. Superorder Dilleniidae

1. Order Dilleniales
1. Dilleniaceae 18 530 tr/o tr/n te/Aus
2. Paeoniaceae 1 40 te/Euras N/Am
3. Crossosomataceae 1 2 W/N/Am

2. Order Theales
1. Ochnaceae 29 400 tr
2. Dipterocarpaceae 22 400 tr/o
3. Caryocaraceae 2 25 tr/n
4. Theaceae 36 580 tr te
5. Stachyuraceae 1 5 E/As
6. Marcgraviaceae 5 120 tr/n
7. Quiinaceae 3 37 tr/n
8. Elatinaceae 2 45 tr te
9. Medusagynaceae 1 1 Seychelles
10. Guttiferae 48 910 tr te

3. Order Malvales
?1. Rhopalocarpaceae 2 14 Mad
?2. Sarcolaenaceae 8 33 Mad
3. Elaeocarpaceae 10 400 tr/Euras n
4. Scytopetalaceae 5 32 tr/Afr
5. Tiliaceae 45 400 tr te
6. Sterculiaceae 60 1000 tr te
7. Bombacaceae 28 190 tr
8. Malvaceae 90 1570 tr te

?4. Order Urticales
1. Ulmaceae 16 150 tr te
2. Barbeyaceae 1 1 Afr Sw/As
3. Moraceae 60 1550 tr te
4. Cannabaceae 2 4 te/Euras N/Am
5. Urticaceae 45 700 tr te

?5. Order Lecythidales
1. Lecythidaceae 24 450 tr

6. Order Violales
1. Flacourtiaceae 82 1270 tr te
2. Peridiscaceae 2 2 tr/n
3. Scyphostegiaceae 1 1 Borneo
4. Violaceae 16 850 ww
5. Turneraceae 8 120 tr n/Afr
6. Passifloraceae 20 650 tr te
7. Malesherbiaceae 2 25 te/S/Am
8. Bixaceae 3 20 tr
9. Cistaceae 8 180 te/Euras N/Am S/Am
10. Tamaricaceae 3 120 Med te/As
11. Frankeniaceae 5 60 te
?12. Dioncophyllaceae 3 3 tr/Afr
?13. Ancistrocladaceae 1 16 tr/As Afr
14. Fouquieriaceae 2 7 W/N/Am
15. Achariaceae 3 3 S/Afr
16. Caricaceae 4 45 tr/n te/S/Am tr/Afr
?17. Loasaceae 15 250 tr/te/n tr/Afr
18. Begoniaceae 5 820 tr
19. Datiscaceae 3 4 te/tr/As Mal N/Guin N/Am
20. Cucurbitaceae 120 1000 tr te

7. Order Salicales
1. Salicaceae 3 340 ww

8. Order Capparales
1. Tovariaceae 1 2 tr/n
2. Capparaceae 47 900 tr te
3. Cruciferae 350 3000 te aa
4. Resedaceae 6 75 Med te/tr/n
5. Moringaceae 1 10 tr/Afr Mad SW/As

9. Order Ericales
?1. Actinidiaceae 3 320 tr/te/As Aus N/Am
2. Cyrillaceae 3 13 tr te n
3. Clethraceae 2 50 te/tr/As N/Am
?4. Grubbiaceae 1 5 S/Afr
5. Ericaceae 80 2500 ww
6. Epacridaceae 30 400 Aus N/Z Mal te/S/Am
7. Empetraceae 3 9 te/aa/Euras N/Am S/Am
8. Pyrolaceae 4 45 te/Euras N/Am
9. Monotropaceae 13 16 te/Euras N/Am

10. Order Diapensiales
1. Diapensiaceae 6 18 te/aa/Euras N/Am

11. Order Ebenales
1. Sapotaceae 60 800 tr
2. Ebenaceae 7 450 tr te
3. Styracaceae 12 150 tr te
4. Lissocarpaceae 1 2 tr/S/Am
5. Symplocaceae 1 400 tr/te/As Aus N/Cal N/Am

12. Order Primulales
1. Theophrastaceae 4 110 tr/n Hawaii
2. Myrsinaceae 38 1000 tr te
3. Primulaceae 30 800 te aa

V. SUPERORDER ROSIDAE

1. Order Rosales
1. Eucryphiaceae 1 5 te/Aus S/Am
2. Cunoniaceae 27 380 tr/Aus N/Guin N/Cal S/te
3. Davidsoniaceae 1 1 te/Aus
4. Pittosporaceae 9 240 Aus N/Z tr/As Afr
5. Byblidaceae 2 4 Aus S/Afr

?6. Droseraceae 4 100 te tr
7. Hydrangeaceae 19 260 te/tr/Euras N/Am
8. Grossulariaceae (incl. Escalloniaceae *et al.*) 27 345 ww
9. Bruniaceae 12 75 S Afr
?10. Alseuosmiaceae 3 11 N/Z N/Cal
11. Crassulaceae 35 1450 ww
12. Cephalotaceae 1 1 SW/Aus
13. Saxifragaceae 42 670 te aa
14. Rosaceae 45 5000 te aa
15. Neuradaceae 3 10 Med S/Afr
16. Chrysobalanaceae 13 530 tr

2. Order Fabales

1. Leguminosae 700 17600 ww

3. Order Podostemales

1. Podostemaceae 43 200 tr

4. Order Haloragales

1. Haloragaceae 7 130 ww
2. Gunneraceae 1 40 S/te Mad Mal N/Guin Pac-/Ids
3. Hippuridaceae 1 1 te/Euras N/Am
?4. Theligonaceae 1 3 Med te/As

5. Order Myrtales

1. Sonneratiaceae 2 7 tr/o
2. Lythraceae 23 500 tr te
3. Rhizophoraceae 16 120 tr
4. Peneaceae 5 20 S/Afr
5. Crypteroniaceae 1 4 tr/As Mal
6. Thymeleaceae 50 650 te tr
7. Trapaceae 1 15 tr/te/Afr Euras
8. Dialypetalanthaceae 1 1 tr/S/Am
9. Myrtaceae 100 3000 tr te
10. Punicaceae 1 2 Med
11. Onagraceae 20 650 ww
12. Oliniaceae 1 8 tr/Afr S/Afr
13. Melastomataceae 200 4000 tr te
14. Combretaceae 18 550 tr

6. Order Cornales

1. Davidiaceae 1 1 te/E/As
2. Nyssaceae 2 9 te/E/As E/N/Am
3. Alangiaceae 1 20 tr/te/o
4. Cornaceae 17 140 ww
5. Garryaceae 1 17 te/tr/N/Am

7. Order Proteales

1. Eleagnaceae 3 65 te tr
2. Proteaceae 62 1400 tr S/te

8. Order Santalales

?1. Medusandraceae 1 1 tr/Afr
?2. Dipentodontaceae 1 1 E As
3. Olacaceae 25 250 tr
4. Opiliaceae 9 60 tr S/te
5. Santalaceae 30 400 tr te
6. Loranthaceae 40 1400 tr te
7. Misodendraceae 1 11 te/S/Am
8. Balanophoraceae 18 110 tr
9. Cynomoriaceae 1 2 Med As

9. Order Rafflesiales

1. Hydnoraceae 2 18 tr/Afr Mad S/Afr
2. Rafflesiaceae 9 55 tr te

10. Order Celastrales

1. Geissolomataceae 1 1 S/Afr
2. Hippocrateaceae 18 300 tr
3. Celastraceae 59 850 tr te
4. Siphonodontaceae 1 5 tr/As Mal Aus
5. Stackhousiaceae 2 22 te/Aus
6. Salvadoraceae 3 12 tr/te/Afr SW/As
7. Aquifoliaceae 4 450 te tr
8. Icacinaceae 45 40 tr
?9. Cardiopteridaceae 1 3 tr/As Mal Aus N/Guin
?10. Dichapetalaceae 4 250 tr
?11. Corynocarpaceae 1 5 Aus N Guin Mal

11. Order Euphorbiales

1. Buxaceae 6 60 te tr
2. Euphorbiaceae 290 7500 tr te
3. Daphniphyllaceae 1 35 E/As Mal
4. Aextoxicaceae 1 1 te/S/Am
5. Pandaceae 3 35 tr/Afr As N/Guin

12. Order Rhamnales

1. Rhamnaceae 60 900 tr te
2. Leeaceae 1 70 tr/As
3. Vitaceae 11 700 tr te

13. Order Sapindales

1. Staphyleaceae 7 50 te/tr/Euras Mal te/N/Am
2. Melianthaceae 2 35 tr/Afr S/Afr
3. Greyiaceae 1 3 S/Afr
4. Connaraceae 24 400 tr

5. Akaniaceae 1 1 Aus
?6. Sabiaceae 4 100 tr/As tr/n
7. Sapindaceae 140 1600 tr te
8. Hippocastaneaceae 2 25 te/tr/Euras N/Am
9. Aceraceae 2 150 te/Euras N/Am
10. Burseraceae 20 600 tr te
11. Anacardiaceae 80 600 tr te
12. Julianaceae 2 5 tr/n
13. Simaroubaceae 30 220 tr te
14. Cneoraceae 2 3 Med Cuba
?15. Coriariaceae 1 10 Med tr/As Mal N/Guin N/Z tr/n
16. Rutaceae 150 1600 tr te
17. Meliaceae 50 1400 tr te
?18. Zygophyllaceae 29 265 te tr

?14. Order Juglandales

1. Rhoipetalaceae 1 1 E/As
2. Juglandaceae 8 70 te/tr/Euras n

15. Order Geraniales

1. Humiriaceae 8 50 tr/n
2. Linaceae 23 370 ww
3. Erythroxylaceae 4 200 tr
4. Geraniaceae 11 830 te tr
5. Oxalidaceae 5 900 te tr
6. Limnanthaceae 2 8 te/N/Am
7. Tropaeolaceae 2 80 tr/N/Am te/tr/S/Am
8. Balsaminaceae 2 450 tr te

16. Order Polygalales

?1. Malpighiaceae 60 870 tr
2. Trigoniaceae 2 40 tr/n tr/As Mad
3. Vochysiaceae 6 200 tr/n Afr
4. Tremandraceae 3 28 te/Aus
5. Polygalaceae 14 900 tr te
6. Krameriaceae 1 20 te tr n

17. Order Umbellales

1. Araliaceae 70 850 tr te
2. Umbelliferae 300 3000 ww

VI. Superorder Asteridae

1. Order Gentianales

1. Loganiaceae 22 460 tr te
2. Gentianaceae 70 1100 ww
3. Apocynaceae (incl. Asclepiadaceae) 400 4000 tr te
?4. Oleaceae 29 600 tr te

2. Order Polemoniales

?1. Nolanaceae 2 83 Chile Peru
?2. Solanaceae 85 2300 tr te
?3. Hoplestigmataceae 1 2 tr/Afr
?4. Convolvulaceae 50 1500 tr te
?5. Cuscutaceae 1 17 tr te
?6. Menyanthaceae 5 40 tr te
7. Polemoniaceae 18 320 te tr N/Am S/Am Euras
8. Hydrophyllaceae 20 275 te tr
?9. Boraginaceae 100 2000 te tr
10. Lennoaceae 3 4 tr/te/n

3. Order Lamiales

1. Verbenaceae 100 2600 tr te
2. Labiatae 200 3500 ww
3. Phrymaceae 1 2 te/E/As N/Am
4. Callitrichaceae 1 25 ww

4. Order Plantaginales

1. Plantaginaceae 3 265 ww

5. Order Scrophulariales

1. Buddlejaceae 10 170 tr te
?2. Collumeliaceae 1 4 Andes
3. Scrophulariaceae 200 3000 ww
4. Myoporaceae 5 180 tr/te/Aus S/Afr
5. Globulariaceae 3 27 Med
6. Gesneriaceae 14 1800 tr te
7. Orobanchaceae 13 160 te/tr/Euras N/Am
8. Bignoniaceae 120 800 tr te
9. Acanthaceae 250 2600 tr te
10. Pedaliaceae 19 72 tr te
11. Lentibulariaceae 5 300 ww
12. Hydrostachyaceae 1 30 tr/Afr S/Afr Mad

6. Order Campanulales

1. Sphenocleaceae 1 2 tr/o
2. Campanulaceae (incl. Lobeliaceae) 75 2000 ww
3. Stylidiaceae 6 140 Aus N/Z tr/As te/S/Am
4. Brunoniaceae 1 1 Aus
5. Goodeniaceae 14 320 S/te tr/As Mal Pac/Ids

7. Order Rubiales

1. Rubiaceae 450 7000 tr te

8. Order Dipsacales

1. Caprifoliaceae 16 440 te/tr/Euras N/Am S/Am
2. Adoxaceae 1 1 te/Euras N/Am
3. Valerianaceae 13 420 te/Euras Afr N/Am

4. Dipsacaceae 10 280 Med te/As S/Afr
?5. Calyceraceae 6 60 S/Am

9. Order Asterales

1. Compositae 1000 20,000 ww

B. SUBCLASS MONOCOTYLEDONES

I. Superorder Alismatidae

1. Order Alismatales

1. Butomaceae 1 1 te/Euras
2. Limnocharitaceae 3 12 tr
3. Alismataceae 13 70 tr te

2. Order Hydrocharitales

1. Hydrocharitaceae 15 100 tr te

3. Order Najadales

1. Aponogetonaceae 1 40 tr o
2. Scheuchzeriaceae 1 1 te/Euras N/Am
3. Juncaginaceae 5 18 te aa
4. Najadaceae 1 35 te tr
5. Potamogetonaceae 2 90 ww
6. Ruppiaceae 1 3 te tr
7. Zannichelliaceae 3 11 ww
8. Zosteraceae 6 28 ww

4. Order Triuridales

1. Petrosaviaceae 1 1 tr/As
2. Triuridaceae 7 80 tr

II. Superorder Commelinidae

1. Order Commelinales

1. Rapateaceae 16 80 tr/n tr/Afr
2. Xyridaceae 4 270 tr te
3. Mayacaceae 1 4 tr/n tr/Afr
4. Commelinaceae 49 600 tr te

2. Order Eriocaulales

1. Eriocaulaceae 13 1200 tr te

3. Order Restionales

1. Flagellariaceae 3 9 tr/As Mal N/Cal Pac/Ids
2. Restionaceae 30 400 tr/As Mad S/te
3. Centrolepidaceae 6 40 tr/As Aus N/Z N/Guin te/S/Am

4. Order Poales

1. Gramineae 700 9000 ww

5. Order Juncales

1. Juncaceae 8 350 te aa
2. Thurniaceae 1 3 tr n

6. Order Cyperales

1. Cyperaceae 95 3800 ww

7. Order Typhales

1. Sparganiaceae 1 20 te/Euras N/Am Aus N/Z
2. Typhaceae 1 15 ww

8. Order Bromeliales

1. Bromeliaceae 50 2000 tr/te/n tr/Afr

9. Order Zingiberales

1. Strelitziaceae 3 7 tr/Afr Mad S/Am
2. Lowiaceae 1 5 tr/As Mal
3. Heliconiaceae 1 150 tr/n
4. Musaceae 2 70 tr/o
5. Zingiberaceae 45 1300 tr
6. Costaceae 5 170 tr
7. Cannaceae 1 50 tr/n
8. Marantaceae 32 350 tr

III. Superorder Arecidae

1. Order Arecales

1. Palmae 240 3400 tr te

2. Order Cyclanthales

1. Cyclanthaceae 11 180 tr/n

3. Order Pandanales

1. Pandanaceae 3 880 tr/te/o

4. Order Arales

1. Araceae 110 2000 tr te
2. Lemnaceae 4 25 ww

IV. Superorder Liliidae

1. Order Liliales

1. Philydraceae 4 5 Tr/As Mal Aus N/Guin
2. Pontederiaceae 7 40 tr te
3. Liliaceae (incl. Amaryllidaceae) 282 4425 ww
4. Iridaceae 70 1500 ww
5. Tecophilaeaceae 6 25 S/Afr W/N/Am Chile
6. Agavaceae 18 560 tr te
7. Xanthorrheaceae 8 60 Aus N/Guin N/Cal
8. Velloziaceae 3 190 tr Am Mad
9. Haemodoraceae 16 100 S/te tr/n te/N/Am

10. Taccaceae 2 35 tr
11. Cyanastraceae 1 5 tr/Afr
12. Stemonaceae 3 30 tr/te/As N/Am
13. Smilacaceae 3 320 tr te
14. Dioscoreaceae 9 650 tr te

2. Order Orchidales

1. Geosiridaceae 1 1 Mad
2. Burmanniaceae 22 130 tr te
3. Corsiaceae 2 9 N/Guin Chile
4. Orchidaceae 800 30,000 ww

	Total numbers		
	Families	Genera	Species
Dicotyledons	288	10,590	166,545
Monocotyledons	61	1,744	64,868
Angiosperms	349	12,334	231,413

Bibliography

Allard, R. W. 1965. "Genetic systems associated with colonizing ability in predominantly self-pollinated species." In H. G. Baker and G. L. Stebbins, eds., *The Genetics of Colonizing Species* (Academic Press, New York), pp. 49–76.

——— 1969. "Breeding systems and population structures: synthesis" (abstr.). Abstracts of the papers presented at the XI International Botanical Congress (Seattle): 2.

——— and A. L. Kahler. 1972. "Patterns of molecular variation in plant populations," in L. M. LeCam, J. Neyman, and E. L. Scott, eds. *Darwinian, Neo-Darwinian and Non-Darwinian Evolution* (Proc. 6th Berkeley Symp. Math. Stat. Prob.; University of California Press, Berkeley and Los Angeles), vol. 5, pp. 237–254.

Alston, R. E., and B. L. Turner. 1963. *Biochemical Systematics* (Prentice-Hall, Englewood Cliffs, N. J.).

Anderson, E. 1937. "Supra-specific variation in nature and in classification, from the viewpoint of botany," *Am. Nat.* 71:223–235.

Andrews, E. C. 1913. "The development of the natural order Myrtales," *Proc. Linn. Soc. New South Wales* 38:529–568.

Andrews, H. N. 1963. "Early seed plants," *Science* 142:925–931.

Antonovics, J. 1971. "The effects of a heterogeneous environment on the genetics of natural populations," *Amer. Scientist* 59:593–599.

Arber, A. 1925. *Monocotyledons, a Morphological Study* (Cambridge University Press, Cambridge, England).

——— 1934. *The Gramineae—A Study of Cereal, Bamboo and Grass* (Cambridge University Press, Cambridge, England).

Arber, E. A. N., and J. Parkin. 1907. "On the origin of angiosperms," *J. Linn. Soc. Bot.* 38:29–80.

Arldt, T. 1938. *Die Entwicklung der Kontinente und ihrer Lebewelt,* vol. I, 2nd ed. (Borntraeger, Berlin).

Arnold, C. A. 1947. *An Introduction to Paleobotany* (McGraw-Hill, New York).

Arnott, H. J. 1962. "The seed, germination and seedling of *Yucca*," *Univ. Calif. Publ. Bot.* 35:1–164.

Ashton, P. S. 1969. "Speciation among tropical forest trees; some deductions in the light of recent evidence," *Biol. J. Linn. Soc.* 1:155–196.

Axelrod, D. I. 1952. "A theory of angiosperm evolution," *Evolution* 6:29–59.

——— 1958. "Evolution of the Madro-Tertiary Geoflora," *Bot. Rev.* 24:433–509.
——— 1959. "Poleward migration of the early Angiosperm flora," *Science* 130:203–207.
——— 1960. "The evolution of flowering plants," in S. Tax, ed., *The Evolution of Life* (University of Chicago Press, Chicago), pp. 227–305.
——— 1966. "Origin of deciduous and evergreen habits in temperate forests," *Evolution* 20:1–15.
——— 1970. "Mesozoic paleogeography and early angiosperm history," *Bot. Rev.* 36:277–319.
——— 1972. "Edaphic aridity as a factor in angiosperm evolution," *Amer. Nat.* 106:311–320.
Babcock, E. B. 1947. *The Genus Crepis*, I. *Univ. Calif. Publ. Bot.*, vols. 21 and 22.
——— 1949. "Supplementary notes on *Crepis*. II. Phylogeny, distribution and Matthew's principle," *Evolution* 3:374–376.
——— 1950. "Supplementary notes on *Crepis*. III. Taproot versus rhizome in phylogeny," *Evolution* 4:358–359.
Bailey, I. W. 1944*a*. "The development of vessels in angiosperms and its significance in morphological research," *Amer. J. Bot.* 31:421–438.
——— 1944*b*. "The comparative morphology of the Winteraceae. III. Wood," *J. Arnold Arb.* 25:97–103.
——— 1949. "Origin of the angiosperms: need for a broadened outlook," *J. Arnold Arb.* 30:64–70.
——— 1956. "Nodal anatomy in retrospect," *J. Arnold Arb.* 37:269–287.
——— 1957. "The potentialities and limitations of wood anatomy in the study of the phylogeny and classification of angiosperms," *J. Arnold Arb.* 38:243–254.
——— and C. G. Nast. 1943*a*. "The comparative morphology of the Winteraceae. I. Pollen and stamens," *J. Arnold Arb.* 24:340–346.
——— and C. G. Nast. 1943*b*. "The comparative morphology of the Winteraceae. II. Carpels," *J. Arnold Arb.* 24:472–481.
——— and C. G. Nast. 1944*a*. "The comparative morphology of the Winteraceae. IV. Anatomy of the node and vascularization of the leaf," *J. Arnold Arb.* 25:215–221.
——— and C. G. Nast. 1944*b*. "The comparative morphology of the Winteraceae. V. Foliar epidermis and sclerenchyma," *J. Arnold Arb.* 25:342–348.
——— and C. G. Nast. 1945*a*. "The comparative morphology of the Winteraceae. VII. Summary and conclusions," *J. Arnold Arb.* 26:37–47.
——— and C. G. Nast. 1945*b*. "Morphology and relationships of *Trochodendron* and *Tetracentron*. I. Stem, root, and leaf," *J. Arnold Arb.* 26:143–153.
——— and C. G. Nast. 1948. "Morphology and relationships of *Illicium, Schisandra*, and *Kadsura*. I. Stem and leaf," *J. Arnold Arb.* 29:77–89.
———, C. G. Nast, and A. C. Smith. 1943. "The family Himantandraceae," *J. Arnold Arb.* 24:190–206.
——— and E. W. Sinnott. 1914. "The origin and dispersal of herbaceous angiosperms," *Ann. Bot.* 28:547–600.
——— and A. C. Smith. 1942. "Degeneriaceae, a new family of flowering plants from Fiji," *J. Arnold Arb.* 23:356–365.
——— and B. G. L. Swamy. 1948. "*Amborella trichopoda* Baill. A new morphological type of vesselless dicotyledon," *J. Arnold Arb.* 29:245–253.
——— and B. G. L. Swamy. 1951. "The conduplicate carpel and its initial trends of specialization," *Amer. J. Bot.* 38:373–379.
Baker, H. G. 1948. "Dimorphism and monomorphism in the Plumbaginaceae. I. A survey of the family," *Ann. Bot.* (n.s.) 12:207–219.
——— 1953*a*. "Dimorphism and monomorphism in the Plumbaginaceae. II. Pollen and stigmata in the genus *Limonium*," *Ann. Bot.* (n.s.) 17:433–445.
——— 1953*b*. "Dimorphism and monomorphism in the Plumbaginaceae. III. Correlation of geographical distribution patterns with dimorphism and monomorphism in *Limonium*," *Ann. Bot.* (n.s.) 17:615–627.
——— 1955. "Self-compatibility and establishment after "long-distance" dispersal," *Evolution* 9:347–348.
——— 1957. "The pollination of *Parkia* by bats and its attendant evolutionary problems," *Evolution* 11:449–460.
——— 1959. "Reproductive methods as factors in speciation in flowering plants," *Cold Spring Harbor Symp. Quant. Biol.* 24:177–191.
——— 1961*a*. "Pollination mechanisms and inbreeders. Rapid speciation in relation to changes in the breeding systems of plants," in *Recent Advances in Botany* (University of Toronto Press, Toronto), pp. 881–885.

——— 1961*b*. "The adaptation of flowering plants to nocturnal and crepuscular pollinators," *Quart. Rev. Biol.* 36:64–73.

——— 1961*c*. "*Ficus* and *Blastophaga*," *Evolution* 15:378–379.

——— 1966. "The evolution, functioning and breakdown of heteromorphic incompatibility systems. I. The Plumbaginaceae," *Evolution* 20:349–368.

——— 1967. "The evolution of weedy taxa in the *Eupatorium microstemon* species aggregate," *Taxon* 16:293–300.

——— and I. Baker. 1968. "Chromosome numbers in the Bombacaceae," *Bot. Gaz.* 129:294–296.

——— and B. J. Harris. 1959. "Bat-pollination of the silk-cotton tree, *Ceiba pentandra* (L.) Gaertn. (sensu lato), in Ghana," *J. West African Sci. Assoc.* 5:1–9.

——— and P. D. Hurd, Jr. 1968. "Intrafloral Ecology," *Ann. Rev. Entomology* 13:385–414.

Ball, E. 1950. "Isolation, removal and attempted transplants of the central portion of the shoot apex of *Lupinus albus* L.," *Amer. J. Bot.* 37:117–136.

Barber, J. T., and F. C. Steward. 1968. "The proteins of *Tulipa* and their relation to morphogenesis," *Devel. Biol.* 17:326–349.

Barker, W. G., and F. C. Steward. 1962. "Growth and development of the banana plant. 1. The growing regions of the vegetative shoot; 2. The transition from the vegetative to the floral shoot in *Musa acuminata* cv. Gros Michel," *Ann. Bot.* 26:389–423.

Barlow, B. A. 1959. "Polyploidy and apomixis in the *Casuarina distyla* species group," *Austral. J. Bot.* 7:238–251.

Barnard, C. 1955. "Histogenesis of the inflorescence and flower of *Triticum aestivum* L.," *Austral. J. Bot.* 3:1–20.

Bateman, A. J. 1951. "The taxonomic discrimination of bees," *Heredity* 5:271–278.

Bate-Smith, E. C. 1962. "The phenolic constituents of plants and their taxonomic significance. I. Dicotyledons," *J. Linn Soc. Bot.* 58:95–173.

——— 1965. "Recent progress in the chemical taxonomy of some phenolic constituents of plants," *Bull. Soc. Bot. France, Mém.* 1965:16–28.

——— 1972. "Chemistry and phylogeny of angiosperms," *Nature* 236:353–354.

——— and T. Swain. 1966. "The asperuloides and the aucubins," in T. Swain, ed., *Comparative Phytochemistry* (Academic Press, New York), pp. 159–174.

Baude, E. 1956. "Die Embryoentwicklung von *Stratiotes aloides* L.," *Planta* 46:649–671.

Baum, H. 1952. "Der Bau des Karpelstiels von *Grevillea thelemanniana* und seine Bedeutung für die Beurteilung der epeltaten Karpelle," *Phytomorphology* 2:191–197.

——— 1953. "Die Karpelle von *Eranthis hiemalis* und *Cimicifuga americana* als weitere Verbindungsglieder zwischen peltaten und epeltaten Karpellen," *Osterr. Bot. Z.* 100:353–357.

Benson, L., E.A. Phillips, and P. A. Wilder. 1967. "Evolutionary sorting of characters in a hybrid swarm. I. Direction of slope," *Amer. J. Bot.* 54:1017–1026.

Berg, R. Y. 1954. "Development and dispersal of the seed of *Pedicularis silvatica*." *Nytt. Mag. Bot.* 2:1–158.

——— 1958. "Seed dispersal, morphology, and phylogeny of *Trillium*," *Norske Vidensk-Akad. Oslo, Mat.-Naturv. Klasse* 1:1–36.

——— 1959. "Seed dispersal, morphology, and taxonomic position of *Scoliopus*, Liliaceae," *Norske Vidensk-Akad. Oslo, Mat.-Naturv. Klasse.* 4:1–56.

——— 1966. "Seed dispersal of *Dendromecon*: its ecologic, evolutionary, and taxonomic significance," *Amer. J. Bot.* 53:61–73.

——— 1972. "Dispersal ecology of *Vancouveria* (Berberidaceae)," *Amer. J. Bot.* 59:109–122.

Bernier, G. 1964. "Etude histophysiologique et histologique de l'évolution du méristème apical de *Sinapis alba* L. cultivé en milieu conditioné et en diverses durées de jours favorables à la mise à fleurs," *Mém. Acad. Roy. Belg. Cl. Sci.* 16:1–150.

Berry, E. W. 1920. "Palaeobotany: a sketch of the origin and evolution of floras," *Ann. Rep. Smithsonian Inst.* 1918:289–407.

Bersillon, G. 1957. "Ontogénie comparée de la tige feuillée et de l'inflorescence chez *Oenothera biennis* L.," *C. R. Acad. Sci. Paris* 245:1455–1458.

Bessey, C. E. 1897. "Phylogeny and taxonomy of angiosperms," *Bot. Gaz.* 24:145–178.

——— 1915. "The phylogenetic taxonomy of flowering plants," *Ann. Missouri Bot. Gard.* 2:109–164.

Bews, J. W. 1927. "Studies in the ecological evolution of angiosperms," *New Phytol.*, Reprint No. 16:2–37.

——— 1929. *The World's Grasses, Their Differentiation,*

Distribution Economics and Geology (Longmans Green, London and New York).
Bhar, D. S., and N. W. Radforth. 1969. "Vegetative and reproductive development of shoot apices of *Pharbitis nil* as influenced by photoperiodism," *Canad. J. Bot.* 47:1403–1406.
Björkman, O., J. E. Boynton, M. A. Nobs, and R. W. Pearcy. 1971. "Physiological ecology investigations," *Carnegie Inst. Yearbook*, No. 69:624–648.
Black, G. A., Th. Dobzhansky, and C. Pavan. 1950. "Some attempts to estimate species diversity and population density of trees in Amazonian forests," *Bot. Gaz.* 111:413–425.
Black, J. N. 1956. "The influence of seed size and depth of sowing on pre-emergence and early vegetative growth of subterranean clover (*Trifolium subterraneum* L.)," *Austral. J. Agr. Res.* 7:98–109.
——— 1957. "The early vegetative growth of three strains of subterranean clover (*Trifolium subterraneum* L.) in relation to size of seed," *Austral. J. Agr. Res.* 8:1–14.
——— 1959. "Seed size in herbage legumes," *Herb. Abstr.* 29:235–241.
Blaser, H. W. 1944. "Studies in the morphology of the Cyperaceae. II. The prophyll," *Amer. J. Bot.* 31:53–64.
Boke, N. H. 1947. "Development of the adult shoot apex and floral initiation in *Vinca rosea* L.," *Amer. J. Bot.* 34:433–439.
——— 1964. "The cactus gynoecium: a new interpretation," *Amer. J. Bot.* 51:598–601.
Bolkhovskikh, Z., V. Grif, T. Matvejeva, and O. Zakharyeva. 1969. *Chromosome Numbers of Flowering Plants* (Academy of Sciences, USSR, Komarov Botanical Institute). In Russian.
Bonnett, O. T. 1937. "The development of the oat panicle," *J. Agr. Res.* 54:927–931.
——— 1961. "The oat plant: its histology and development," *Bull. 672. Univ. Ill. Agr. Exper. Sta.*
Borthwick, H. A., and M. W. Parker. 1938. "Influence of photoperiods upon the differentiation of meristems and the blossoming of Biloxi soy beans," *Bot. Gaz.* 99:825–839.
Boulter, D., D. A. Thurman, and E. Derbyshire. 1967. "A disc electrophoretic study of globulin proteins of legume seeds with reference to their systematics," *New Phytol.* 66:27–36.
——— E. W. Thompson, J. A. M. Ramshaw, and M. Richardson. 1970*a*. "Higher plant cytochrome c," *Nature* 228:552–554.
——— M. V. Laycock, A. M. Ramshaw, and E. W. Thompson. 1970*b*. "Amino acid sequence studies of plant cytochromes c," *Taxon* 19:561–564.
Boyd, L. 1932. "Monocotyledonous seedlings. Morphological studies in the post-seminal development of the embryo," *Trans. Bot. Soc. Edinburgh* 31:1–224.
Bradshaw, A. D. 1965. "Evolutionary significance of phenotypic plasticity in plants," *Advances in Genetics* 13:115–155.
——— 1971. "Plant evolution in extreme environments," in R. Creed, ed., *Ecological Genetics and Evolution* (Blackwell Scientific Publications, Oxford and Edinburgh), pp. 20–50.
Brandza, M. 1891. "Développement des téguments de la graine," *Rev. Gén. Bot.* 3:1–32, 105–117, 150–165, 229–240.
Brewbaker, J. L. 1967. "The distribution and phylogenetic significance of binucleate and trinucleate pollen grains in the angiosperms," *Amer. J. Bot.* 54:1069–1083.
Briggs, B. G. 1964. "The control of interspecific hybridization in *Darwinia*," *Evolution* 18:292–303.
Britten, R. J., and E. H. Davidson. 1969. "Gene regulation for higher cells: a theory," *Science* 165: 349–357.
——— and D. E. Kohne. 1968. "Repeated sequences in DNA," *Science* 161:529–540.
Brown, R. W. 1956. "Palmlike plants from the Dolores formation (Triassic) Southwestern Colorado," *U. S. Geol. Surv. Prof. Paper* 274H:205–209.
Burtt, B. L. 1960. "Compositae and the study of functional evolution," *Trans. Proc. Bot. Soc. Edinburgh* 39:216–232.
——— 1961. "Interpretive morphology," *Notes Roy. Bot. Garden Edinburgh* 23:569–572.
——— 1970. "Studies in the Gesneriaceae of the Old World. XXXI. Some aspects of functional evolution," *Notes Roy. Bot. Garden Edinburgh* 30:1–10.
Butzin, F. 1965. "Neue Untersuchungen über die Blüte der Gramineae," Inaug. Diss., *Math. Nat. Fakultät der Freien Univ. Berlin*:3–183.
Buvat, R. 1951. "Transformation du point végétatif de *Myosurus minimus* L. en méristème floral," *C. R. Acad. Sci. Paris* 232:2466–2468.
——— 1952. "Structure, évolution, et fonctionnement

du méristème apical de quelques dicotylédones," *Ann. Sci. Nat.*, sér. 11, 13:198–300.

Buxbaum, F. 1961. "Vorläufige Untersuchungen über Umfang, systematische Stellung und Gliederung der Caryophyllales (Centrospermae)," *Beitr. Biol. Pflanz.* 36:3–56.

Cain, A. J. 1969. "Speciation in tropical environments: summing up," *Biol. J. Linn. Soc.* 1:233–236.

Cammerloher, H. 1923. "Zur Biologie der Blüte von *Aristolochia grandiflora* Swartz," *Oesterr. Bot. Z.* 72:180–198.

Camp, W. H. 1947. "Distribution patterns in modern plants and problems of ancient dispersals," *Ecol. Mon.* 17:159–183.

de Candolle, C. 1891. "Recherches sur les inflorescences épiphylles," *Mém. Soc. Phys. Hist. Nat. Genève Suppl.* 6:1–37.

Canright, J. E. 1952. "The comparative morphology and relationships of the Magnoliaceae. I. Trends of specialization in the stamens," *Amer. J. Bot.* 39:484–497.

——— 1955. "The comparative morphology and relationships of the Magnoliaceae. IV. Wood and nodal anatomy," *J. Arnold Arb.* 36:119–140.

——— 1960. "The comparative morphology and relationships of the Magnoliaceae. III. Carpels," *Amer. J. Bot.* 47:145–155.

Carlquist, S. 1962. "A theory of paedomorphosis in dicotyledonous woods," *Phytomorphology* 12:30–45.

——— 1966*a*. "The biota of long-distance dispersal. I. Principles of dispersal and evolution," *Quart. Rev. Biol.* 41:247–270.

——— 1966*b*. "Wood anatomy of Compositae: a summary, with comments on factors controlling wood evolution," *Aliso* 6:25–44.

——— 1967. "Anatomy and systematics of *Dendroseris* (sensu lato)," *Brittonia* 19:99–121.

——— 1969*a*. "Wood anatomy of Goodeniaceae and the problem of insular woodiness," *Ann. Missouri Bot. Gard.* 56:358–390.

——— 1969*b*. "Wood anatomy of Lobelioideae (Campanulaceae)," *Biotropica* 1:47–72.

——— 1969*c*. "Toward acceptable evolutionary interpretations of floral anatomy," *Phytomorphology* 19:332–362.

——— 1970. "Wood anatomy of *Echium* (Boraginaceae)," *Aliso* 7:183–199.

Carr, S. G. M., and D. J. Carr. 1961. "The functional significance of syncarpy," *Phytomorphology* 11:249–256.

Cave, M. S., H. J. Arnott, and S. A. Cook. 1961. "Embryogeny in the California peonies with reference to their taxonomic position," *Amer. J. Bot.* 48:397–404.

Chambers, K. L. 1955. "A biosystematic study of the annual species of *Microseris*," *Contr. Dudley Herb. Stanford Univ.* 4:207–312.

Cheadle, V. I. 1942. "The occurrence and types of vessels in various organs of the plant in the Monocotyledoneae," *Amer. J. Bot.* 29:441–450.

——— 1943. "Vessel specialization in the late metaxylem of the various organs in the Monocotyledoneae," *Amer. J. Bot.* 30:484–490.

——— 1953. "Independent origin of vessels in the monocotyledons and dicotyledons," *Phytomorphology* 3:23–44.

Clausen, J. 1951. *Stages in the Evolution of Plant Species* (Cornell University Press, Ithaca, New York).

——— and W. M. Hiesey. 1958. "Experimental studies on the nature of species. IV. Genetic structure of ecological races," *Carnegie Inst. Wash. Publ.* No. 615.

——— D. D. Keck, and W. M. Hiesey. 1940. "Experimental studies on the nature of species. I. The effect of varied environments on western North American plants," *Carnegie Inst. Wash. Publ.* No. 520.

——— D. D. Keck, and W. M. Hiesey. 1948. "Experimental studies on the nature of species. III. Environmental responses of climatic races of *Achillea*," *Carnegie Inst. Wash. Publ.* No. 581.

Cockayne, L. 1899. "An inquiry into the seedling forms of New Zealand phanerogams and their development," *Trans. New Zealand Inst.* 31:354–398.

Constance, L. 1964. "Systematic botany—an unending synthesis," *Taxon* 13:257–273.

Cook, S. A. 1968. "Adaptation to heterogeneous environments. I. Variation in heterophylly in *Ranunculus flammula* L.," *Evolution* 22:496–516.

Corner, E. J. H. 1946. "Centrifugal stamens," *J. Arnold Arb.* 27:423–437.

——— 1949. "The durian theory or the origin of the modern tree," *Ann. Bot.* 52:367–414.

——— 1953. "The durian theory extended. I," *Phytomorphology* 3:465–476.

——— 1954*a*. "The durian theory extended. II. The arillate fruit and the compound leaf," *Phytomorphology* 4:152–165.

——— 1954*b*. "The durian theory extended. III. Pachycauly and megaspermy. Conclusion," *Phytomorphology* 4:263–274.

——— 1954*c*. *The Evolution of Tropical Forest: Evolution as a Process* (Allen and Unwin, London), pp. 34–46

——— 1958. "Transference of function," *J. Linn. Soc. (Bot.-Zool.)* 56:33–40.

Correll, D. S. 1950. *Native Orchids of North America, North of Mexico* (Chronica Botanica, Waltham, Mass.).

Corson, G. E., Jr., and E. M. Gifford, Jr. 1969. "Histochemical studies of the shoot apex of *Datura stramonium* during transition to flowering," *Phytomorphology* 19:189–196.

Cowan, R. S. 1968. "*Swartzia* (Leguminosae, Caesalpinoideae, Swartzieae)," *Flora Neotropical Monograph* 1:1–228.

Craig-Holmes, A. P., and M. W. Shaw. 1971. "Polymorphism of human constitutive heterochromatin," *Science* 174:702–704.

Cronquist, A. 1955. "Phylogeny and taxonomy of the Compositae," *Amer. Midl. Nat.* 53:478–511.

——— 1968. *The Evolution and Classification of Flowering Plants* (Houghton Mifflin, Boston).

Crosby, J. L. 1940. "High proportions of homostyle plants in populations of *Primula vulgaris*," *Nature* 145:672.

——— 1959. "Outcrossing on homostyle primroses," *Heredity* 13:127–131.

Crosswhite, F. S., and C. D. Crosswhite. 1966. "Insect pollinators of *Penstemon* series *Graciles* (Scrophulariaceae) with notes on *Osmia* and other Megachilidae," *Amer. Midl. Nat.* 76:450–467.

da Cunha, A. B., and Th. Dobzhansky. 1954. "A further study of chromosomal polymorphism in *Drosophila willistoni* in its relation to the environment," *Evolution* 8:119–134.

Cusick, F. 1956. "Studies of floral morphogenesis. I. Median bisections of flower primordia in *Primula Bulleyana* Forrest," *Trans. Roy. Soc. Edinburgh* 63:153–166.

Cutter, E. 1971. *Plant Anatomy: Experiment and Interpretation. Part 2. Organs* (Arnold, London).

Darlington, C. D., and K. Mather. 1949. *The Elements of Genetics* (Allen and Unwin, London).

Dandy, J. E. 1927. "The genera of Magnoliaceae," *Kew Bull.* 1927:257–264.

Darrah, W. C. 1960. *Principles of Paleobotany* (Ronald Press, New York).

Darwin, C. 1877. *The Effects of Cross and Self-Fertilization in the Vegetable Kingdom* (Appleton, New York).

——— 1880. *The Different Forms of Flowers on Plants of the Same Species* (Murray, London).

Datta, S. C., M. Evenari, and Y. Gutterman. 1970. "The heteroblasty of *Aegilops ovata* L.," *Israel J. Bot.* 19:463–483.

Davis, G. L. 1966. *Systematic Embryology of the Angiosperms* (Wiley, New York).

Davis, P. H., and V. H. Heywood. 1963. *Principles of Angiosperm Taxonomy* (Van Nostrand, Princeton, N. J.).

De Beer, G. R. 1951. *Embryos and Ancestors*, (rev. ed.; Oxford University Press, New York).

Delevoryas, T. 1962. *Morphology and Evolution of Fossil Plants* (Holt, Rinehart, and Winston, New York).

Dempster, L. T., and G. L. Stebbins. 1968. "A cytotaxonomic revision of the fleshy-fruited *Galium* species of the Californias and Southern Oregon (Rubiaceae)," *Univ. Calif. Publ.* 46:1–57.

Denffer, D. V. 1952. "Durch die Behandlung mit 2,3,5 Trijodbenzoessigsäure hervorgerufene Gamophyllien," *Ber. Deu. Bot. Ges.* 64:269–274.

DeVlaming. V., and V. W. Proctor. 1968. "Dispersal of aquatic organisms: Viability of seeds recovered from the droppings of captive killdeer and mallard ducks," *Amer. J. Bot.* 55:20–26.

Dickison, W. C. 1967. "Comparative morphological studies in Dilleniaceae. I. Wood anatomy," *J. Arnold Arb.* 48:1–23.

——— 1968. "Comparative studies in the Dilleniaceae. II. The carpels," *J. Arnold Arb.* 49:317–329.

——— 1969. "Comparative morphological studies in Dilleniaceae. IV. Anatomy of the node and vascularization of the leaf," *J. Arnold Arb.* 50:384–400.

Dickson, J. 1936. "Studies in floral anatomy. III. An interpretation of the gynoecium in the Primulaceae," *Amer. J. Bot.* 23:385–393.

Diels, L. 1908. *Pflanzengeographie* (Borntraeger, Berlin).

——— 1910. "Käferblumen bei den Ranales und ihre Bedeutung für die Phylogenie der Angiospermen," *Ber. Deu. Bot. Ges.* 34:758–774.

Diettert, R. A. 1938. "The morphology of *Artemisia tridentata* Nutt.," *Lloydia* 1:3–74.
Diomauito-Bonnand, J. 1966. "Morphologie et ontogénie de quelques espèces de *Nicotiana.* Etude de l'inflorescence," *Rev. Cytol. Biol. Vég.* 29:1–172.
Dobzhansky, Th. 1951. *Genetics and the Origin of Species* (third ed.; Columbia University Press, New York).
——— 1968*a*. "On some fundamental concepts of Darwinian biology," Th. Dobzhansky, M. K. Hecht, and W. C. Steere, eds., *Evolutionary Biology,* vol 2:1–34.
——— 1968*b*. "Are they compatible? On Cartesian and Darwinian aspects of biology," *Grad. J.* 8:99–117.
———1970. *Genetics of the Evolutionary Process* (Columbia University Press, New York).
Dodson, C. H., R. L. Dressler, H. G. Hills, R. M. Adams, and N. H. Williams. 1969. "Biological active compounds in Orchid fragrances," *Science* 164:1243–1249.
Dormer, K. J., and E. G. Cutter. 1959. "On the arrangement of flowers on the rhizomes of some Nymphaeaceae," *New Phytol.* 58:176–181.
Douglas, A. G., and G. Eglinton. 1966. "The distribution of alkanes," in T. Swain, ed., *Comparative Phytochemistry* (Academic Press, New York), pp. 57–77.
Douglas, G. E. 1957. "The inferior ovary. II," *Bot. Rev.* 23:1–46.
Dowrick, P. J. 1956. "Heterostyly and homostyly in *Primula obconica,*" *Heredity* 10:219–236.
Doyle, J. A. 1969. "Cretaceous angiosperm pollen of the Atlantic coastal plain and its evolutionary significance," *J. Arnold Arb.* 50:1–35.
Dressler, R. L. 1968. "Pollination by Euglossine bees," *Evolution* 22:202–210.
Dronamraju, K. R. 1958. "The visits of insects to different coloured flowers of *Lantana camara* L.," *Current Sci.* 27:452–453.
——— 1960. "Selective visits of butterflies to flowers: a possible factor in sympatric speciation," *Nature* 186:178.
——— and H. Spurway. 1960. "Constancy to horticultural varieties shown by butterflies and its possible evolutionary significance," *J. Bombay Nat. Hist. Soc.* 57:136–143.
Eames, A. J. 1931. "The vascular anatomy of the flower, with refutation of the theory of carpel polymorphism," *Amer. J. Bot.* 18:147–188.
——— 1961. *Morphology of the Angiosperms* (McGraw-Hill, New York).
——— and L. H. MacDaniels. 1947. *An Introduction to Plant Anatomy* (2nd. ed.; McGraw-Hill, New York).
Eckardt, T. 1964. "Die natürliche Verwandtschaft bei den Blütenpflanzen," *Umschau* 64:496–502.
Ehrendorfer, F. 1951. "Rassengliederung, Variabilitätszentren und geographische Merkmalsprogressionen als Ausdruck der raum-zeitlichen Entfaltung des Formenkreises *Galium incanum* S. S.," *Oesterr. Bot. Z.* 98:428–490.
——— 1965. "Dispersal mechanisms, genetic systems, and colonizing abilities in some flowering plant families," in H. G. Baker and G. L. Stebbins, eds., *The Genetics of Colonizing Species* (Academic Press, New York), pp. 331–352.
——— F. Krendl, E. Habeler, and W. Sauer. 1968. "Chromosome number and evolution in primitive Angiosperms," *Taxon* 17:337–353.
Ehrlich, P. R., and P. H. Raven. 1969. "Differentiation of populations," *Science* 165:1228–1232.
Eiseley, L. 1958. *Darwin's Century* (Doubleday, New York).
Emberger, L. 1950. "La valeur morphologique et l'origine de la fleur (à propos d'une théorie nouvelle)," *Ann. Biol.* III 54:279–296.
——— 1960. "Les végétaux vasculaires," in M. Chadefaud and L. Emberger *Traité de Botanique,* II (Masson, Paris).
Endress, P. K. 1967. "Systematische Studie über die verwandtschaftlichen Beziehungen zwischen den Hamamelidaceen und Betulaceen," *Bot. Jahrb.* 87:431–525.
Engard, C. J. 1944. "Organogenesis in *Rubus,*" *Univ. Hawaii Res. Publ.* 21.
Engler, A. 1928. "Saxifragaceae," in A. Engler and K. Prantl, *Die Natürlichen Pflanzenfamilien* (2nd ed.; Duncker and Humboldt, Berlin), 18a:74–226.
Erdtman, G. 1966. *Pollen Morphology and Plant Taxonomy* (Hafner, New York, London).
Ertl, O. 1932. "Vergleichende Untersuchungen über die Entwicklung der Blattnervatur der Araceen," *Flora* 26:116–248.
Evans, M. W., and F. O. Grover. 1940. "Developmental morphology of the growing point of the

shoot and the inflorescence in grasses," *J. Agr. Res.* 61:481–520.
Eyde, R. H. 1969. "Flower of *Tetraplasandra gymnocarpa:* hypogyny with epigynous ancestry," *Science* 166:506–508.
Faegri, K., and L. van der Pijl. 1966. *The Principles of Pollination Ecology* (Pergamon, London).
Fagerlind, F. 1946. "Strobilus und Blüte von *Gnetum* und die Möglichkeit, aus ihrer Struktur den Blütenbau der Angiospermen zu deuten," *Ark. Bot.* 33A(8):1–57.
Fairbrothers, D. E. 1966. "Comparative serological studies in plant systematics," *Bull. Serol Mus., New Brunswick* 35:2–6.
——— and M. A. Johnson. 1961. "The precipitin reaction as an indicator of relationships in some grasses," *Recent Advances in Botany* (University of Toronto Press, Toronto), vol. 1, pp. 116–120.
——— and M. A. Johnson. 1964. "Comparative serological studies within the families Cornaceae (dogwood) and Nyssaceae (sour gum)," in C. A. Leone, ed., *Taxonomy, Biochemistry and Serology* (Ronald Press, New York), pp. 305–318.
Fedorov, A. A. 1966. "The structure of the tropical rain forest and speciation in the humid tropics," *J. Ecol.* 54:1–11.
Ferguson, I. K., and D. A. Webb. 1970. "A pollen morphology in the genus *Saxifraga* and its taxonomic significance," *Bot. J. Linn. Soc.* 63:295–311.
Fernald, M. L. 1950. *Gray's Manual of Botany* (8th ed.; American Book, New York).
Fiori, A. 1921. *Flora Italiana Illustrata* (Sasciano Val di Pesa, Stab. Tipo-Litografico Fratelli Santi).
Fisher, R. A. 1930. *The Genetical Theory of Natural Selection* (Clarendon Press, Oxford).
van Fleet, D. S. 1959. "Analysis of the histochemical localization of peroxidase related to the differentiation of plant tissues." *Canad. J. Bot.* 37:449–458.
Florin, R. 1944. "Die Koniferen des Oberkarbons und des unteren Perms," *Paleontographica* 85:365–654.
——— 1950. "Upper Carboniferous and lower Permian conifers," *Bot. Rev.* 16:258–282.
——— 1951. "Evolution in Cordaites and Conifers," *Acta Hort. Berg.* 15:285–388.
——— 1953. "The female reproductive organs of conifers and taxads," *Biol. Rev.* 29:367–389.
Ford, E. B. 1964. *Ecological Genetics* (Wiley, New York).
Forde, M. B., and M. M. Blight. 1964. "Geographical variation in the turpentine of Bishop Pine," *New Zealand J. Bot.* 2:44–52.
Foster, A. S. 1950. "Venation and histology of the leaflets in *Touroulia guianensis* Aubl. and *Froesia tricarpa* Pires," *Amer. J. Bot.* 37:848–862.
——— 1951. "Heterophylly and foliar variation in *Lacunaria,*" *Bull. Torr. Bot. Club* 78:382–400.
——— 1952. "Foliar venation in angiosperms from an ontogenetic standpoint," *Amer. J. Bot.* 39:752–766.
Fraenkel, G. S. 1959. "The raison d'être of secondary plant substances," *Science* 129:1466–1470.
Frankel, O. H., and J. B. Hair. 1937. "Studies on the cytology, genetics, and taxonomy of New Zealand *Hebe* and *Veronica* (Part 1)," *New Zealand J. Sci. Tech.* 18:669–687.
Fries, R. E. 1919. "Studien über die Blütenstandverhaltnis bei der Familie Annonaceae," *Acta Horti. Berg.* 6(6):48.
——— 1959. "Annonaceae," in A. Engler and K. Prantl, *Die Natürlichen Pflanzenfamilien* (2nd ed.; Duncker and Humboldt, Berlin), 17a II:1–171.
Fritsche, E. 1955. "Quelques observations concernant la biologie de *Colchicum autumnale,*" *Bull. Acad. Roy. Belgique* (Cl. des Sci.) 41:238–258.
Fryxell, P. A. 1957. "Mode of reproduction of higher plants," *Bot. Rev.* 23:135–233.
——— 1965. "Stages in the evolution of *Gossypium* L.," *Adv. Front. Plant Sci.* 10:31–56.
Fukuda, I. 1961. "On insects visiting *Trillium* flowers," *Essays and Studies* (Tokyo Women's Christian College), 12:23–34.
——— 1968. "Chromosomal compositions of natural populations in *Trillium ovatum.,*" *Sci. Rep. Tokyo Women's Christian Coll.* 10:110–137.
Gajewski, W. 1959. "Evolution in the genus *Geum,*" *Evolution* 13:378–388.
——— 1964. "The heredity of seed-dispersing mechanisms in *Geum,*" *Genetics Today, Proc. XI Int. Congr. Genet.*, pp. 423–430.
Galil, J. 1968. "Biological studies on the seedling of *Colchicum steveni* Kunth.," *Beitr. Biol. Pfl.* 45:243–256.
Galston, A. W., and P. J. Davies. 1969. "Hormonal regulation in higher plants," *Science* 163:1288–1297.
Galun, E., J. Gressel, and A. Keynan. 1964. "Suppression of floral induction by actinomycin D

—an inhibitor of "messenger" RNA synthesis," *Life Sci.* 3:911–915.
Ganong, W. F. 1901. "The cardinal principles of morphology," *Bot. Gaz.* 31:426–434.
Gates, D. M., R. Alderfer, and E. Taylor. 1968. "Leaf temperatures of desert plants," *Science* 159: 994–995.
Gatin, C. L. 1906. "Recherches anatomiques et chimiques sur la germination des palmiers," *Ann Sci. Nat. Bot.*, ser. 9, III:191–315.
Gaussen, H. 1946. *Les Gymnospermes, Actuelles et Fossiles* (Trav. Lab. Forest. Toulouse), vol. 2, sec. 1.
Gelius, L. 1967. "Studien zur Entwicklungsgeschichte an Blüten der Saxifragales sensu lato mit besonderer Berücksichtigung des Androeceums," *Bot. Jahrb.* 87:253–303.
Gifford, E. M., Jr. 1951. "Early ontogeny of the foliage leaf in *Drimys Winteri* var. *chilensis*," *Amer. J. Bot.* 38:93–104.
——— 1969. "Initiation and early development of the inflorescence in pineapple (*Ananas comosus*, var. Smooth Cayenne) treated with acetylene," *Amer. J. Bot.* 56:892–897.
——— and G. E. Corson, Jr. 1971. "The shoot apex in seed plants," *Bot. Rev.* 37:143–229.
——— and K. D. Stewart. 1965. "Ultrastructure of vegetative and reproductive apices of *Chenopodium album*," *Science* 149:75–77.
——— and H. B. Tepper. 1961. "Ontogeny of the inflorescence in *Chenopodium album*," *Amer. J. Bot.* 48:657–667.
——— and H. B. Tepper. 1962. "Ontogenetic and histochemical changes in the vegetative shoot tip of *Chenopodium album*," *Amer. J. Bot.* 49:902–911.
Gilg, E., and E. Werdermann. 1925. "Dilleniaceae," in A. Engler and K. Prantl, *Die Natürliche Pflanzenfamilien* (2nd ed.; Duncker and Humboldt, Berlin), 21:7–36.
Goebel, K. 1931. "Blütenbildung und Sprossgestaltung," *Org. der Pflanzen*, Suppl. 2. 242 pp. (Fischer, Jena).
Goldschmidt, R. 1940. *The Material Basis of Evolution* (Yale University Press, New Haven, Conn.).
Good, R. 1956. *Features of Evolution in the Flowering Plants* (Longman's, London).
Gorter, C. J. 1951. "The influence of 2,3,5-triiodobenzoic acid on the growing points of tomatoes. II. The initiation of ring fasciations," *Proc. Kon. Nederl. Akad. Wetens.* 54:181–190.
Gottlieb, L. D. 1971. "Evolutionary relationships in the outcrossing diploid annual species of *Stephanomeria* (Compositae)," *Evolution* 25:312–329.
Gottsberger, G. 1970. "Beiträge zur Biologie von Annonaceen-Blüten," *Oesterr. Bot. Z.* 118:237–279.
Gottschalk, W. 1971. *Die Bedeutung der Genmutationen für die Evolution der Pflanzen* (Fischer, Stuttgart).
Grant, K. A. 1966. "A hypothesis concerning the prevalence of red coloration in California hummingbird flowers," *Amer. Nat.* 100:85–97.
——— and V. Grant. 1964. "Mechanical isolation of *Salvia apiana* and *Salvia mellifera*," *Evolution* 18:196–212.
——— and V. Grant. 1967. "Effects of hummingbird migration on plant speciation in the California flora," *Evolution* 21:457–465.
Grant, V. 1949. "Pollination systems as isolating mechanisms in angiosperms," *Evolution* 3:82–97.
——— 1950*a*. "The pollination of *Calycanthus occidentalis*," *Amer. J. Bot.* 37:294–297.
——— 1950*b*. "The protection of the ovules in flowering plants," *Evolution* 4:179–201.
——— 1952. "Isolation and hybridization between *Aquilegia formosa* and *A. pubescens*," *Aliso* 2:341–360.
——— 1963. *The Origin of Adaptations* (Columbia University Press, New York).
——— 1964. "The biological composition of a taxonomic species in *Gilia*," *Adv. Genet.* 12:281–328.
——— 1971. *Plant Speciation* (Columbia University Press, New York).
——— and K. A. Grant. 1965. *Flower Pollination in the Phlox Family* (Columbia University Press, New York).
——— and K. A. Grant. 1966. "Records of hummingbird pollination in the Western American flora. I. Some California species," *Aliso* 6:51–66.
Gregory, M. P. 1956. "A phyletic rearrangement in the *Aristolochiaceae*," *Amer. J. Bot.* 43:110–122.
Gregory, W. K. 1936. "On the meaning and limits of irreversibility of evolution," *Amer. Nat.* 70:517–528.
Gut, B. J. 1966. "Beiträge zur Morphologie des Gynaeciums und der Blütenachse einiger Rutaceen," *Bot. Jahrb.* 85:151–247.
Haccius, B. 1952. "Die Embryoentwicklung bei *Ottelia alismoides* und das Problem des terminalen Monokotylen-Keimblattes," *Planta* 40:433–460.

——— 1954. "Embryologische und histologische Studien an 'monokotylen Dikotylen.' I. *Claytonia virginica* L.," *Oesterr. Bot. Z.* 101:285–303.

——— 1960. "Experimentell induzierte Einkeimblättrigkeit bei *Eranthis hiemalis.* II. Monokotylie durch Phenylborsäure," *Planta* 54:482–497.

——— and G. Trompeter. 1960. "Experimentell induzierte Einkeimblättrigkeit bei *Eranthis hiemalis.* I. Synkotylie durch 2,4-dichlorphenoxessigsäure," *Planta* 54:466–481.

Hagemann, W. 1960. "Kritische Untersuchungen über die Organisation des Sprossscheitels dikotyler Pflanzen," *Oesterr. Bot. Z.* 107:366–402.

——— 1963. "Die morphologische Sprossdifferenzierung und die Anordnung des Leitgewebes," *Ber. Deu. Bot. Ges.* 76. Sondernummer. I. Generalversammlungsheft:113–120.

Hagerup, O. 1934. "Zur Abstammung einiger Angiospermen durch Gnetales und Coniferae," *Biol. Medd.* 11(4):1–82.

——— 1936. "Zur Abstammung einiger Angiospermen durch Gnetales und Coniferae. II. *Centrospermae,*" *Biol. Medd.* 13.

——— 1938. "On the origin of some angiosperms through the Gnetales and the Coniferae. III. The gynoecium of *Salix cinerea,*" *Biol. Medd.* 14(4):1–34.

——— 1939. "On the origin of some Angiosperms through the Gnetales and the Coniferae. IV. The gynoecium of Personatae," *Biol. Medd.* 15(2):1–39.

——— 1950. "Rain-pollination," *Kgl. Danske Vidensk. Selsk. Biol. Medd.* 18:1–19

——— 1951. "Pollination in the Faroes—in spite of rain and poverty of insects," *Kgl. Danske Vidensk. Selsk. Biol. Medd.* 18:3–48.

Hair, J. B., and E. J. Beuzenberg. 1958. "Chromosomal evolution in the Podocarpaceae," *Nature* 181:1584–1586.

Hall, H. M., and F. E. Clements. 1923. *The Phylogenetic Method in Taxonomy: the North American Species of Artemisia, Chrysothamnus and Atriplex* (Carnegie Inst. Wash. Publ. No. 326).

Hall, M. T. 1961. "Teratology in *Trillium grandiflorum,*" *Amer. J. Bot.* 48:803.

Hallier, H. 1912. "L'Origine et la système phylétique des angiospermes exposés a l'aide de leur arbre généalogique," *Archives Néerl.*, ser. IIIB, 1:146–234.

Halvorson, H. O. 1965. "Sequential expression of biochemical events during intracellular differentiation," *Symp. Soc. General Microbiol.* 343–368.

Hammond, D. 1941. "The expression of genes for leaf shape in *Gossypium hirsutum* L. and *Gossypium arboreum* L. I and II.," *Amer. J. Bot.* 28:124–150.

Handel-Mazzetti, H. von. 1907. *Monographie der Gattung Taraxacum* (Leipzig and Vienna).

Harborne, J. S. 1966. "The evolution of flavonoid pigments in plants," in T. Swain, ed., *Comparative Phytochemistry* (Academic Press, New York), pp. 271–295.

Harder, R., and A. Oppermann. 1952. "Einfluss von 2,3,5-Trijodbenzoesäure auf die Blütenbildung und die vegetative Gestaltung von *Kalanchoë Blossfeldiana,*" *Planta* 41:1–24.

Hardy, A. C. 1954. "Escape from specialization," in J. Huxley, A. C. Hardy, and E. B. Ford, eds., *Evolution as a Process* (Allen and Unwin, London), pp. 122–142.

Harlan, J. R. 1945*a*. "Cleistogamy and chasmogamy in *Bromus carinatus* Hook and Arn.," *Amer. J. Bot.* 32:66–72.

——— 1945*b*. "Natural breeding structure in the *Bromus carinatus* complex as determined by population analyses," *Amer. J. Bot.* 32:142–148.

Harling, G. 1958. "Monograph of the Cyclanthaceae," *Acta Hort. Berg.* 18(1):1–428.

Harms, H. 1917. "Ueber eine Meliacee mit blattbürtigen Blüten," *Ber. Deu. Bot. Ges.* 35:338–348.

Harper, J. L. 1965. "Establishment, aggression and cohabitation in weedy species," in H. G. Baker and G. L. Stebbins, eds., *The Genetics of Colonizing Species* (Academic Press, New York), pp. 243–265.

——— and A. P. Chancellor. 1959. "The comparative biology of closely related species living in the same area. IV. *Rumex:* Interference between individuals in populations of one and two species," *J. Ecol.* 47:679–695.

——— and J. N. Clatworthy. 1963. "The comparative biology of closely related species. VI. Analysis of the growth of *Trifolium repens* and *T. fragiferum* in pure and mixed populations," *J. Exp. Bot.* 14:172–190.

——— P. H. Lovell, and K. G. Moore. 1970. "The shapes and sizes of seeds," *Ann. Rev. Ecol. Syst.* 1:327–356.

——— and M. Obeid. 1967. "Influence of seed size and depth of sowing on the establishment and growth of varieties of fiber and oil seed flax," *Crop Sci.* 7:527–532.

——— and J. Ogden. 1970. "The reproductive strategy of higher plants. I. The concept of strategy with special reference to *Senecio vulgaris* L.," *J. Ecol.* 58:681–698.

——— J. T. Williams, and G. R. Sagar. 1965. "The behavior of seeds in soil. I. The heterogeneity of soil surfaces and its role in determining the establishment of plants from seeds," *J. Ecol.* 53:273–286.

Harris, B. J., and H. G. Baker. 1958. "Pollination in *Kigelia africana* Benth." *J. West African Sci. Assoc.* 4:25–30.

——— and H. G. Baker. 1960. "Pollination of flowers by bats in Ghana," *Nigerian Field* 24:151–159.

Harris, T. M. 1932. "The fossil flora of Scoresby Sound, East Greenland, Part 2. Description of seed plants *incertae sedis* together with a discussion of certain cycadophyte cuticles," *Medd. om Grønland,* 85; 3:1–112.

——— 1951. "The relationships of the Caytoniales," *Phytomorphology* 1:29–33.

——— 1964. "The Yorkshire Jurassic Flora. II. Caytoniales, Cycadales and Pteridosperms," *Publ. British Museum (Natural History) London.*

Haskell, G. 1949. "Variation in the number of stamens in the common chickweed," *J. Genet.* 49:291–301.

——— 1954. "Pleiocotyly and differentiation within angiosperms," *Phytomorphology* 4:140–152.

Hatch, M. D., C. R. Slack, and H. S. Johnson. 1967. "Further studies on a new pathway of photosynthetic carbon dioxide fixation in sugar cane and its occurrence in other plant species," *Biochem. J.* 102:417–422.

Hawkes, J. G., and P. Smith. 1965. "Continental drift and the age of angiosperm genera," *Nature* 207:48–50.

——— and W. G. Tucker. 1968. "Serological assessment of relationships in a flowering plant family (Solanaceae)," in J. G. Hawkes, ed., *Chemotaxonomy and Serotaxonomy* (Systematics Assoc. special vol. No. 2; Academic Press, New York), pp. 77–88.

Hayata, B. 1913. *Icones Plantarum Formosanarum,* vol. III, plate IV.

Hedberg, Olov. 1968. "Taxonomic and ecological studies on the afroalpine flora of Mt. Kenya," *Hochgebirgsforschung* 1:171–194.

van Heel, W. A. 1966. "Morphology of the androecium in the Malvales," *Blumea* 13:177–394.

——— 1970. "Distally lobed integuments in some angiosperm ovules," *Blumea* 18:67–70.

Hegi, G. 1906. *Illustrierte Flora von Mittel-Europa* (Lehmanns, Munich), vol. V, pt. 3.

Hegnauer, R. 1966. "Comparative phytochemistry of alkaloids," in T. Swain, ed., *Comparative Phytochemistry* (Academic Press, New York), pp. 211–230.

Heinsbroek, P. G., and W. A. van Heel. 1969. "Note on the bearing of the pattern of vascular bundles on the morphology of the stamens of *Victoria amazonica* (Poepp.) Sowerby," *Proc. Kgl. Ned. Akad. Wetens.* 72:431–444.

Heiser, C. B., Jr. 1962. "Some observations on pollination and compatibility in *Magnolia,*" *Proc. Indiana Acad. Sci.* 72:259–265.

Henslow, G. 1911. "The origin of monocotyledons from dicotyledons through self-adaptation to a moist or aquatic habitat," *Ann. Bot.* 26:717–744.

Heslop-Harrison, J. 1961. "The function of the glume pit and the control of cleistogamy in *Bothriochloa decipiens* (Hack.)," *Phytomorphology* 11(4):378–383.

——— 1967. "Differentiation," *Ann. Rev. Plant Physiology* 18:325–348.

——— 1971. *Pollen: Development and Physiology* (Butterworth's, London), 338 pp.

——— and Y. Heslop-Harrison. 1957. "Studies on flowering-plant growth and organogenesis. I. Morphogenetic effects of 2,3,5-triiodobenzoic acid on *Cannabis sativa,*" *Proc. Royal Soc. Edinburgh,* Ser. B, 66:409–434.

Hicks, G. S., and I. M. Sussex. 1971. "Organ regeneration in sterile culture after median bisection of the flower primordia of *Nicotiana tabacum,*" *Bot. Gaz.* 132:350–363.

Hiepko, P. 1965*a*. "Das zentrifugale Androeceum von *Paeonia,*" *Ber. Deu. Bot. Ges.* 77:427–435.

——— 1965*b*. "Vergleichend-morphologische und entwicklungsgeschichtliche Untersuchungen über das Perianth bei den Polycarpicae," *Bot. Jahrb.* 84:359–508.

Hirmer, M. 1917. "Beiträge zur Morphologie polyandrische Blüten," *Flora* 110:140–192.

Hitchcock, A. S., and A. Chase. 1950. *Manual of the Grasses of the United States* (2nd ed.; U. S. Dept. Agr. Misc. Publ. No. 200).
Hjelmquist, H. 1964. "Variations in embryo sac development," *Phytomorphology* 14:186–196.
Holden, H. S. 1920. "Observations on the anatomy of teratological seedlings. III. On the anatomy of some atypical seedlings of *Impatiens Roylei* Walp.," *Ann. Bot.* 34:321–344.
——— and M. E. Daniels. 1921. "Observations on the anatomy of teratological seedlings of *Impatiens Roylei* Walp.," *Ann. Bot.* 35:461–492.
Holm, R. W. 1950. "The American species of *Sarcostemma* R. Br. (Asclepiadaceae)," *Ann. Missouri Bot. Gard.* 37:477–560.
Holmgren, P. 1968. "Leaf factors affecting light-saturated photosynthesis in ecotypes of *Solidago virgaurea* from exposed and shaded habitats," *Physiol. Pl.* 21:676–698.
——— P. G. Jarvis, and M. S. Jarvis. 1965. "Resistance to carbon dioxide and water vapour transfer in leaves of different plant species," *Physiol. Pl.* 18:557–573.
Holtorp, H. E. 1944. "Tricotyledony," *Nature* 153:13–14.
Holttum, R. E. 1955. "Growth-habits of Monocotyledons—variation on a theme," *Phytomorphology* 5:399–413.
Hoogland, R. D. 1952. "A revision of the genus *Dillenia*," *Blumea* 7:1–145.
——— 1953. "The genus *Tetracera* (Dilleniaceae) in the eastern Old World," *Reinwardtia* 2:185–225.
Hotchkiss, A. T. 1958. "Pollen and pollination in the Eupomatiaceae," *Proc. Linn. Soc. New South Wales* 83:86–91.
Houghtaling, H. 1935. "A devlopmental analysis of size and shape in tomato fruits," *Bull. Torr. Bot. Club* 62:243–252.
Howell, J. T. 1949. *Marin Flora* (University of California Press, Berkeley).
Hoyer, B. H., E. T. Bolton, B. J. McCarthy, and R. B. Roberts. 1965. "The evolution of polynucleotides," in V. Bryson and H. G. Vogel, eds., *Evolving Genes and Proteins* (Academic Press, New York), pp. 581–590.
——— B. J. McCarthy, and E. T. Bolton. 1964. "Molecular approach in the systematics of higher organisms," *Science* 144:959–967.
Hu, S-Y. 1954–1956. "A monograph of the genus *Philadelphus*," *J. Arnold Arb.* 35:275–333; 36:52–109, 325–368; 37:15–90.
Huether, C. A., Jr. 1966. "The extent of variability for a canalized character (corolla lobe number) in natural populations of *Linanthus* (Benth.)," unpub. Ph. D. thesis, Univeristy of California, Davis.
——— 1968. "Exposure of natural genetic variability underlying the pentamerous corolla constancy in *Linanthus androsaceus* ssp. *androsaceus*," *Genetics* 60:123–146.
——— 1969. "Constancy of the pentamerous corolla phenotype in natural populations of *Linanthus*," *Evolution* 23:572–588.
Hunziker, J. H. 1969. "Molecular data in plant systematics," *Systematic Biology* (Nat. Acad. Sci. Publ. 1692), pp. 280–312.
Hutchinson, J. 1926. *The Families of Flowering Plants. I. Dicotyledons* (Macmillan, London).
——— 1934. *The Families of Flowering Plants.* II. *Monocotyledons* (Macmillan, London).
——— 1959. *The Families of Flowering Plants* (2nd ed., 2 vols.; Oxford University Press, New York).
——— 1967. *The Genera of Flowering Plants.* II (Clarendon Press, Oxford).
——— 1969. *Evolution and Phylogeny of Flowering Plants: Dicotyledons; Facts and Theory* (Academic Press, London, New York).
Hutchinson, J. B., and S. G. Stephens. 1947. *The Evolution of Gossypium,* (Oxford University Press, London, New York).
Irwin, H. S., and B. L. Turner. 1960. "Chromosomal relationships and taxonomic considerations in the genus *Cassia*," *Amer. J. Bot.* 47:309–318.
van Iterson, G. 1907. *Mathematische und mikroskopische-anatomische Studien über Blattstellung* (Fischer, Jena).
Jain, S. K., and D. R. Marshall. 1967. "Population studies in predominantly self-pollinating species. X. Variation in natural populations of *Avena fatua* and *A. barbata*," *Amer. Nat.* 101:19–32.
Janchen, E. 1938. "Der morphologischen Wert der Gramineen-Vorspelze," *Oesterr. Bot. Z.* 87:51–61.
Janzen, D. H. 1966. "Coevolution of mutualism between ants and acacias in Central America," *Evolution* 20:249–275.
——— 1967. "Interaction of the bull's horn acacia (*Acacia cornigera* L.) with an ant inhabitant

(*Pseudomyrmex ferruginea* F. Smith) in eastern Mexico," *Univ. Kansas Sci. Bull.* 47:315–558.
——— 1969. "Seed-eaters versus seed size, number, toxicity and dispersal," *Evolution* 23:1–27.
——— 1970. "Herbivores and the number of tree species in tropical forests," *Amer. Nat.* 104:501–528.
——— 1971*a*. "Euglossine bees as long-distance pollinators of tropical plants," *Science* 171:203–205.
——— 1971*b*. "Seed predation by animals," *Ann. Rev. Ecol. Syst.* 2:465–492.
Jeffrey, C. 1966. "Notes on Compositae: I. The Cichorieae in East Tropical Africa," *Kew Bull.* 18:427–486.
Jeffrey, E. C. 1916. *The Anatomy of Woody Plants* (University of Chicago Press, Chicago).
Jenkin, T. J. 1954. "Interspecific and intergeneric hybrids in herbage grasses. V. *Lolium rigidum* sens. ampl. with other *Lolium* species," *J. Genetics* 52:252–281.
Jensen, U. 1968*a*. "Serologische Beiträge zur Systematik der Ranunculaceae," *Bot. Jahrb.* 88:269–310.
——— 1968*b*. "Serologische Beiträge zur Frage der Verwandtschaft zwischen Ranunculaceen und Papaveraceen," *Ber. Deu. Bot. Ges.* 80:621–624.
Johansen, D. A. 1945. "A critical survey of the present status of plant embryology," *Bot. Rev.* 11:87–107.
——— 1950. *Plant Embryology* (Chronica Botanica, Waltham, Mass.).
Johnson, M. A. 1954. "The precipitin reaction as an index of relationship in the Magnoliaceae," *Serol. Mus. Bull.* 13:1–5.
——— and D. E. Fairbrothers. 1965. "Comparison and interpretation of serological data in the Magnoliaceae," *Bot. Gaz.* 126:260–269.
Johnson, M. P., and S. A. Cook. 1968. "'Clutch size' in buttercups," *Amer. Nat.* 102:405–411.
Jukes, T. H. 1966. *Molecules and Evolution* (Columbia University Press, New York).
Just, T. 1948. "Gymnosperms and the origin of angiosperms," *Bot. Gaz.* 110:91–103.
Kaeiser, M., and S. G. Boyce. 1962. "Embryology of *Liriodendron tulipifera* L.," *Phytomorphology* 12:103–109.
Kaplan, D. R. 1967. "Floral morphology, organogenesis and interpretation of the inferior ovary in *Downingia bacigalupii*," *Amer. J. Bot.* 54:1274–1290.
——— 1970. "Comparative foliar histogenesis in *Acorus calamus* and its bearing on the phyllode theory of monocotyledonous leaves," *Amer. J. Bot.* 57:331–361.
Kaplan, S. M., and D. L. Mulcahy. 1971. "Mode of pollination and floral sexuality in *Thalictrum*," *Evolution* 25:659–668.
Kaul, R. B. 1968. "Floral morphology and phylogeny in the Hydrocharitaceae," *Phytomorphology* 18:13–35.
——— 1969. "Morphology and development of the flowers of *Boottia cordata*, *Ottelia alismoides*, and their synthetic hybrid (Hydrocharitaceae)," *Amer. J. Bot.* 56:951–959.
——— 1970. "Evolution and adaptation of inflorescence in the Hydrocharitaceae," *Amer. J. Bot.* 57:708–715.
Khan, R. 1950. "A case of twin ovules in *Isomeris arborea*," *Current Sci.* 19:326.
Khoshoo, T. N., and U. Sachdeva. 1961. "Cytogenetics of Punjab weeds. I. Causes of polymorphicity in *Convolvulus arvensis*," *Indian J. Agric. Sci.* 31:13–17.
Kiermayer, O. 1960. "Vergleichende Untersuchungen über die formative Wirksamkeit von 5-Chloropyridazin-6-glycolsäure (CPGS) und 2,3,5-Triiodbenzoesäure (TIBA) bei *Lycopersicum esculentum*," *Ber. Deu. Bot. Ges.* 73:157–166.
Kimura, M. 1968. "Genetic variability maintained in a finite population due to mutational production of neutral and nearly neutral isoalleles," *Genet. Res.* 11:247–269.
——— and T. Ohta. 1972. "Population genetics, molecular biology and evolution," in L. M. LeCam, J. Neyman, and E. L. Scott, eds. *Darwinian, Neo-Darwinian and Non-Darwinian Evolution* (Proc. 6th Berkeley Symp. Math. Stat. Prob.; University of California Press, Berkeley and Los Angeles), vol. 5, pp. 43–68.
King, J. L., and T. H. Jukes. 1969. "Non-Darwinian evolution," *Science* 164:788–798.
Kloz, J., V. Turkova, and G. Klosova. 1960. "Serological investigations of taxonomic specificity of proteins in various plant organs in some taxons of the family Viciaceae," *Biol. Plant* 2:126–137.
Kneebone, W. R. 1972. "Breeding for seedling vigor," in V. B. Youngner and C. M. McKell, eds., *The Biology and Utilization of Grasses* (Academic Press, New York, London), pp. 90–100.

Knuth, P. 1895–1905. *Handbuch der Blütenbiologie.* I–III (Engelmann, Leipzig).
Koltzov, N. K. 1936. *Organizatsiia Kletki* (*The Organization of Cells;* Publ. Akad. Nauk, Moscow-Leningrad). In Russian.
Kratzer, J. 1918. "Die verwandtschäftlichen Beziehungen der Cucurbitaceen auf Grund ihrer Samenentwicklung," *Flora* 110:275–343.
Kruckeberg, A. R. 1951. "Intraspecific variability in the response of certain native plant species to serpentine soil," *Amer. J. Bot.* 38:408–419.
Kubitzki, K. 1968. "Flavonoide und Systematik der Dilleniaceen," *Ber. Deu. Bot. Ges.* 81:238–251.
——— 1969. "Chemosystematische Betrachtungen zur Grossgliederung der Dicotylen," *Taxon* 18:360–368.
Kugler, H. 1955. *Einführung in die Blütenökologie* (Fischer, Stuttgart).
——— 1963. "UV-Musterung auf Blüten und ihre Zustandekommen," *Planta* 59:296–329.
Kullenberg, B. 1956*a*. "On the scents and colours of *Ophrys* flowers and their specific pollinators among the aculeate *Hymenoptera*," *Svensk Bot. Tidsk.* 50:25–46.
——— 1956*b*. "Field experiments with chemical sexual attractants on aculeate *Hymenoptera* males. I," *Zool. Bidr. Uppsala* 31:254–354.
——— 1961. "Studies in *Ophrys* pollination," *Zool. Bidr. Uppsala* 34:1–340.
Laetsch, W. M. 1968. "Chloroplast specialization in dicotyledons possessing the C_4-dicarboxylic acid pathway of photosynthetic CO_2 fixation," *Amer. J. Bot.* 55:875–883.
Laing, R. M., and H. W. Gourley. 1935. "The small-leaved species of the genus *Pittosporum* occurring in New Zealand, with descriptions of new forms," *Trans. Roy. Soc. New Zealand* 65:44–62.
Lam, H. J. 1952. "L'Evolution des plantes vasculaires," *Ann. Biol.* 28 (5–6):57–88.
——— 1959. "Some fundamental considerations on the new morphology," *Trans. Bot. Soc. Edinburgh* 38:100–134.
——— 1961. "Reflections on angiosperm phylogeny. I and II. Facts and theories," *Proc. Kgl. Nederl Akad. Wetens.*, Ser. C, 64:251–276.
Lance, A. 1957. "Recherches cytologiques sur l'évolution de quelques méristèmes apicaux et sur les variations provoquées par des traitements photopériodiques," *Ann. Sci. Nat. Bot.* sér. 11, 18:91–422.
——— and P. Rondet. 1958. "Sur le fonctionnement du méristème apical de *Beta vulgaris* L. (variété Cérès sucrière) de depuis la phase adulte jusqu'à la fleur terminale," *C. R. Acad. Sci. Paris* 246:3177–3180.
Lance-Nougarède, A. 1961. "Comparaison du fonctionnement reproducteur chez deux plantes vivaces construissant des inflorescences en grappe indéfinie sans fleur terminale, *Teucrium scorodonia* (Labiées) et *Alyssum maritimum* Lamk (Crucifères)," *C. R. Acad. Sci. Paris* 252:924–926.
Lankester, E. R. 1870. "On the use of the term homology," *Ann. Nat. Hist.* 6:34–43.
Lawton, J. R. S., and J. R. Lawton. 1967. "The morphology of the dormant embryo and young seedling of five species of *Dioscorea* from Nigeria," *Proc. Linn. Soc. London* 178:153–160.
LeCam, L., J. Neyman, and E. L. Scott, eds. 1972. *Darwinian, Neo-Darwinian and Non-Darwinian Evolution* (Proc. 6th Berkeley Symp. Math. Stat. Prob.; University of California Press, Berkeley and Los Angeles), vol. 5.
Lee, J. A. 1960. "A study of plant competition in relation to development." *Evolution* 14:18–28.
Leinfellner, W. 1950. "Der Bauplan des synkarpen Gynözeums," *Oesterr. Bot. Z.* 97:403–436.
Leins, P. 1964*a*. "Die frühe Blütenentwicklung von *Hypericum hookerianum* Wright et Arn. und *H. aegypticum* L.," *Ber. Deu. Bot. Ges.* 77:112–123.
——— 1964*b*. "Das zentripetale und zentrifugale Androecium," *Ber. Deu. Bot. Ges.* 77, suppl., 22–26.
——— 1965. "Die Infloreszens und frühe Blütenentwicklung von *Melaleuca nesophila* F. Muell. (Myrtaceae)," *Planta* 65:195–204.
——— 1967. "Die frühe Blütenentwicklung von *Aegle marmelos* (Rutaceae)," *Ber. Deu. Bot. Ges.* 80:320–325.
——— 1971. "Das Androecium der Dicotylen," *Ber. Deu. Bot. Ges.* 84:191–193.
Leopold, A. C. 1955. *Auxins and Plant Growth* (University of California Press, Berkeley, Los Angeles).
Lerner, I. M. 1954. *Genetic Homeostasis* (Wiley, New York).
Leppik, E. E. 1953. "The ability of insects to distinguish number," *Amer. Nat.* 87:229–236.
——— 1955. "*Dichromena ciliata,* a noteworthy entomophilous plant among Cyperaceae," *Amer. J. Bot.* 42:455–458.

——— 1956. "The form and function of numeral patterns in flowers," *Amer. J. Bot.* 43:445–455.

——— 1957. "Evolutionary relationship between entomophilous plants and anthophilous insects," *Evolution* 11:466–481.

——— 1960. "Early evolution of flower types," *Lloydia* 3:72–92.

——— 1964. "Floral evolution in the Ranunculaceae," *Iowa State J. Sci.* 39:1–101.

Levin, D. A. 1971. "Plant phenolics: an ecological perspective," *Amer. Nat.* 105:157–181.

Lewin, R. A. 1954. "Mutants of *Chlamydomonas moewusii* with impaired motility," *J. Gen. Microbiol.* 11:358–363.

Lewis, D. 1955. "Sexual incompatibility," *Sci. Prog.* 172:593–605.

Lewis, H. 1966. "Speciation in flowering plants," *Science* 152:167–172.

——— and C. Epling. 1959. "*Delphinium gysophilum*, a diploid species of hybrid origin," *Evolution* 13:511–525.

——— and P. H. Raven. "Rapid evolution in *Clarkia*," *Evolution* 12:319–336.

——— and J. Szweykowski. 1964. "The genus *Gayophytum* (Onagraceae)," *Brittonia* 16:343–391.

Lewis, M. C. 1969. "Genecological differentiation of leaf morphology in *Geranium sanguineum* L.," *New Phytol.* 68:481–503.

Lewis, W. H., R. L. Oliver, and Y. Suda. 1967. "Cytogeography of *Claytonia virginica* and its allies," *Ann. Missouri Bot. Gard.* 54:153–171.

Li, H.-L. 1951. "Evolution in the flowers of *Pedicularis*," *Evolution* 5:158–164.

——— 1960. "A theory on the ancestry of angiosperms," *Acta Biotheor.* 13:185–202.

Lindner, E. 1928. "*Aristolochia Lindneri* Berger und ihre Befruchtung durch Fliegen," *Biol. Centralbl.* 48:93–101.

Lindsey, A. A. 1940. "Floral anatomy in the Gentianaceae," *Amer. J. Bot.* 27:640–652.

Linsley, E. G., J. W. MacSwain, and P. H. Raven. 1963. "Comparative behavior of bees and Onagraceae. I. *Oenothera* bees of the Colorado Desert. II. *Oenothera* bees of the Great Basin," *Univ. Calif. Publ. Entom.* 33:1–58.

——— J. W. MacSwain, and P. H. Raven. 1964. "Comparative behavior of bees and Onagraceae. III. *Oenothera* bees of the Mojave Desert, California," *Univ. Calif. Publ. Entom.* 33:59–98.

Loiseau, J.-E. 1959. "Observations et expérimentation sur la phyllotaxie et le functionnement du sommet végétatif chez quelques Balsaminacées," *Ann. Sci. Nat. Bot.* 11 sér. 20:1–214.

——— 1960. "Application des techniques de microchirurgie à l'étude expérimentale des méristèmes caulinaires," *Ann. Biol.* 36:249–304.

Long, A. G. 1966. "Some lower Carboniferous fructifications from Berwickshire, together with a theoretical account of the evolution of ovules, cupules and carpels," *Trans. Roy. Soc. Edinburgh* 66:345–375.

Lowe-McConnell, R. H. 1969. "Speciation in tropical freshwater fishes," *Biol. J. Linn. Soc.* 1:51–76.

Lubbock, J. 1892. *A Contribution to Our Knowledge of Seedlings,* vol. 1 (London).

Luckwill, L. C. 1943. "The Genus *Lycopersicon*," *Aberdeen University Studies,* No. 20, 44 pp.

Ludwig, J. W., and J. L. Harper. 1958. "The influence of the environment on seed and seedling mortality. VIII. The influence of soil colour," *J. Ecol.* 46:381–389.

Lutz, F. E. 1933. "'Invisible' colors of flowers and butterflies," *Nat. Hist.* 33:565–576.

Mabry, T. J. 1964. "The betacyanins, a new class of red-violet pigments, and their phylogenetic significance," in C. A. Leone, ed., *Taxonomic Biochemistry and Serology* (Ronald Press, New York), pp. 239–254.

——— 1966. "The betacyanins and betaxanthins," in T. Swain, ed., *Comparative Phytochemistry* (Academic Press, New York), pp. 231–244.

Macior, L. W. 1966. "Foraging behavior of *Bombus* (*Hymenoptera: Apidae*) in relation to *Aquilegia* pollination," *Amer. J. Bot.* 53:302–309.

——— 1968. "Pollination adaptation in *Pedicularis groenlandica*," *Amer. J. Bot.* 55:927–932.

Madison, J. H. 1970. "An appreciation of monocotyledons," *Notes Roy. Bot. Garden Edinburgh* 30:377–390.

Maekawa, F. 1960. "A new attempt in phylogenetic classification of plant kingdom," *J. Fac. Sci. Univ. Tokyo, Sec. III, Bot.* 7:543–569.

Maheshwari, P. 1950. *The Embryology of Angiosperms* (McGraw-Hill, New York).

Mahlberg, P. G. 1959. "Development of the non-articulated laticifer in proliferated embryos of *Euphorbia marginata* Pursh.," *Phytomorphology* 9:156–162.

——— and P. S. Sabharwal. 1967. "Mitosis in

the non-articulated laticifer of *Euphorbia marginata*," *Amer. J. Bot.* 54:465–472.
Malyshev, S. I. 1964. "Stanovlenie tsvetkovykh rastenii v aspekte evoliutsii povedenniia osoobraznykh predkov pchelynykh," *Usp. Sovr. Biol.* 57:1.
Maneval, W. E. 1914. "The development of *Magnolia* and *Liriodendron*, including a discussion of the primitiveness of the Magnoliaceae," *Bot. Gaz.* 57:1–31.
Manning, A. 1956. "Bees and flowers," *New Biology* (Penguin Books) 21:59–73.
Marchant, C. J. 1967. "Chromosome evolution in the *Bromeliaceae*," *Kew Bull.* 21:161–168.
Margoliash, E., and W. M. Fitch. 1968. "Evolutionary variability of cytochrome C primary structures," *Ann. New York Acad. Sci.* 151(1):359–381.
Marsden, M. P. F., and I. W. Bailey. 1955. "A fourth type of nodal anatomy in Dicotyledons, illustrated by *Clerodendron trichotomum* Thunb.," *J. Arnold Arb.* 36:1–50.
Martin, F. W. 1967. "Distyly, self-incompatibility, and evolution in *Melochia*," *Evolution* 21:493–499.
Marushige, Y. 1965. "Ontogeny of the vegetative and the reproductive apices in *Pharbitis nil* Chois. II. Development of the terminal flower bud," *Bot. Mag. Tokyo* 78:397–406.
Mathiessen, A. 1962. "A contribution to the embryology of *Paeonia*," *Acta Hort. Berg.* 20:57–61.
Mayer, A. M., and A. Poljakoff-Mayber. 1963. *The Germination of Seeds* (Macmillan, New York).
Mayr, B. 1969. "Ontogenetische Studien an Myrtales-Blüten," *Bot. Jahrb.* 89:210–271.
Mayr, E. 1960. "The emergence of evolutionary novelties," in S. Tax, ed., *Evolution after Darwin* 1: *The Evolution of Life* (University of Chicago Press, Chicago, Ill.), pp. 349–380.
——— 1963. *Animal Species and Evolution* (Harvard University Press, Cambridge, Mass.).
——— 1970. *Populations, Species, and Evolution* (Harvard University Press, Cambridge, Mass.).
McKell, C. M. 1972. "Seedling vigor and seedling establishment," in V. B. Youngner and C. M. McKell, eds., *The Biology and Utilization of Grasses* (Academic Press, New York, London), pp. 74–89.
McMinn, H. E. 1942. "A systematic study of the genus *Ceanothus*," in *Ceanothus: Publ. Santa Barbara Botanic Garden*, pp. 131–279.
McNeilly, T., and J. Antonovics. 1968. "Evolution in closely adjacent plant populations. IV. Barriers to gene flow," *Heredity* 23:205–218.
McPherson, J. K., and C. H. Muller. 1969. "Allelopathic effects of *Adenostoma fasciculatum*, 'chamise,' in the California chaparral," *Ecological Mon.* 39:177–198.
Meeuse, A. D. J. 1965. *Fundamentals of Phytomorphology* (Ronald Press, New York).
——— 1967. "Again: the growth habit of the early angiosperms," *Acta Bot. Neerl.* 16:33–41.
——— 1971. "Interpretative gynoecial morphology of the Lactoridaceae and Winteraceae—a reassessment," *Acta Bot. Neerl.* 20:221–238.
Meier, K. I. 1960. "K embriologii *Nuphar luteum* Sm.," *Biull. Mosk. obsh. ispyt. prir. Otd. Biol.* 65(6):48–58.
Melville, R. 1960. "A new theory of the angiosperm flower," *Nature* 188:14–18.
——— 1962. "A new theory of the angiosperm flower. I. The gynoecium," *Kew Bull.* 16:1–50.
——— 1963. "A new theory of the angiosperm flower. II. The androecium," *Kew Bull.* 17:1–63.
Merrill, E. D. 1923. *An Enumeration of Philippine Flowering Plants* (4 vols.; Bureau of Science, Manila).
Merxmüller, Hermann. 1967. "Chemotaxonomie?" *Ber. Deu. Bot. Ges.* 80:608–620.
Metcalfe, C. R. 1936. "An interpretation of the morphology of the single cotyledon of *Ranunculus ficaria* based on embryology and seedling anatomy," *Ann. Bot.* 50:103–120.
——— and L. Chalk. 1950. *Anatomy of the Dicotyledons* (Clarendon Press, Oxford).
Meusel, H. 1965. "Die Evolution der Pflanzen in pflanzengeographischer-ökologischer Sicht," *Beitr. Abstammungsl.* 2:7–39.
Meyer, F. J. 1932. "Die Verwandschaftsbeziehungen der Alismataceen zu den Ranales im Lichte der Anatomie," *Bot. Jahrb.* 65:53–59.
Mez, C., and H. Siegenspeck. 1926. "Der Königsberger serodiagnostische Stammbaum," *Bot. Archiv* 12:163–202.
Michaux, N. 1964. "Structure et évolution du méristème apical, du *Jussieva grandiflora* Michx. durant les phases végétative et reproductrice," *Rev. Gén. Bot.* 71:91–170.
Michener, C. D. 1965. "A classification of the bees of the Australian and South Pacific regions," *Bull. Amer. Mus. Nat. Hist.* 139:1–362.
Miki, S. 1941. "On the change of flora in eastern

Asia since Tertiary Period I. The clay or lignite beds flora in Japan with special reference to the *Pinus trifolia* beds in central Hondo," *Jap. J. Bot.* 11:237–303.

Miksche, J. P., and J. A. M. Brown. 1965. "Development of vegetative and floral meristems of *Arabidopsis thaliana*," *Amer. J. Bot.* 52:533–537.

Mirov, N. T. 1967. *The Genus Pinus* (Ronald Press, New York).

Monod, J. 1971. *Chance and Necessity* (Knopf, New York).

Mooney, H. A., and E. L. Dunn. 1970. "Convergent evolution of Mediterranean climate evergreen sclerophyll shrubs," *Evolution* 24:292–303.

Morf, E. 1950. "Vergleichend morphologische Untersuchungen am Gynoecium der Saxifragaceen," *Ber. Schweiz. Bot. Ges.* 60:516–590.

Morison, C. G. T., A. C. Hoyle, and J. F. Hope-Simpson. 1948. "Tropical soil-vegetation catenas and mosaics. A study in the southwestern part of the Anglo-Egyptian Sudan," *J. Ecol.* 36:1–84.

Moss, E. H. 1940. "Interxylary cork in *Artemisia* with a reference to its taxonomic significance," *Amer. J. Bot.* 27:762–768.

Mulcahy, D. L. 1964. "The reproductive biology of *Oxalis Priceae*," *Amer. J. Bot.* 51:1045–1050.

Muller, J. 1970. "Palynological evidence on early differentiation of angiosperms," *Biol. Rev.* 45:417–450.

Munz, P. A., and D. D. Keck. 1959. *A California Flora* (University of California Press, Berkeley).

Müntzing, A. 1961. *Genetic Research* (LTs. Forlag, Stockholm).

Mürbeck, S. 1912. "Untersuchungen über den Blütenbau der Papaveraceen," *Kgl. Svenska Vetensk. Akad. Handlingar* 50:1–168.

——— 1914. "Uber die Baumechanik bei Anderungen im Zahlenverhaltnis der Blüte," *Lunds Univ. Arsskr.* (n.f.) *Avd.* 2, 11(3):1–36.

——— 1941. "Das Androecium der Rosaceen," *Lunds Univ. Arsskr.* (n.f.) *Avd.* 2, 37(7):1–56.

Narayana, H. S. 1962. "Studies in the Capparidaceae. I. The embryology of *Capparis decidua* (Forsk.) Pax.," *Phytomorphology* 12:167–177.

Nemejc, F. 1956. "On the problem of the origin and phylogenetic development of the angiosperms," *Act. Mus. Nat. Pragae* 12B:59–143.

Newman, I. V. 1936. "Studies in the Australian acacias. VI. The meristematic activity of the floral apex of *Acacia longifolia* and *Acacia suaveolens* as a histogenetic study of the ontogeny of the carpel," *Proc. Linn. Soc. New South Wales* 61:56–88.

Nitsch, C. 1968. "Induction *in vitro* de la floraison chez une plante de jours courtes: *Plumbago indica*," *Ann. Sci. Nat. Bot.* Sér. 12, 9:1–92.

Nitsch, J. P. 1965. "Physiology of flower and fruit development," *Encyclopedia of Plant Physiology*, W. Ruhland ed. 15(1):1537–1647.

Nobs, M. A. 1963. "Experimental studies on species relationships in *Ceanothus*," *Carnegie Inst. Wash. Publ.* No. 623.

Nordhagen, R. 1932. "Zur Morphologie und Verbreitungsbiologie der Gattung *Roscoea* Sm.," *Bergens Mus. Arb.* 1932, *Natur.-Vitensk. Rekke* 4:1–57.

——— 1959. "Remarks on some new or little known myrmecochorous plants from North America and East Asia," *Bull. Res. Counc. Israel* 7D:184–201.

Nougarède, A. 1965. "Organisation et fonctionnement du méristème apical des végétaux vasculaires," in *Travaux dédiés à Lucien Plantefol* (Masson, Paris), pp. 171–340.

Nougarède, A. R. B., G. Bernier, and P. Rondet. 1964. "Comportement du méristème apical du *Perilla nankingensis* (Lour.) Decne. en relation avec les conditions photopériodiques," *Rev. Gén. Bot.* 71:205–238.

Olson, K. C., T. W. Tibbits, and B. E. Struckmeyer. 1969. "Leaf histogenesis in *Lactuca sativa* with emphasis upon laticifer ontogeny," *Amer. J. Bot.* 56:1212–1216.

Ornduff, R. 1966. "The origin of dioecism from heterostyly in *Nymphoides*," *Evolution* 20:309–314.

——— 1969. "Reproductive biology in relation to systematics," *Taxon* 18:121–133.

——— and T. J. Crovello. 1968. "Numerical taxonomy of Limnanthaceae," *Amer. J. Bot.* 55:173–182.

Ozenda, P. 1949. "Recherches sur les dicotylédones apocarpiques; contribution a l'étude des angiospermes dites primitives," *Ecole Normale Supér. Publ. Lab. Biol. Paris.*

Page, V. M. 1951. "Morphology of the spikelet of *Streptochaeta*," *Bull. Torr. Bot. Club* 78:22–37.

Parkin, J. 1914. "The evolution of the inflorescence," *J. Linn. Soc. Bot.* 42:511–563.

——— 1952. The unisexual flower—a criticism," *Phytomorphology* 2:75–79.

——— 1955. "A plea for a simpler gynoecium," *Phytomorphology* 5:46–57.

Payer, J. B. 1857. *Traité d'Organogénie Comparée de la Fleur* (Paris).

Payne, W. W. 1963. "The morphology of the inflorescence of ragweeds (*Ambrosia-Franseria:* Compositae)," *Amer. J. Bot.* 50:872–879.

——— 1964. "A re-evaluation of the composite genus *Ambrosia* (*Compositae*)," *J. Arnold Arb.* 45:401–438.

Pearson, H. H. W. 1929. *Gnetales* (Cambridge University Press, London).

Pervukhina, N. V. 1967. "Opylenie pervikhnykh pokrytosemmenikh i evoliutsiia sposobov opyleniia," *Bot. Zhurn.* 52:157–188.

Peterson, R. L., and E. G. Cutter. 1969*a*. "The fertile spike of *Ophioglossum petiolatum.* I. Mechanism of elongation," *Amer. J. Bot.* 56:473–483.

——— and E. G. Cutter. 1969*b*. "The fertile spike of *Ophioglossum petiolatum.* II. Control of spike elongation and a study of aborted spikes," *Amer. J. Bot.* 56:484–491.

Philipson, W. R. 1946. "Studies in the development of the inflorescence. I. The capitulum of *Bellis perennis,*" *Ann. Bot.* (n.s.) 10:257–270.

——— 1947*a*. "Studies in the development of the inflorescence. II. The capitulum of *Succisa pratensis* Moench. and *Dipsacus fullonum* L.," *Ann. Bot.* (n.s.) 11:285–297.

——— 1947*b*. "Studies in the development of the inflorescence. III. The thyrse of *Valeriana officinalis,*" *Ann. Bot.* (n.s.) 11:409–416.

Pijl, L. van der. 1953. "On the flower biology of some plants from Java. With general remarks on fly-traps (species of *Annona, Artocarpus, Typhonium, Gnetum, Arisaema* and *Abroma*)," *Ann. Bogorienses* 1:77–99.

——— 1954. "*Xylocopa* and flowers in the tropics. I and II," *Proc. Kgl. Neder. Akad. Wetens.* 57:414–562.

——— 1956. "Remarks on pollination by bats in the genera *Freycinetia, Duabanga* and *Haplophragma,* and on chiropterophily in general," *Acta Bot. Néerl.* 5:135–144.

——— 1960. "Ecological aspects of flower evolution. I. Phyletic evolution," *Evolution* 14:403–416.

——— 1961. "Ecological aspects of flower evolution. II. Zoophilous flower classes," *Evolution* 15:44–59.

——— 1969. *Principles of Dispersal in Higher Plants* (Springer, Berlin).

——— and C. Dodson. 1966. *Orchid Flowers: Their Pollination and Evolution* (University of Miami Press, Coral Gables, Florida).

Pittendrigh, C. S. 1948. "The Bromeliad-*Anopheles*-malaria complex in Trinidad. I. The Bromeliad flora," *Evolution* 2:58–89.

Plantefol. L. 1948. "Fondements d'une théorie florale nouvelle. L'Ontogenie de la fleur," *Ann. Sci. Nat. Bot.*, ser. 11, 9:33–186.

Polyakov, P. P. 1961. "*Artemisia,*" in B. K. Shishkin and E. G. Bobrov, eds., *Flora USSR,* (Publ. Akad. Nauk, Leningrad), vol. 26:425–631.

Poore, M. E. D. 1968. "Studies in Malaysian rainforest. I. The forest on Triassic sediments in Jengka Forest Reserve," *J. Ecol.* 56:143–196.

Popham, R. A., and A. P. Chan. 1952. "Origin and development of the receptacle of *Chrysanthemum morifolium,*" *Amer. J. Bot.* 39:329–339.

Porsch, O. 1934-1935. "Säugetiere als Blumenausbeuter und die Frage der Säugetierblume. I, II," *Biol. Gen.* 10:657–685; 11:171–188.

——— 1942. "Eine neuer Typus Fledermausblumen," *Biol. Gen.* 15:283–294.

Prakash, S., R. C. Lewontin, and J. L. Hubby. 1969. "A molecular approach to the study of genic heterozygosity in natural populations. IV. Patterns of genic variation in central, marginal and isolated populations of *Drosophila pseudoobscura,*" *Genetics* 61:841–858.

Pray, T. R. 1955*a*. "Foliar venation in angiosperms. II. Histogenesis of the venation of *Liriodendron,*" *Amer. J. Bot.* 42:18–27.

——— 1955*b*. "Foliar venation of angiosperms. IV. Histogenesis of the venation of *Hosta,*" *Amer. J. Bot.* 42:698–706.

Puri, V. 1951. "The role of floral anatomy in the solution of morphological problems," *Bot. Rev.* 17:471–553.

——— 1952. "Placentation in angiosperms," *Bot. Rev.* 18:603–651.

——— 1961. "The classical concept of angiosperm carpel: a reassessment," *J. Indian Bot. Soc.* 60:511–524.

Rao, V. S. 1938. "Studies on the Capparidaceae. III. Genus *Capparis,*" *J. Indian Bot. Soc.* 17:69–80.

Raschke, K. 1960. "Heat transfer between the plant and the environment," *Ann. Rev. Pl. Physiol.* 11:111–126.

Rauh, W., and H. Reznik. 1951. "Histogenetische Untersuchungen an Blüten- und Infloreszenachsen. I. Teil. Die Histogenese becherformigen

Blüten- und Infloreszenachsen, sowie der Blütenachsen einigen Rosoideen," *Sitzber. Heidelberg Akad. Wiss. Math. -naturw. Klasse* 1951:139–207.

——— and H. Reznik. 1953. "Histogenetische Untersuchungen an Blüten- und Infloreszenzachsen. II. Die Histogenese der kopfchenförmiger Infloreszenzen," *Beitr. Biol. Pflanzen* 29:233–296.

Raunkiaer, C. 1934. *The Life Forms of Plants and Statistical Plant Geography* (Clarendon Press, Oxford).

Raven, P. H. 1969. "A revision of the genus *Camissonia* (Onagraceae)," *Contr. U. S. Nat. Herb.* 37:161–396.

——— Kyhos, D. W., and Cave, M. S. 1971. "Chromosome numbers and relationships in Anno-niflorae," *Taxon* 20:479–483.

Ray, P. M., and F. Chisaki. 1957*a*. "Studies on *Amsinckia*. I. Heterostyly," *Amer. J. Bot.* 44:529–536.

——— and F. Chisaki. 1957*b*. "Studies on *Amsinckia*. II. Relationships among the primitive species," *Amer. J. Bot.* 44:537–544.

Reeve, R. M. 1943. "Comparative ontogeny of the inflorescence and the axillary vegetative shoot in *Garrya elliptica*," *Amer. J. Bot.* 30:608–619.

Rensch, B. 1960. *Evolution Above the Species Level* (Columbia University Press, New York).

Reveal, J. L. 1969. "The subgeneric concept in *Eriogonum* (Polygonaceae)," in J. E. Gunckel, ed., *Current Topics in Plant Science* (Academic Press, New York), pp. 229–249.

Reznik, H. 1955. "Die Pigmente der Centrospermen als systematisches Element," *Z. Bot.* 43:499–530.

——— 1957. "Die Pigmente des Centrospermen als systematisches Element. II. Untersuchungen über das ionophoretische Verhalten," *Planta* 49:406–434.

Richards, P. 1952. *The Tropical Rain Forest; an Ecological Study* (Cambridge University Press, Cambridge, England).

——— 1969. "Speciation in the tropical rain forest and the concept of the niche," *Biol. J. Linn. Soc.* 1:149–154.

Rick, C. M. 1950. "Pollination relations of *Lycopersicon esculentum* in native and foreign regions," *Evolution* 4:110–122.

——— and R. I. Bowman. 1961. "Galapagos tomatoes and tortoises," *Evolution* 15:407–417.

Rickett, H. W. 1944. "The classification of inflorescences," *Bot. Rev.* 10:187–231.

Ridley, H. N. 1930. *The Dispersal of Plants Throughout the World* (L. Reeve, Ashford, Kent, England).

Rizzini, C. de T. 1963. "A flora do cerrado—Análise floristica das savanas centrais," in Editora da Universidade de São Paulo, *Simposio Sobre o Cerrado* (São Paulo, Brazil), pp. 125–177.

Robson, N. K. B. 1972. "Evolutionary recall in *Hypericum* (Guttiferae)?" *Trans. Bot. Soc. Edinburgh* 41:365–383.

Rodriguez, C. L. R. 1956. "A graphic representation of Hutchinson's phylogenetic system," *Rev. Biol. Trop.* 4:35–40.

Rohweder, O. 1967. "Karpelbau und Synkarpie bei Ranunkulazeen," *Ber. Schweiz. Bot. Ges.* 77:376–432.

Ruijgrok, H. W. L. 1966. "The distribution of ranunculin and cyanogenetic compounds in the Ranunculaceae," in T. Swain, ed., *Comparative Phytochemistry* (Academic Press, New York), pp. 175–186.

Rüter, E. 1918. "Ueber Vorblattbildung bei Monokotylen," *Flora* 10:193–261.

Sachar, R. C. 1955. "The embryology of *Argemone mexicana*: a reinvestigation," *Phytomorphology* 5:200–218.

Sagar, G. R., and J. L. Harper. 1961. "Factors affecting the germination and early establishment of plantains (*Plantago lanceolata, P. media* and *P. major*)," in J. L. Harper, ed., *Biology of Weeds* (Symp. Brit. Ecol. Soc.; Blackwell, Oxford).

Salisbury, E. L. 1942. *The Reproductive Capacity of Plants* (Bell, London).

Sargant, E. 1903. "A theory of the origin of monocotyledons, founded on the structure of their seedlings," *Ann. Bot.* 17:1–92.

Sastri, R. L. N. 1958. "Floral morphology and embryology of some Dilleniaceae," *Bot. Notiser* 111:495–511.

Sattler, Rolf. 1962. "Zur frühen Infloreszenz- und Blütenentwicklung der Primulales sensu lato mit besonderer Berücksichtigung der Stamen-Petalum-Entwicklung," *Bot. Jahrb.* 81:358–396.

——— 1966. "Towards a more adequate approach to comparative morphology," *Phytomorphology* 16:417–429.

Sauer, W., and F. Ehrendorfer. 1970. "Chromosomen, Verwandtschaft und Evolution tropischer Holzplanzen. II. *Himantandraceae,*" *Oesterr. Bot. Z.* 118:38–54.

Saunders, A. P., and L. Stebbins. 1938. "Cytogenetic studies in *Paeonia.* I. The compatibility of the species and the appearance of the hybrids," *Genetics* 28:65–82.

Saunders, E. R. 1928. "*Matthiola,*" *Bibl. Genet.* 4:141–170.

Scandalios, J. 1969. "Genetic control of multiple molecular forms of enzymes in plants: a review," *Biochem. Genet.* 3:37–79.

Schlising, R. A. 1969. "Seedling morphology in *Marah* (Cucurbitaceae) related to the Californian Mediterranean climate," *Amer. J. Bot.* 56:552–560.

Schnarf, K. 1933. "Embryologie der Gymnospermen," in K. Linsbauer ed., *Handbuch der Pflanzenanatomie* (Borntraeger, Berlin), vol. 2, p. 2.

Schuster, J. 1910. "Ueber die Morphologie der Grasblüte," *Flora* 100:213–266.

Schuster, W. 1910. "Zur Kenntnis der Aderung des Monocotyledonblattes," *Ber. Deu. Bot. Ges.* 28: 268–278.

Scott, D. 1924. *Extinct Plants and Problems of Evolution* (Macmillan, London).

Scott, R. A., E. S. Barghoorn, and E. B. Leopold. 1960. "How old are the Angiosperms?" *Amer. J. Sci.* 258A:284–299.

Selander, R. K., S. Y. Yang, and W. G. Hunt. 1969. "Polymorphism in esterases and hemoglobin in wild populations of the house mouse (*Mus musculus*)," University of Texas Publ. No. 6918, *Studies in Genetics* V:271–338.

———, S. Y. Yang, R. C. Lewontin, and W. E. Johnson. 1970. "Genetic variation in the horseshoe crab (*Limulus polyphemus*), a phylogenetic 'relic,'" *Evolution* 24:402–414.

Serebryakova, T. I. 1971. *Morfogenez pobegov i evoliutsiia zhiznennikh form zlakov* (Publ. Akad. Nauk, Moscow).

Sernander, R. 1906. "Entwurf einer Monographie der Europäischen Myrmekochoren," *Kgl. Svenska Vet. Handl.* 41:1–410.

Seward, A. C. 1931. *Plant Life Through the Ages* (Cambridge University Press, Cambridge, England).

Shaw, C. R. 1965. "Electrophoretic variation in enzymes" *Science* 149:936–943.

Simpson, G. G. 1953. *The Major Features of Evolution* (Columbia University Press, New York).

——— 1970. "Uniformitarianism. An inquiry into principle, theory and method in geohistory and biohistory," in M. K. Hecht and W. C. Steere, eds., *Essays in Honor of Theodosius Dobzhansky* (Evolutionary Biology Suppl.), pp. 43–96.

Sing, C. F., and G. J. Brewer. 1969. "Isozymes of a polyploid series of wheat," *Genetics* 61:391–398.

Sinnott, E. W. 1960. *Plant Morphogenesis* (McGraw-Hill, New York).

——— and Bailey, I. W. 1914. "Investigations on the phylogeny of the angiosperms. IV. The origin and dispersal of herbaceous angiosperms," *Ann. Bot.* 28:547–600.

——— and S. Kaiser. 1934. "Two types of genetic control over the development of shape," *Bull. Torr. Bot. Club* 61:1–7.

Skipworth, J. P., and W. R. Philipson. 1966. "The cortical vascular system and the interpretation of the *Magnolia* flower," *Phytomorphology* 16:463–469.

Slatyer, R. O. 1967. *Plant-Water Relationships* (Academic Press, New York, London).

Small, J., and I. K. Johnston. 1937. "Quantitative evolution in Compositae," *Proc. Roy. Soc. Edinburgh* 57:26–54.

Smith, A. C. 1970. "The Pacific as a key to flowering plant history," *University of Hawaii, Harold L. Lyon Arboretum Lecture* No. 1:1–26.

Smith, D. L. 1966. "Development of the inflorescence in *Carex,*" *Ann. Bot.* (n.s.) 30:475–486.

Smith, P. 1969. "Serological relationships and taxonomy in certain tribes of the Gramineae," *Ann. Bot.* 33:591–613.

——— 1972. "Serology and species relationships in annual bromes (*Bromus* L. sect. *Bromus*)," *Ann. Bot.* 36:1–30.

Smith-White, S. 1955. "Chromosome numbers and pollen types in the Epacridaceae," *Austr. J. Bot.* 3:48–67.

——— 1960. "Cytological evolution in the Australian flora," *Cold Spring Harbor Symp. Quant. Biol.* 24:273–289.

——— C. R. Carter, and H. M. Stace. 1970. "The cytology of *Brachycome.* I. The subgenus *Eubrachycome:* A general survey," *Austr. J. Bot.* 18:99–125.

Smyth, N. 1970. "Relationships between fruiting

seasons and seed dispersal methods in a neotropical forest," *Amer. Nat.* 104:25–35.

Snaydon, R. W. 1962. "The growth and competitive ability of contrasting natural populations of *Trifolium repens* L. on calcareous and acid soils," *J. Ecol.* 50:439–447.

——— and A. D. Bradshaw. 1961. "Differential response to calcium within the species *Festuca ovina* L.," *New Phytol.* 60:219–234.

Snow, D. W. 1966. "A possible selective factor in the evolution of flowering seasons in tropical forests," *Oikos* 15:274–281.

Snow, M., and R. Snow. 1934. "The interpretation of phyllotaxis," *Biol. Rev.*, 9:132–137.

——— and R. Snow. 1948. "On the determination of leaves," *Symp. Soc. Exp. Biol., II, Cambridge:* 263–275.

Snow, R. 1955. "Problems of phyllotaxis and leaf determination," *Endeavour* 14:190–199.

Snyder, L. A. 1950. "Morphological variability and hybrid development in *Elymus glaucus*," *Amer. J. Bot.* 37:628–636.

Soderstrom, T. R., and C. E. Calderón. 1971. "Insect pollination in tropical rain forest grasses," *Biotropica* 3:1–16.

Soertiato, S., and E. Ball. 1969. "Ontogenetical and experimental studies of the floral apex of *Portulaca grandiflora.* 2. Bisection of the meristem in successive stages," *Canad. J. Bot.* 47:1067–1076.

Souèges, R. 1954. "L'Origine du cone végétatif de la tige, et la question de la 'terminalité' du cotylédon des Monocotylédones," *Ann. Sci. Nat. Bot.* XI, 15:1–20.

Sparks, R. S., and D. T. Arakaki. 1971. "Chromosomal polymorphism in inbred subspecies of *Peromyscus maniculatus* (Deer mouse)," *Canad. J. Genet. Cytol.* 13:277–282.

Sporne, K. R. 1948. "Correlation and classification in dicotyledons," *Proc. Linn. Soc. London* 160:40–47.

——— 1954. "Statistics and the evolution of dicotyledons," *Evolution* 8:55–64.

——— 1956. "The phylogenetic classification of the angiosperms," *Biol. Rev.* 31:1–29.

——— 1959. "On the phylogenetic classification of plants," *Amer. J. Bot.* 46:385–394.

——— 1960. "Correlation of biological characters," *Proc. Linn. Soc. London* 171:83–88.

——— 1969. "The ovule as an indicator of evolutionary status in angiosperms," *New Phytol.* 68:555–566.

Sprague, E. F. 1962. Pollination and evolution in *Pedicularis* (Scrophulariaceae)," *Aliso* 5:181–209.

Stahl, E. 1888. "Pflanzen und Schnecken. Biologische Studie über die Schutzmittel der Pflanzen gegen Schneckenfrass," *Jenaische Z. Naturw.* 22: 557–684.

Standley, P. C. 1928. "Flora of the Panama Canal Zone," *Contr. U. S. Nat. Herbarium* 27:1–417.

Stant, M. Y. 1964. "Anatomy of the Alismataceae," *J. Linn. Soc. Bot.* 59:1–42.

——— 1970. "Anatomy of *Petrosavia stellaris* Becc., a saprophytic monocotyledon," in N. K. B. Robson, D. F. Cutler, and M. Gregory, eds., *New Research in Plant Anatomy* (Academic Press, New York), pp. 147–161.

Stebbins, G. L., Jr. 1932*a*. "Cytology of *Antennaria.* I. Normal species," *Bot. Gaz.* 94:134–151.

——— 1932*b*. "Cytology of *Antennaria.* II. Parthenogenetic species," *Bot. Gaz.* 94:322–345.

——— 1939. "Notes on the systematic relationships of the Old World species and of some horticultural forms of *Paeonia*," *Univ. Calif. Publ. Bot.* 19:245–266.

——— 1949. "Speciation, evolutionary trends and distribution patterns in *Crepis*," *Evolution* 3:188–193.

——— 1950. *Variation and Evolution in Plants* (Columbia University Press, New York).

——— 1951. "Natural selection and the differentiation of angiosperm families," *Evolution* 5:299–324.

——— 1952. "Aridity as a stimulus to plant evolution," *Amer. Nat.* 86:33–44.

——— 1953. "A new classification of the tribe *Cichorieae*, family *Compositae*," *Madroño* 12:65–80.

——— 1956*a*. "Cytogenetics and evolution of the grass family," *Amer. J. Bot.* 43:890–905.

——— 1956*b*. "Taxonomy and the evolution of genera, with special reference to the family Gramineae," *Evolution* 10:235–245.

——— 1957*a*. "Self-fertilization and population variability in the higher plants," *Amer. Nat.* 91:337–354.

——— 1957*b*. "The hybrid origin of microspecies in the *Elymus glaucus* complex," *Proc. Int. Genet. Symp. Cytologia, Suppl. Vol.* pp. 336–340.

——— 1958. "Longevity, habitat and release of genetic variability in the higher plants," *Cold Spring Harbor Symp. Quant. Biol.* 23:365–378.

——— 1959*a*. "The role of hybridization in evolution," *Proc. Amer. Phil. Soc.* 103:231–251.

——— 1959*b*. "Seedling heterophylly in the Califor-

nia flora," *Bull. Res. Counc. of Israel* 7D(3–4):248–255.
——— 1965. "The probable growth habit of the earliest flowering plants," *Ann. Missouri Bot. Gard.* 52:457–468.
——— 1966. "Chromosome variation and evolution," *Science* 152:1463–1469.
——— 1967. "Adaptive radiation and trends of evolution in higher plants," in Th. Dobzhansky, M. K. Hecht, and W. C. Steere, eds., *Evolutionary Biology* (Appleton-Century-Crofts, New York), vol. 2, pp. 101–142.
——— 1969. *The Basis of Progressive Evolution* (University of North Carolina Press, Chapel Hill).
——— 1970*a*. "Adaptive radiation in Angiosperms. I. Pollination mechanisms," *Ann. Rev. Ecol. Syst.* 1:307–326.
——— 1970*b*. "Transference of function as a factor in the evolution of seeds and their accessory structure," *Israel J. Bot.* 19:59–70.
——— 1971*a*. *Chromosomal Variation in Higher Plants* (Arnold, London).
——— 1971*b*. "Adaptive radiation of reproductive characters of angiosperms, II: Seeds and seedlings," *Ann. Rev. Ecol. Syst.* 2:237–260.
——— 1972. "Ecological distribution of centers of major adaptive radiation in angiosperms," in D. Valentine, ed., *Taxonomy, Phytogeography and Evolution* (Academic Press, New York, London), pp. 7–34.
——— J. A. Jenkins, and M. S. Walters. 1953. "Chromosomes and phylogeny in the Compositae, tribe Cichorieae," *Univ. Calif. Publ. Bot.* 26: 401–430.
——— and G. S. Khush. 1961. "Variation in the organization of the stomatal complex in the leaf epidermis of monocotyledons and its bearing on their phylogeny," *Amer. J. Bot.* 48:51–59.
——— and S. S. Shah. 1960. "Developmental studies of cell differentiation in the epidermis of monocotyledons. II. Cytological features of stomatal development in the Gramineae," *Dev. Biol.* 2:477–500.
——— and J. Major. 1965. "Endemism and speciation in the California flora," *Ecol. Mon.* 35:1–35.
——— J. I. Valencia, and R. M. Valencia. 1946. "Artificial and natural hybrids in the Gramineae, tribe Hordeae. I. *Elymus, Sitanion,* and *Agropyron,*" *Amer. J. Bot.* 33:338–351.
——— and E. Yagil. 1966. "The morphogenetic effects of the hooded gene in barley. I. The course of development in hooded and awned genotypes," *Genetics* 54:727–741.
van Steenis, C. G. G. J. 1969. "Plant speciation in Malesia, with special reference to the theory of non-adaptive saltatory evolution," *Biol. J. Linn. Soc.* 1:97–133.
Stephens, S. G. 1958. "Salt water tolerance of seeds of *Gossypium* species as a possible factor in seed dispersal," *Amer. Nat.* 92:83–93.
Sterling, C. 1964. "Comparative morphology of the carpel in the Rosaceae. II. Prunoideae: *Maddenia, Pygaeum, Osmaronia,*" *Amer. J. Bot.* 51:354–360.
——— 1966*a*. "Comparative morphology of the carpel in the Rosaceae. VII. Pomoideae: *Chaenomeles, Cydonia, Docynia,*" *Amer. J. Bot.* 53:225–231.
——— 1966*b*. "Comparative morphology of the carpel in the Rosaceae. IX. Spiraeoideae: Quillajeae, Sorbarieae," *Amer. J. Bot.* 53:951–960.
Stone, D. E. 1959. "A unique balanced breeding system in the vernal pool mouse-tails," *Evolution* 13:151–174.
Stopp, K. 1958. "Die verbreitungshemmenden Einrichtungen in der Sudafrikanischen Flora," *Botan. Studien* (Fischer, Jena), vol. 8.
Straw, R. M. 1955. "Hybridization, homogamy, and sympatric speciation," *Evolution* 9:441–444.
——— 1956*a*. "Adaptive morphology of the *Penstemon* flower," *Phytomorphology* 6:112–119.
——— 1956*b*. "Floral isolation in *Penstemon,*" *Amer. Nat.* 90:47–53.
Stubbe, H. 1963. "Ueber die Stabilisierung des sich Variabel manifestierenden Merkmals 'Polycotylie' von *Antirrhinum majus* L.," *Kulturpflanze* 11:250–263.
Suessenguth, K. 1920. "Beiträge zur Frage des systematischen Anschlusses der Monokotylen," *Beih. Bot. Centr.* 38:1–79.
Suneson, C. A. 1949. "Survival of four barley varieties in a mixture," *Agron. J.* 41:459–461.
Swain, T., ed. 1966. *Comparative Phytochemistry* (Academic Press, New York).
Swamy, B. G. L. 1949. "Further contributions to the anatomy of the Degeneriaceae," *J. Arnold Arb.* 30:10–38.
——— and I. W. Bailey. 1949. "The morphology and relationships of *Cercidiphyllum,*" *J. Arnold Arb. 30:187–210.*
——— and I. W. Bailey. 1950. "*Sarcandra,* a

vesselless genus of Chloranthaceae," *J. Arnold Arb.* 34:375–408.

Swanson, J. R. 1964. "*Claytonia* (*Montia*) *perfoliata:* A genecological and evolutionary study," unpub. Ph. D. thesis, University of California, Berkeley.

Takhtajan, A. 1954. Proiskhozhdenie Pokrytosemenykh Rastenii (*Origin of Angiospermous Plants;* Publ. Akad. Nauk, Moscow).

——— 1959. *Die Evolution der Angiospermen* (Fischer, Jena).

——— 1966. *Sistema i Filogeniia Tsvetovikh Rastenii* (Izdatelstvo Nauk, Moscow, Leningrad).

——— 1969. *Flowering Plants: Origin and Dispersal* (Oliver and Boyd, Edinburgh).

——— 1970. *Proiskhozhdenie i Rasselenie Tsvetkovykh Rastenii* (Akademia Nauk, Bot. Inst. im. V. L. Komarova, Leningrad).

Taubert, P. 1894. "Leguminosae," in A. Engler and K. Prantl, eds., *Die Natürliche Pflanzenfamilien,* pp. 70–385.

Taylor, R. J., and D. Campbell. 1969. "Biochemical systematics and phylogenetic interpretations in the genus *Aquilegia,*" *Evolution* 23:153–162.

Tepfer, S. 1953. "Floral anatomy and ontogeny in *Aquilegia formosa* v. *truncata* and *Ranunculus repens,*" *Univ. Calif. Publ. Bot.* 25:513–648.

Thomas, H. H. 1925. "The Caytoniales, a new group of angiospermous plants from the Jurassic rocks of Yorkshire," *Phil. Trans. Roy. Soc. B* 213:299–313.

——— 1933. "On some pteridospermous plants from the Mesozoic rocks of South Africa," *Phil. Trans. Roy. Soc. B* 222:193–265.

——— 1934. "The nature and origin of the stigma," *New Phytol.* 33:173–198.

——— 1947. "The history of plant form," *Adv. Sci.* 4:243–254.

——— 1955. "Mesozoic pteridosperms," *Phytomorphology* 5:177–185.

——— 1958. "*Lidgettonia,* a new type of fertile *Glossopteris,*" *Bull. British Mus. Nat. Hist.* 3:179–189.

Thompson, J. M. 1929. "A study in advancing sterility with special reference to the Leguminous gynaecium," *Publ. Hartley Bot. Lab. Liverpool,* No. 6. 47 pp.

Thomson, B. F. 1942. "The floral morphology of the Caryophyllaceae," *Amer. J. Bot.* 29:333–349.

Thorne, R. F. 1958. "Some guiding principles of angiosperm phylogeny," *Brittonia* 10:72–77.

——— 1963. "Some problems and guiding principles of angiosperm phylogeny," *Amer. Nat.* 97:287–305.

——— 1968. "Synopsis of a putatively phylogenetic classification of the flowering plants," *Aliso* 6:57–66.

Tomlinson, P. B. 1970. "Monocotyledons—toward an understanding of their morphology and anatomy," *Adv. Bot. Res.* 3:207–292.

——— and H. E. Moore, Jr. 1968. "Inflorescence in *Nannorrhops Ritchiana* (Palmae)," *J. Arnold Arb.* 49:16–34.

Torrey, J. G. 1955. "On the determination of vascular patterns during tissue differentiation in excised pea roots," *Amer. J. Bot.* 42:183–198.

——— 1957. "Auxin control of vascular pattern formation in regenerating pea root meristems grown in vitro," *Amer. J. Bot.* 44:859–870.

Townrow, J. A. 1962. "On *Pteruchus,* a microsporophyll of the Corystospermaceae," *Bull. Brit. Mus. Nat. Hist. Geology* 6:289–320.

——— 1965. A new member of the Corystospermaceae Thomas. *Ann. Bot.* (n.s.) 29:495–511.

Troll, W. 1932. "Morphologie der schildförmigen Blätter," *Planta* 17:153–314.

——— 1964. *Die Infloreszenzen. Typologie und Stellung im Aufbau des Vegetationskörpers* (Fischer, Stuttgart), pt. 1.

Tucker, S. C. 1959. "Ontogeny of the inflorescence and the flower in *Drimys winteri* var. *chilensis,*" *Univ. Calif. Publ. Bot.* 30:257–336.

——— 1960. "Ontogeny of the floral apex of *Michelia fuscata,*" *Amer. J. Bot.* 47:266–277.

——— 1966. "The gynoecial vascular supply in *Caltha,*" *Phytomorphology* 16:339–342.

——— and E. M. Gifford. 1966. "Carpel development in *Drimys lanceolata,*" *Amer. J. Bot.* 53:671–678.

Tzveliov, N. N. 1969. "Some problems of the evolution of *Poaceae,*" *Bot. Zhurn.* 54:361–373.

Valentine, J. W. 1967. "The influence of climatic fluctuations on species diversity within the tethyan provincial system," in C. G. Adams and D. V. Ager, eds., *Aspects of Tethyan Biogeography* (Systematics Assoc. Pub. No. 7), pp. 153–166.

Van Overbeek, J. 1962. "Endogenous factors of fruit growth," *Plant Science Symposium* (Campbell Soup Corp., Camden, N. J.), pp. 37–56.

Vasek, F. C. 1964. "The evolution of *Clarkia*

unguiculata derivatives adapted to relatively xeric environments," *Evolution* 18:26–42.

——— 1968. "The relationships of two ecologically marginal, sympatric *Clarkia* populations," *Amer. Nat.* 103:25–40.

Vink, W. 1970. "The Winteraceae of the Old World. I. *Pseudowintera* and *Drimys*—morphology and taxonomy," *Blumea* 18:225–234.

Vogel, S. 1963. "Blüten-Oekotypen und die Gliederung systematischer Einheiten," *Ber. Deu. Bot. Ges.* 76, Sondernummer:98–101.

——— 1968. "Chiropterophilie in der neotropischen Flora, Neue Mitteilungen. I," *Flora Abtl.* B 157: 562–602.

——— 1969. "Chiropterophilie in der neotropischen Flora, Neue Mitteilungen. II. und III," *Flora Abtl.* B 158:185–222, 289–323.

Vuilleumier, B. S. 1967. "The origin and evolutionary development of heterostyly in the angiosperms," *Evolution* 21:210–226.

Waddington, C. H. 1962. *New Patterns in Genetics and Development* (Columbia University Press, New York).

Wagner, W. H. 1961. "Problems in the classification of ferns," *Recent Advances in Botany* (University of Toronto Press, Toronto), pp. 841–844.

Waisel, Y. 1959. "Ecotypic variation in *Nigella arvensis* L.," *Evolution* 13:469–475.

Walker, J. W. 1971. "Pollen morphology, phytogeography and phylogeny of the Annonaceae," *Contr. Gray Herb. Harvard Univ.* No. 202.

Walter, H. 1962. *Vegetation der Erde. in Oekologischer Betrachtung* (Fischer, Jena).

Walton, J. 1953. "The evolution of the ovule in Pteridosperms," *Advancement of Sci.* 10:223–230.

Wardlaw, C. W. 1953. "Action of tri-iodobenzoic and trichlorobenzoic acids in morphogenesis," *New Phytol.* 52:210–217.

——— 1957*a*. "The floral meristem as a reaction system," *Proc. Bot. Soc. Edinburgh* 66:394–408.

——— 1957*b*. "On the organization and reactivity of the shoot apex in vascular plants," *Amer. J. Bot.* 44:176–185.

——— 1965. "The organization of the shoot apex," in *Encyclopedia of Plant Physiology* (Springer, Berlin), XV, 1:966–1076.

Wareing, P. F., and G. Ryback. 1970. "Abscisic acid: a newly discovered growth-regulating substance in plants," *Endeavour* 29:84–88.

Waterbury, A. 1972. "Clinal variation in the karyotype of *Peromyscus maniculatus* Wagner from California," unpub. Ph. D. thesis, University of California, Davis.

Watson, J. D. 1970. *Molecular Biology of the Gene* (2nd. ed.; Benjamin, New York.).

Weber, H. 1938. "Gramineenstudien. I. Ueber das Verhalten des Gramineenvegetationskegel beim Uebergang zur Infloreszenzbildung," *Planta* 28:275–289.

——— 1939. "Gramineenstudien. II. Ueber Entwicklungsgeschichte und Symmetrie einiger Grasinfloreszenzen," *Planta* 29:427–449.

Weberling, F. 1964. "Homologien im Infloreszenzbereich und ihre systematischer Wert," *Ber. Deu. Bot. Ges.* 76:102–112.

——— 1965. "Typology of inflorescences," *J. Linn. Soc. Bot.* 59:215–221.

Weibel, R. 1945. "La placentation chez les Tiliacées," *Candollea* 10:155–177.

Weier, T. E., K. Stocking, and K. L. Shumway. 1966. "The photosynthetic apparatus of higher plants," *Brookhaven Symp. Biol.* 19:353–374.

Weimarck, H. 1941. "Phytogeographical groups, centres, and intervals within the Cape Flora," *Lunds Univ. Arsk.* (n.f.) *Avd.* 2, 37:nr. 5.

Weissman, G. 1966. "The distribution of terpenoids," in T. Swain, ed., *Comparative Phytochemistry* (Academic Press, New York), pp. 97–120.

Wells, P. V. 1969. "The relation between mode of reproduction and extent of speciation in woody genera of the California chaparral," *Evolution* 23:264–267.

Werth, E. 1956. *Bau und Leben der Blumen* (Enke Verlag, Stuttgart).

Wetmore, R. H., and J. P. Rier. 1963. "Experimental induction of vascular tissues in callus of angiosperms," *Amer. J. Bot.* 50:418–429.

White, F. 1962. "Geographic variation and speciation in Africa with particular reference to *Diospyros*," *Publ. Syst. Assoc.* 4:71–103.

Whittaker, R. H., and P. P. Feeny. 1971. "Allelochemics: Chemical interactions between species," *Science* 171:757–770.

Whyte, L. L. 1965. *Internal Factors in Evolution* (Braziller, New York).

Wiebes, J. T. 1963. "Taxonomy and host preferences of Indo-Australian fig wasps of the genus *Ceratosolen* (Agaonidae)," *Tijdschr. Entomol.* 106:1–112.

Wieland, G. R. 1933. "Origin of angiosperms," *Nature* 131:360–361.

Willis, J. C. 1922. *Age and Area* (Cambridge University Press, Cambridge, England).

——— 1940. *The Course of Evolution* (Cambridge University Press, Cambridge, England).

Wilson, C. L. 1924. "Medullary bundle in relation to primary vascular system in Chenopodiaceae and Amaranthaceae," *Bot. Gaz.* 78:175–199.

——— 1937. "The phylogeny of the stamen," *Amer. J. Bot.* 24:686–699.

——— 1942. "The telome theory and the origin of the stamen," *Amer. J. Bot.* 29:759–764.

——— 1950. "Vascularization of the stamen in the Melastomaceae, with some phyletic implications," *Amer. J. Bot.* 37:431–444.

——— 1965. "The floral anatomy of the Dilleniaceae. I. *Hibbertia* Andr.," *Phytomorphology* 15:248–274.

Wilson, E. O. 1961. "The nature of the taxon cycle in the Melanesian ant fauna," *Amer. Nat.* 95:169–193.

Wilson, T. K. 1964. "Comparative morphology of the Canellaceae. III. Pollen," *Bot. Gaz.* 125:192–197.

Wodehouse, R. P. 1935. *Pollen Grains* (McGraw-Hill, New York).

Wohlpart, A., and T. J. Mabry. 1968. "The distribution and phylogenetic significance of the betalains with respect to the Centrospermae," *Taxon* 17:148–152.

Woodson, R. E. 1935. "Observations on the inflorescence of Apocynaceae," *Ann. Missouri Bot. Gard.* 22:1–48.

——— 1954. "The North American species of *Asclepias* L.," *Ann. Missouri Bot. Gard.* 41:1–211.

Wright, S. 1934. "Genetics of abnormal growth in the guinea pig," *Cold Spring Harbor Symp. Quant. Biol.* 2:137–147.

——— 1940. "Breeding structure of populations in relation to evolution," *Amer. Nat.* 74:232–248.

Wulff, E. V. 1943. *Historical Plant Geography* (Chronica Botanica, Waltham, Mass.).

Wunderlich, R. 1959. "Zur Frage der Phylogenie der Endospermtypen bei den Angiospermen," *Oesterr. Bot. Z.* 106:203–293.

Yagil, E., and G. L. Stebbins. 1969. "The morphogenetic effects of the hooded gene in barley. II. Cytological and environmental factors affecting gene expression," *Genetics* 62:307–319.

Yakovlev, M. S., and M. D. Yoffe. 1957. "On the embryogenesis in *Paeonia* L.," *Bot. Zh.* 42:1503–1506.

Zeevart, J. A. D. 1962. "Physiology of flowering," *Science* 137:723–731.

Zeiger, E., and G. L. Stebbins. 1972. "Developmental genetics in barley: a mutant for stomatal development," *Amer. J. Bot.* 59:143–148.

Zimmermann, W. 1959. *Die Phylogenie der Pflanzen* (2nd ed.; Fischer, Stuttgart).

——— 1964. "Wie entstehen neue Organe?" *Natur* 72(3):100–106.

——— 1965. "Die Blütenstände, ihr System und ihre Phylogenie," *Ber. Deu. Bot. Ges.* 78:3–12.

Zohary, D. 1959. "Is *Hordeum agriocrithon* the ancestor of six-rowed cultivated barley?" *Evolution* 13:279–280.

——— 1960. "Studies on the origin of cultivated barley," *Bull. Res. Counc. Israel* 9D:21–42.

——— 1963. "Spontaneous brittle six-row barleys, their nature and origin," *Proc. 1st Int. Barley Genet. Symp. Wageningen* 1963, *Barley Genetics* I:27–31.

——— 1965. "Colonizer species in the wheat group," in H. G. Baker and G. L. Stebbins, eds., *The Genetics of Colonizing Species* (Academic Press, New York), pp. 403–419.

Zohary, M. 1937. "Die verbreitungsökologischen Verhältnisse der Pflanzen Palästinas. I," *Beih. Bot. Zentralbl.* A 56:1–155.

Index

NOTE: An asterisk following a page number indicates that an illustration of the subject entry appears on the page.